LEICESTER SOUTH FIELDS COLLEGE
LIBRARY
Andromeda
06127844
061278

Chemistry and Technology of Water Based Inks

Chemistry and Technology of Water Based Inks

Edited by

P. LADEN

BLACKIE ACADEMIC & PROFESSIONAL

An Imprint of Chapman & Hall

London · Weinheim · New York · Tokyo · Melbourne · Madras

Published by Blackie Academic and Professional, an imprint of Chapman & Hall, 2–6 Boundary Row, London SE1 8HN, UK

Chapman & Hall, 2–6 Boundary Row, London SE1 8HN, UK

Chapman & Hall, GmbH, Pappelallee 3, 69469 Weinheim, Germany

Chapman & Hall USA, 115 Fifth Avenue, New York, NY 10003, USA

Chapman & Hall Japan, ITP-Japan, Kyowa Building, 3F, 2-2-1 Hirakawacho, Chiyoda-ku, Tokyo 102, Japan

DA Book (Aust.) Pty Ltd, 648 Whitehorse Road, Mitcham 3132, Victoria, Australia

Chapman & Hall India, R. Seshadri, 32 Second Main Road, CIT East, Madras 600 035, India

---

First edition 1997

Typeset in 10/12 Times by Blackpool Typesetting Services Limited, UK
Printed in Great Britain by Hartnolls Ltd, Bodmin, Cornwall

ISBN 0 7514 0165 X

A catalogue record for this book is available from the British Library

Library of Congress Catalog Card Number: 96-83829

∞ Printed on acid-free text paper, manufactured in accordance with ANSI/NISO Z39.48-1992 (Permanence of Paper)

Preface

This book has been a long time in the making. Since its beginning the concept has been refined many times. This is a first attempt at a technical book for me and fortunately the goals I have set have been achieved.

I have been involved in water based ink evaluation since its unclear beginnings in the early 1970s. This book is fashioned much like a loose-leaf binder I had put together for early reference and guidance. The format has worked for me over the years; I trust it will work for you.

I would like to thank the many people who made this book possible, particularly Blackie Academic & Professional for their saint-like patience. Thanks again to W.B. Thiele (Thiele-Engdahl), to Lucille, my wife, and to James and Frank, my two boys. A final and special thank you to Richard Bach who taught me there are no limits.

Patrick Laden
Anna Maria Island, Florida

Contributors

G.H. Cabiro	SC Johnson Polymers, 1525 Howe Street, Racine, WI 53403, USA
A.C.D. Cowley	Zeneca Ltd, Wilmington, DE, USA
M.A. Dalton	SC Johnson Polymers, 1525 Howe Street, Racine, WI 53403, USA
C. Durgan	Business Manager (Fluid Inks), Pigments Div., Sun Chemical Corp., 411 Sun Avenue, Cincinnati, OH 45232, USA
S. Fingerman	President, Garln Graphics, PO Box 1621, Cranford, NJ 07016, USA
H. Gaines	American Ink Maker, Melville, NY, USA
R.P. Grandke	SC Johnson Polymers, 1525 Howe Street, Racine, WI 53403, USA
H.J. Hartschuh	SC Johnson Polymers, 1525 Howe Street, Racine, WI 53403, USA
T.R. Holshue	President, WN Stevenson Co., 246 Rock Hill Road, Bala Cynwyd, PA 19004-2134, USA
P. Laden	Kohl & Madden Printing Ink Corp., 9025 Junction Drive, Annapolis Junction, MD 20701, USA
G. Oliver	SC Johnson Polymers, 1525 Howe Street, Racine, WI 53403, USA
E. Reich	Reich Corporation, 3146 Marion Street, Reading, PA 19405, USA
D. Sagraves	Technical Marketing Manager, Ink Division, General Printing, PO Box 2037, NC 28241, USA
J.A. Schak	Kady International, Pleasant Hill Road, Scarboro, ME, USA
W. Schumacher	Jakob-Altmaier-Straße 15, D-63457 Hanau, Germany
F. Shapiro	President, PF Technical Services, 210 Stephen Street, Levittown, NY 11756, USA

T. Smith	SC Johnson Polymers, 1525 Howe Street, Racine, WI 53403, USA
D. Tuttle†	Flexographic Technical Association, USA
T. Walsh	Ultra Additives, Park Station, Patterson, NJ, USA
H.W. Way	Dispersion Equipment Division, Netzsch Inc., 119 Pickering Way, Exton, PA 19341-1393, USA

Contents

3 Flexographic printing presses 43
D. TUTTLE

1 Introduction

F. SHAPIRO and D. SAGRAVES

1.1 Environmental considerations for water-based inks

The overriding concern of inkmakers and printers since the early 1980s has been the global movement to save the environment. Spurred by popular writers such as Rachel Carson and inflamed by such massive evidence of neglect as the poisonous gas leak in Bhopal, India and the pollution of the Love Canal in New York, United States by hazardous waste in landfills, the public has demanded and the government has responded with, increasingly tougher measures to control pollutants in all media.

In the wake of the Bhopal tragedy in 1982, the United States Congress became acutely aware of the environmental and safety hazards posed by the storage and handling of toxic chemicals. According to Congressional sources at the time, the Bhopal disaster focused public attention on the fact that extremely hazardous chemicals are present at chemical manufacturing plants and other facilities in communities all across America.

The thrust of legislative measures dealt with control of the wastes of the process, coping with emissions, discharges and hazardous waste through varying levels of control technology. With the passage of time, the limitations and drawbacks of this approach became evident. Considerable attention at both the State and Federal levels is now focusing on restricting the use of potentially hazardous chemicals. The prevention of pollution at the start of a process rather than the end-of-the-pipe handling and disposal will be the focus of legislative initiatives over the next few years.

Public groups around the globe have placed considerable pressure on the political process to move this agenda forward at a very fast pace. Frequently, emotions win out over facts. The stakes are too high for the political machines to put the environmental agenda at a lower priority. This is most evident now in business circles with the move to manage the environment under similar restraints as have been applied to quality management. ISO14000 seeks to codify business responsibility for the environment using methodology that is now common under the ISO9000 quality management standards.

1.1.1 *Effect on the ink and printing industry*

The basic liquid ink for flexography and gravure contains upward of 60% volatile organic compounds when applied at the press. The advantages of this liquidity have become the focus of efforts to save the environment.

The printing industry, and the ink industry, have been under constant pressure to replace the various hydrocarbon solvents and the heavy pigments used in ink and coatings formulations. In addition, the solvents used as diluents or for cleaning agents for the processes are under attack.

Regulations were passed in the United States, and subsequently in many countries around the world, that dealt with emission of fumes to the atmosphere, hazardous wastes to the landfills and incinerators, process waters discharging into the water supply and ground, as well as the safety regulations for in-plant handling and use. To a great extent, each of these laws operates in its own sphere of influence. To correct the ills of one set of parameters, it was common to seek a solution that led to a transference of pollution from one area to another. Each agency sought to enforce its rules, albeit at the expense of the others.

The decade of the 1990s has evidenced a change in this self-defeating philosophy. The perception has changed from this tunnel vision to a broad scale evaluation of the risks in all media. The resulting philosophy has been termed Multi Media Pollution Prevention (M2P2). Compromises to accomplish radical changes in one arena may be made at the expense of small changes in other areas of concern. This can be done with foresight and scientific judgment to minimize the ramifications of such compromises. The perception has been identified as an approach to the optimum reduction of pollutants in the environment.

Accompanying the total perception of the effects of pollutants is the need to avoid creating the pollutants by design. This effort has been described as Pollution Prevention, Waste Reduction, and Waste Minimization. The driving force is to reduce the potential to generate wastes at the source, by design and within the basic management objectives for the conduct of business. Water based ink and coatings technology has been a primary vehicle in this drive to replace hydrocarbons and other pollutants in the workplace and the community.

1.1.2 *Printing inks as a source of pollution*

Consider the typical printing plant and plot the course of all materials that enter the front doors. One can see that pollutants (wastes) are generated at each and every process. Some of the pollutants can be categorized as hazardous due to the effect on the ecology of the community; others are simply burdens on the community because disposal uses valuable resources to store and destruct vast amounts of these wastes. Air emissions of volatile organic compounds generate ground level ozone; the add-on controls can generate nitrogen oxides (NO_x) and CO_2.

Most solvents that enter the solid waste stream as residues of printing processes or solvent recovery systems have known hazardous characteristics. Disposition to landfills has been limited to prevent leaching of the agents into the ground.

Disposal of these hazardous wastes through incineration, cement kilns or as alternative fuels requires air emission controls to assure high efficiency and clean destruction. The resulting ash must be free of heavy metals in significant quantities.

Release of inks and solvents through water drains from the process areas was common prior to the realization of the affects of these chemicals on the quality of the water system. Such practices have been brought to a halt and are strictly regulated under a variety of clean water regulations.

Solid wastes such as paper or film trim, substandard products, cleaning rags, and rejected product all enter the waste stream that is destined to wind up in landfills or as fuel in incinerators. There are limits to the amounts that can be absorbed into the system without spilling over into the ecological balance in a community. Space is a premium and the potential for abuse of the system has proven that efforts are necessary to control and reduce this waste as well.

Under existing regulations and with the demand for control of the wastes as they are generated, ink makers and printers have found a convenient haven in the use of add-on controls. If wastes are made, burn them or recover what can be reused. Incineration and solvent recovery have become the ideal solutions for government; the quick fix and burden of the printing industry.

However, this end of the pipe option has fallen into question, both as to the cost of such efforts and the real benefits of the approach to the solution to the problem of the proliferation of hazardous chemicals. Pollution prevention efforts propose to get the pollutants out of the system before they are generated, then firms do not have to worry about the cost, performance, and security of the add-on controls.

1.1.3 *Clean air initiatives in the United States*

The legislative attack on air pollution started in the early 1970s in the United States. The early efforts were directed at major sources, with 100 tons of volatile organic compounds as the threshold requiring reduction activities for compliance. Wth the passage of the Clean Air Act Amendments of 1990, the coverage of the regulations was broadened considerably.

The primary target in printing inks is the volatile organic compound (VOCs). Volatile organic compounds are a group of chemicals that react in the atmosphere with nitrogen oxides in the presence of heat and sunlight to form ozone. The EPA does not include methane and other compounds which are deemed to have negligible photochemical reactivity. Typical VOCs in inks are alcohols, acetates, ketones, and glycols.

The enforcement target was lowered from the early threshold of 100 tons of VOCs per year to include firms with annual emissions of 25 tons of VOCs in severe areas or 10 tons of identified Hazardous Air Pollutants (HAPs). The country was divided into categories of severity with varying levels of air emissions qualifying as targets for federally enforceable permits.

A list of 189 chemicals was designated as hazardous air pollutants (HAPs) under the National Emission Standards for Hazardous Air Pollutants (NESHAPs). Among the solvents on this list are toluene, methanol, hexane, and MEK; all common in gravure inks or used cleaning agents for gravure, flexography and publication lithography.

Maximum available control technology (MACT) standards are to be established to reduce the amount of HAPs emitted to the atmosphere. There are federally enforceable laws permitting the use of 'major' facilities, in an effort to standardize handling of air emission sources throughout the US. Permits for major facilities and enforceable reporting and control requirements will bring printing plants throughout the US under similar restraints. Ozone depleting chemicals are banned and a schedule to phase out each group of these chemicals was placed in the law. Pollution prevention efforts and assistance to small businesses are key measures in the amended clean air legislation. Of consequence to the development of water based inks was the inclusion on the list of HAPs of ethylene glycol and all glycol ethers based on ethylene. The formulations of many ink makers utilized these listed hazardous glycols.

1.1.4 *The disposal of hazardous waste*

The Resource Conservation and Recovery Act (RCRA) places restrictions on the handling and disposal of hazardous wastes, the use of underground storage tanks, and the disposal of non-hazardous solid wastes. The major impact on printing inks comes from the enforcement of hazardous waste standards and regulations. The goal of the RCRA is to protect human health and the environment by regulating the management of hazardous waste facilities. Anyone who generates, transports or disposes of hazardous wastes must notify the USEPA of his activities and comply with the provisions of the Act. These measures are designed to prevent future hazardous releases and superfund sites.

The responsibility of the inkmaker and the printer as generator of the wastes is mandated from 'cradle to grave'. Tracking of the hazardous waste from the generator to its final disposition is accomplished by a system manifest, reports and periodic testing. Wastes are classified with regard to general characteristics, or listed for specific hazardous physical traits. The most typical solvent printing ink category is ignitable. Such solvents as toluene, MEK, MIBK, ethyl acetate, and 1,1,1-trichloroethane are listed in specific categories.

Generators are categorized by size. The typical printer generates in excess of 1000 kilograms a month, and is categorized as a large quantity generator. Under this designation, a number of waste management procedures and training are required. Solvent recovery in small distillation units has become common at printing facilities. Reducing the amount of spent solvent for shipment lowers the cost and the liability of off-site disposition of hazardous wastes. Recovering the solvents for reuse reduces the amount purchased as virgin materials.

Storage tanks, particularly those underground, come under the control of the RCRA regulations. Rigid testing, engineering controls and construction specifications have been developed to ensure that the contents of the tanks will not leak into the surrounding ground and groundwater. Waste minimization, an earlier and more limited form of pollution prevention, is mandated for generators of hazardous waste.

1.1.5 *CERCLA and the Superfund Amendments and Reauthorization Act*

The build-up of toxic burial sites around the country led to the legislation, the Comprehensive Environmental, Response, Compensation and Liability Act (CERCLA). Intended as a means of providing the legal and financial backing to clean-up of the waste sites, the Act was amended in 1986 to include the Emergency Planning and Community Right to Know Act, also known as SARA Title III. The purpose of the Title III provision is to provide citizens with the right to know about the chemicals present in their communities, what they are, where they are, and how much is present. Reports are mandated under specific threshold for both inventory and usage.

The statistical inventory of toxic substances is mapped on an annual basis, with trends developed to illustrate the reduced usage of targeted chemicals. The Title III provision for pollution prevention programs has accelerated the drive for multi-media efforts to reduce the release of all types of waste into the environment. Such programs are mandated and closely watched by both federal and state agencies.

1.1.6 *Water pollution control*

The federal government passed legislation in 1972 to protect the quality of the navigable waterways. The intent of the clean water law was to prevent, reduce and eliminate water pollution. National effluent limitations and performance standards were developed for industrial and public treatment plants. The development of standards and the enforcement of legislation for all but the specific industries (known as 'categorical') was left to local authorities, and water or sewer management agencies for the cities, towns, counties or states.

Regulations limit the discharge amounts of various metals, solvents, solids, oils and greases, and biological and chemical contaminants that pollute water. In recent years, this approach has been extended to the effects of stormwater washing residues of chemicals stored or used outdoors into the stormwater drains and into the drinking supplies of the community. Permits are now mandated by the federal government for all facilities with potential to discharge such chemicals into the sewer or drainage systems. The parameters for stormwater permits have been subject to local interpretation; in some areas, a catalytic oxidizer on a roof or pad alongside a building must be permitted under the regulations.

1.1.7 *Chemicals in the workplace*

Safety and health considerations in the workplace have been a driving force for the search for cleaner technologies. This is a global effort. In Europe, the initial regulations that spurred the search for alternatives to solvent based inks were workplace related. In the United States, OSHA regulations for indoor air contaminants preceded the work of the EPA.

Perhaps the biggest impact on safety and health came with the Hazard Communication Standard, commonly known as the Right To Know Laws. The premise for working safely with chemicals became the transfer of information relative to the nature and potential hazards of a chemical (pure or mixture). The doctrine is, that everyone who purchases, transports and uses chemicals has a right to know what they are working with, how to use the chemical, and how to protect themselves in the event of exposure.

The major tools for accomplishing this body of knowledge consists of six major components:

1. Every firm must have an inventory of chemicals in its facility
2. A material safety data sheet (MSDS) must be sent for each product
3. All containers, original and transfer in-house, must be properly labeled
4. A written program must be developed to cover the safe use of chemicals
5. There should be ongoing training of all employees
6. Proprietary chemicals must be approved; this information must be available to appropriate health/environmental officers

The major item that has revolutionized the relationship between the inkmaker and printer has been the MSDS. The document advises the printer of the exact content (of those hazardous materials only) of the product that is being purchased. This enables the printer to request the inkmaker to make changes when specific chemicals in a formulation are not desired. These considerations may be for safety or for compliance with regulations.

Labels are no longer merely the means of identification. They now serve as miniature MSDSs providing basic information as to the nature and practices required when using a specific formulation. In some localities, the label must carry a breakdown of the hazardous ingredients of the formulation.

With the maturation of the concept of the MSDS, there has been a movement to formalize more of the information and standardize the appearance of the documents. The eight part format suggested by the OSHA regulations has been supplemented by an American National Standards Insitute (ANSI) version of the MSDS, which has 16 sections. The intent of these information based regulations is to institute documentation of the safety activities and protection of the employees by industry, and to foster a vehicle that will bring the potential dangers of working with chemicals in the specific workplace to the employees. Included in such information are the physical and health hazards, and the precautions to be taken when handling and using these chemicals.

Among the items which are driving forces to promote the use of water based ink technology are the avoidance of flammability, explosion, and toxicity of the typical solvents used in ink making. When this information was proprietary, the printer had no way of knowing what the hazards were in the inks and solvents he purchased. Now, these items are under a magnifying glass, not only for the printer, but for the community in which the plant is located.

1.1.8 *Regional impact on regulations*

The source of most legislation and standards for complicance come from the federal levels of government. The basic documents set the tone and goals of the particular area of concern. They may also put into place the bureaucracy to implement the law by developing standards, guidelines, reporting structures, and levels of authority. No lower body may exceed the standards set forth by the federal agencies. The implementation of the enforcement of these regulations can be delegated to the states and other regional bodies. These entities can maintain the standards of the federal mandates, or they can rewrite them with more severe limits. This is a factor that enters into the decisions made when formulating chemical products for a broad based market. How does one satisfy all of the varying requirements of the local regulations? This is one of the dilemmas of the water based ink maker. One must contend with the varying water discharge and ground contamination protection regulations that vary from state to state, and country to country. Just how does one write a label which must comply with varying, sometimes conflicting rules of different states? In some cases, can an inkmaker specify or sell the inks to a firm which is not properly permitted? Where does the liability end?

1.1.9 *Why water based ink formulations?*

Given the vast number of regulations that impact on the inkmaker and printer, what are the advantages environmentally of using water technology?

1. Reduction of VOCs emitted to the atmosphere
2. Reduced potential for discharging toxic substances into the water system
3. Reduces the potential for fire and explosion, thereby eliminating some of the costly provisions and regulations required for the flammable solvents and inks
4. Reduces the amount of hazardous wastes
5. Improves the working conditions in the plant
6. Brings the firms into compliance with almost every regulation

1.1.10 *Concerns with the transfer to water based inks*

Notice that we used the term reduces, not eliminate. To be effective, the water based inks will still require a number of components that will qualify as VOCs and will have their own physical and health hazards. There will be a reduction in most areas, a transference in others.

One of the critical environmental areas to be addressed with the use of water based inks, other than their performance as printing inks, will be the disposition of the water wastes. The treatment or recovery of clean water will be a major question for all printers utilizing this alternative.

A secondary concern for larger printers will be the treatment of air emissions of those VOCs which will be present in the water formulations. As thresholds for air compliance are lowered, the larger firms may have to use an add-on control to suit. This need may be answered by work that is currently in progress with the bioremediation of VOCs in the exhaust vapors of the printing process.

Training will be vital. Old methods and habits must be replaced with new ones. Troubleshooting guidelines and practices must change to suit the new chemistry. There will be a learning curve, from top management attitudes and practices to those in the pressroom.

Water technology can be the salve for the problems of the environment.

1.2 Factors affecting the change to water based inks

Changing from solvent based printing to water based printing is becoming both an environmental and economical necessity for many printers. Many printers have experienced past difficulties, or heard of others' difficulties, and have serious doubts about their capabilities for printing water-based inks. That so many have successfully made the change attests to the fact that it can be done with just the right amount of preparation. With a little knowledge of water-based inks, their physical requirements, and an honest evaluation of the press capabilities, most printers have the means to successfully make the change. This article is intended as a starting point for the printer wishing to make a successful change to water-based printing.

Many factors influence the drying rate of water based inks on a particular printing press; how fast the press can run with solvent based inks and what is the targeted press speed for water based inks; what is the trap sequence and how many colors are trapped; what is the volume of the anilox rolls; the amount of fountain roll to anilox squeeze; the amount of printing plate to substrate squeeze; the indicated run-out of the plate cylinders and the anilox cylinders; and of considerable significance, how much air capacity can be delivered to the between color decks. To get maximum performance from water based inks it is absolutely necessary to optimize all these factors and more.

There is no magic water based ink formulation that can overcome all the limitations of some printing press configurations. Furthermore, I do not think it is realistic to believe that water based inks are going to dry significantly faster than the more recent formulations used today, at least not in the next 3 to 5 years. Due to environmental regulations ink companies are under constant pressure to reduce the solvent content of the water based inks to even lower and lower levels—to levels way below the alcohol content of 3.2% beer. The replacement for the most part is water and, as you know, water has a set rate of evaporation under a given temperature and pressure that we just are not going to change.

Water dries approximately four and a half times slower than the solvents commonly used in solvent based printing inks, ethyl alcohol and *n*-propyl acetate. Depending on the press configuration this means that a printer could experience a decrease in press speeds of as much as 40%–50% by switching to water-based inks without making any mechanical modifications to the press metering train and drying capacity.

1.2.1 *What can a printer do to maintain press speeds with water based inks?*

Changing to water based inks and maintaining or increasing press speeds may require the following type of approach:

(a) *Check between color dryers.* Increase the velocity and turbulence of air in the between color dryers, especially in the first deck. Air velocity should be at least 7500 fpm and 12 000 fpm is recommended for multicolor traps. If a job requires trapped colors, perhaps the most significant thing that can done to achieve faster printing speeds is to achieve complete or better drying of the first down color.

Print the thinnest first down ink film possible while still maintaining color density. Apply maximum air turbulence (velocity) in the between color dryer.

Maintain a kiss impression on the second down color and run a slightly wetter ink film. All driers should be set to within $\frac{3}{32}''$ of the web, or to the manufacturer's specifications, and be set perfectly parallel to the drum to prevent loss of nozzle velocity. It is imperative that the correct ratio between supply and exhaust is maintained to prevent the hot air exhaust from blowing onto the print deck. Excessive air blowing down on the print decks will result in the ink drying too fast on the plate and the press operator will fight dirty printing throughout the print run. Proper air balance is critical for clean printing.

All supply and exhaust ductwork must be inspected for leaks and repaired where necessary. Leaks in the ductwork results in loss of supply and/or exhaust air volume that leads to poor ink drying. While this is important when printing solvent based inks it is absolutely critical when printing water based inks due to the slower drying nature of water.

(b) *Check air supply.* Supply dry air, or dehumidified air to the press. On rainy and humid days the press is supplied with moisture laden air that may likely decrease press speeds significantly. As one printer commented, 'Moist air can't dry anything.' One quick way to supply more dry air under humid conditions is to slightly increase the air temperature in the dryer. This can help offset the web cooling effect caused by evaporation of the humid ambient air. Just remember to turn down the oven temperature when conditions change back to normal.

(c) *Deck components.* All printing decks must be inspected and brought to proper mechanical specifications and all decks must repeat to zero when thrown on and off impression. Uneven impression, from rolls or plates, can result in excess squeeze that causes dirty printing, especially with water based inks.

(d) *Check rolls.* Maintain roll uniformity (TIR = total indicated runout) of the anilox rolls and plate cylinders to a mximum TIR of 0.001″ (0.0005″ for the fine process work). High/lows, cylinder bounce, and resultant plate squeeze can result in alternating bands of light and heavy ink application. The heavy bead-line and heavy ink bands may track-off on idler rolls, may ghost, or offset. The same thing applies to printing plate uniformity; the plates should be uniform to 0.0005″. This will help prevent excess plate squeeze and lead to cleaner printing. The stickyback, or plate adhesive, should be checked for uniform thickness to within 0.0005″ to 0.001″ for optimum results.

(e) *Check ink metering.* Change to a finer ink metering system, shallower anilox cell volume and/or doctor blade metering. Two-roll metering can be used if the between color dryers are adequate and it includes shallow anilox rolls, high strength water-based inks, and 70–80 durometer fountain rolls. A harder fountain roll will transfer less ink than a softer durometer roll.

On a two-roll press it is possible to increase the press speed with a decrease in anilox cell volume. Anilox cell volume must be balanced with the different press speeds and the amount of between color air capacity needed to maintain color density and proper ink drying. For instance, maintaining a costant color density at a press speed of 450 fpm might be possible with an anilox cell volume of 3.2–3.9 bcm while the same color density might be attained at 600 fpm with an anilox cell volume of 2.7–3.0 bcm.

(f) *Check ink type.* Change to an ink system with higher emulsion content to achieve faster drying. (The drawback to this approach is poor printability because the inks will also dry faster on the plates and anilox rolls.)

(g) *Use corona treaters.* When printing films, such as polyethylene and polypropylene, a corona treater is highly recommended wherever possible. Properly treated films can result in better ink adhesion, better water resistance, and faster ink drying. Many smaller day-to-day problems with water based inks can be drastically reduced or eliminated with freshly treated films.

(h) *Learn about pH.* Adjusting the pH balance of a water based ink is not that much different from adding a slow, rich solvent to a solvent based ink to keep it printing clean. Most press operators have learned over the years how to adjust the ink solvents, the faster drying and slower drying solvents, for optimum printability. Adjusting the pH balance of a water-based ink simply means adding a solvent to the water based inks. It is an important tool for maintaining water solubility of the ink components for optimum printing.

(i) *Putting it together.* Run a qualifying trial with one or two proven water based inks to evaluate the drying performance of water based inks on the current press equipment. Use that qualifying trial as a measure of current capabilities and as a guideline for future improvements. Then use the information gained as a chance to test the above recommendations.

Without knowing what speeds a printer may be currently achieving, experience would suggest that consistent production speeds on a two-roll press printing polyethylene film could be attained in a two-color trap sequence with the following press configuration:

1. Anilox rolls	360 line, 2.7–3.5 bcm
2. Fountain roll durometer	70–80
3. Printing plates	natural rubber/photopolymer
4. Air capacity; between deck	12 000 ft/min; 7500 ft/min, minimum
5. TIR tolerance	to 0.001″ (0.0005″ for fine process printing)

Given that the evaporation rate of water will not change, there are several methods to achieve better, or faster, drying with water based inks. These methods, as outlined above, include testing between color driers to learn how fast the equipment can blow off the water vapors and making improvements where necessary. Depending on the drying speeds required, these changes may include changes in ink metering by eliminating two-roll metering and changing to doctor blade metering, using higher viscosity, higher solid inks and printing with thinner ink films.

Many printers have already made the successful change to water based ink printing. They have found that water based inks can be as capable of producing high quality flexographic printing as solvent based inks. Critical to their success is the commitment and care given to understanding water based inks and their press requirements. The change to water based printing can be achieved smoothly and successfully and with less downtime with the right amount of proper introduction and training of the press employees. That is not really much different from what was done many years ago with solvent based inks. In fact, it is done by a different printer almost every day.

2 Colorimetry and the calculation of color difference

P.J. LADEN

2.1 Introduction

This chapter was prepared to aid the attendees of the courses or seminars given by the author. It represents an overview of the important elements of colorimetry and the calculation of color differences from an industrial application viewpoint. Examples of each phase of the calculations needed to determine color differences are included.

The readers will find it very useful to refer to the papers listed in the reference section. Some of the figures used in this chapter were taken from those papers. (See References 1–4 for details.)

2.2 Elements needed to describe color

For this particular discussion, a large collection of colored samples arranged in some orderly manner, like the Ostwald or Munsell System, is not being considered.

In order to obtain an objective description of a colored sample, we must consider the characteristics of the sample itself, the light source under which the sample is being viewed, and the human observer.

2.2.1 *The sample*

The sample or object is described by its interaction with light. This description is called a spectrophotometric or spectral curve. The instrument used to make the measurement is called a spectrophotometer. For opaque samples like paint panels, textile headliner material or plastic side panels, the curve is a reflectance curve. For a clear tail light lens, or printed polycarbonate dashboard components, the curve could be a transmittance curve. These curves show the fraction of light reflected or transmitted at each wavelength, from the sample compared to a standard. For reflectance measurements, the standard is a suitable white material. For transmission measurements, the standard is often air. For the remainder of this chapter, all the examples will asume reflectance measurements of opaque samples.

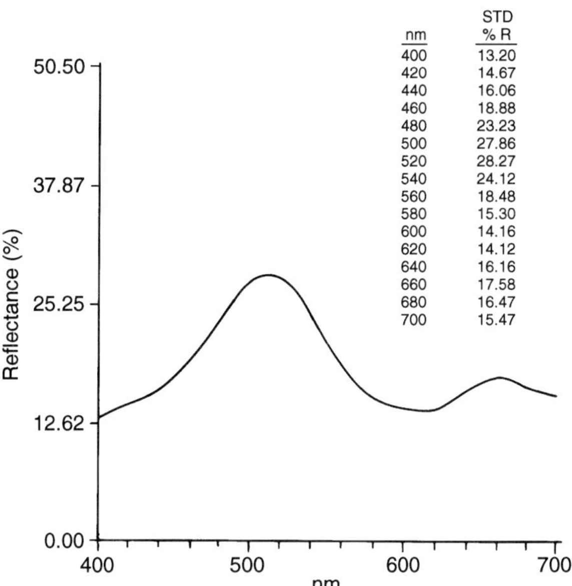

Figure 2.1 Reflectance curve of a green sample.

A spectral measurement is very useful for studying a variety of industrial color problems, but it does not, by itself, represent color as we see it.

Figure 2.1 shows the reflectance curve of a green sample along with the digital data printed out at 20 nanometer (nm) intervals from 400 to 700. The nanometer is the unit of length used to describe the wavelength scale. A nanometer is equal to one billionth of a meter. The wavelength range of 400–700 nm is shown because that represents the approximate range of the sensitivity of the human eye. Sometimes reflectance curves will be given at a range which starts slightly below 400 nm and extends a little beyond 700 nm. From the digital data we can see that the reflectance for the green sample at 500 nm is 27.86%. This means that when compared to the white standard at 500 nm, 27.86% of the light is reflected from the green sample.

2.2.2 *The light source*

We do not see color without light. The color we see depends on the characteristics of the light source under which the sample is viewed. The color of a sample can appear to change considerably when viewed under incandescent light (yellowish) and daylight (bluish). Since the color of the light source affects the color of the sample being viewed, we must have some way of describing the influence of the light source numerically. The light from any source can be described in terms of the relative amount of light (power) emitted at each

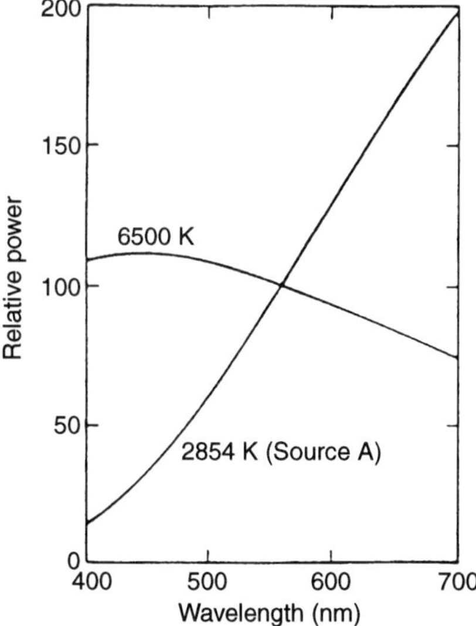

Figure 2.2 Spectral power distribution of two different sources.

wavelength. If we plot the relative amount of light as a function of wavelength, we obtain a curve (which is called a spectral power distribution curve) of the light source.

Figure 2.2 shows an example of the spectral power distribution of two different sources.

2.2.3 *The observer*

Of the three elements needed to describe the color of a sample, the observer is most difficult to describe numerically. It is not my purpose to go into a lot of detail at this time but a general idea of the method used is illustrated by Figure 2.3.

Light from a test lamp shines on a white screen and can be viewed by an observer. Right next to the spot of light from the test lamp, a spot of light from any combination of a red, green and a blue light can also be viewed (at the same time) by the observer. Let us call these red, green and blue lights—primaries. By adjusting the intensities of these primaries, the observer can make their combined color on the screen match the color from the test lamp. Therefore, the color on the screen produced by the test lamp can be described by the amounts of the three primary lamps needed to match it. These three amounts (numbers) are called the tristimulus values of the test lamp. If we wanted the procedure described above to be more precise and much broader in its application, we could perhaps replace the test lamp with a device (prism and slit type) so that white light could be broken up into the visible spectrum one wavelength at a

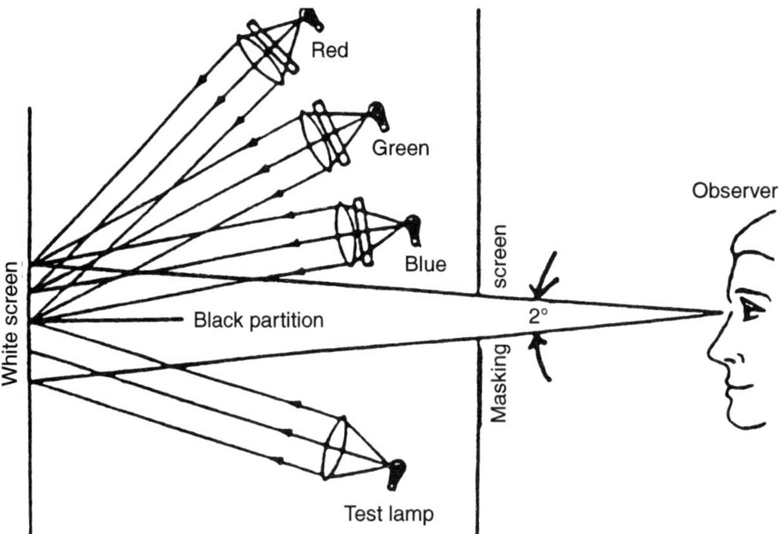

Figure 2.3 Method used to obtain a numerical description of the observer.

time, and each wavelength of light allowed to shine on the test spot. The primary lights could also be made by dispersing white light into the visible spectrum and isolating three single wavelengths of light, a saturated red, green and blue as the primaries. Then using the same procedure described above, each of the colored wavelengths (which when combined make up white light) could be matched one at a time with the three primary lights (each one a single wavelength). If this exercise was carried out by a number of normal observers, we could then calculate the average amount of the three primaries needed to match each wavelength. These amounts are the tristimulus values for the spectrum colors. When this experiment was carried out, it was found that most of the individual wavelengths of light could not be matched exactly with the three primaries. This problem could be overcome by adding light from one of the primaries to the test light, then matching this new color with the remaining primaries. We can think of this as subtracting some light from the other primaries. In these cases, the test light can be described by combinations of positive and negative amounts of the primary lights. By using a transformation equation (like converting temperature from degrees Fahrenheit to degrees Celsius), the values obtained from real primaries could be converted to values obtained from imaginary primaries. These new primaries were determined so that all the spectrum colors (each wavelength) could be matched with positive amounts of the three primaries. Figure 2.4 shows what these curves might look like. In Figure 2.4, the color produced by light at 600 nm could be matched with approximately 1.10 units of the red primary ($\bar{x}$) and 0.75 units of the green primary ($\bar{y}$) and 0.0 units of the blue primary ($\bar{z}$).

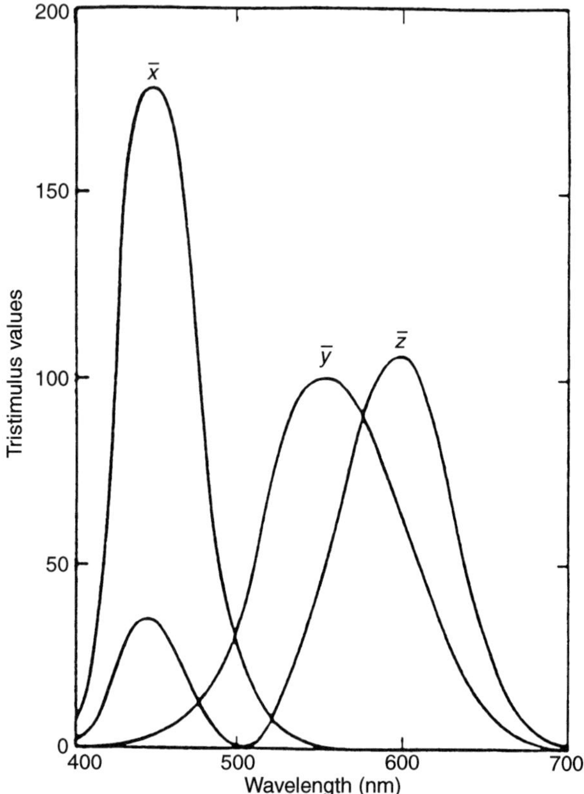

Figure 2.4 Curves obtained when all the spectrum colors are matched with positive amounts of the three primaries.

2.3 The CIE system

In 1931, the International Commission on Illumination, CIE (initials for the French name Commission International de l'Eclairage) put forth a system for the objective description of color by the proper combination of the elements described in section 2.2; i.e. sample (object), light source and observer. The CIE also provided the method to obtain the numbers that yield a measure of the color of a sample as seen under a standard source of illumination by a standard observer. These numbers are called tristimulus values *X*, *Y*, and *Z*.

In 1931, the CIE recommended the use of several sources which were then defined as standard illuminates when their spectral power distibutions were measured. In 1965, the CIE recommended some additional standard illuminants. The most important of these new illuminants (at least for industrial applications) was D65. D65 is an illuminant representing the spectral power distribution of typical daylight having a correlated color temperature of 6500°K (blue white). For industrial color work illuminants D65 and A (2854°K) are most often used.

2.3.1 *Source and illuminant*

In the previous paragraph you will note that I used the terms 'source' and 'illuminant'. It may appear that they represent the same thing; in fact, they may or may not. In CIE terminology, a source is a real light which can be turned on and off and can be used when looking at real samples. An illuminant is light defined by a spectral power distribution which may or may not be realizable as a source. When a standard illuminant can be made into a physical form, it becomes a standard source.

2.3.2 *2° Observer–10° observer*

When the experiments leading to the 1931 CIE standard observer were performed, the conditions were such that only the central area of the retina was used. This area is called the fovea. The area covers about a 2° angle of vision (Figure 2.3). This 2° angle is equivalent to looking at a sample the size of a dime from a distance of about 18″. Scientists have determined that the central area of the retina (fovea) has slightly different characteristics than the rest of the retina. In 1964 the CIE recommended the use of a slightly different observer whenever more accurate correlation to visual perception for larger samples is required. This supplementary standard observer is called the 10° observer. The values for the 10° observer were determined using similar procedures as the 2° observer except that the size of the colored area was considerably larger. The 10° angle is equivalent (approximate) to looking at a sample 3″ in diameter from a distance of about 18″. The 10° observer is recommended for most industrial applications. Figure 2.5 shows the difference between the 1931 2° observer and the 1964 10° observer.

2.3.3 *Computation of CIE tristimulus values, X, Y, Z*

The tristimulus values X, Y, and Z of a colored sample are obtained by multiplying together the reflectance (R) of the sample (Figure 2.1), the relative power (P) of a standard illuminant (Figure 2.2) and the standard observer functions $\bar{x}$, $\bar{y}$, and $\bar{z}$ (Figure 2.4) at each wavelength. You now have three columns: $RP\bar{x}$, $RP\bar{y}$, and $RP\bar{z}$. The products in each column are summed for all the wavelengths to give the tristimulus values X, Y, and Z for the colored samples (calculated for a particular standard illuminant and standard observer).

There is no need to worry too much about the calculation procedures at this point. Detailed examples of all the necessary calculations are given near the end of the chapter.

The values of $P\bar{x}$, $P\bar{y}$, and $P\bar{z}$ for various CIE illuminants and either the 2° or 10° observer are given in a number of publications. In some publications the products have been normalized so that the sum of $P\bar{y}$ is 1.00; in other cases they may be normalized to 100. Tables 2.1, 2.2 and 2.3 show the tristimulus

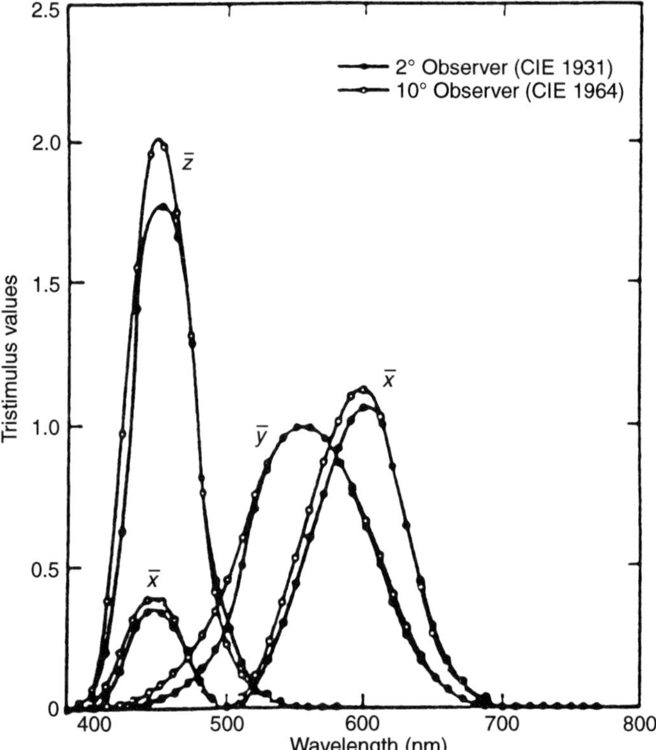

Figure 2.5 Difference between the 1931 2° observer and the 1964 10° observer.

weighting factors for CIE 1964 10° observer and illuminants D65, A, and CWF respectively. Illuminant CWF represents data for a particular cool white fluorescent lamp (source).

2.3.4 *Chromaticity coordinates and the chromaticity diagram*

The tristimulus values *X*, *Y*, and *Z* (as is) are of somewhat limited value as color specifications because they do not correlate very well with visual attributes. They do not define terms such as hue or saturation. The CIE recommended, in 1931, the use of the chromaticity coordinates *x*, *y*, and *z* to define a color specification. The chromaticity coordinate is simply the amount of any one of the tristimulus values divided by the sum of all three. The following relationships define the chromaticity coordinates:

$$x = \frac{X}{X+Y+Z}, \qquad y = \frac{Y}{X+Y+Z}, \qquad z = \frac{Z}{X+Y+Z}$$

Since $x + y + z = 1$, we only need two coordinates to specify chromaticity. The CIE recommended the use of x and y to set up a two-dimensional graph to

Table 2.1 10° 1964 CIE observer, illuminant D65 tristimulus weighting factors

Wavelength (nm)	16 Point data		
	$P\bar{x}$	$P\bar{y}$	$P\bar{z}$
400	0.251	0.023	1.090
420	3.232	0.330	15.383
440	6.679	1.106	34.376
460	6.096	2.620	35.355
480	1.721	4.938	15.897
500	0.059	8.668	3.997
520	2.184	13.846	1.046
540	6.810	17.355	0.237
560	12.165	17.157	0.002
580	16.467	14.148	−0.002
600	17.233	10.105	0.
620	12.894	6.020	0.
640	6.226	2.587	0.
660	2.111	0.827	0.
680	0.573	0.222	0.
700	0.120	0.047	0.
		100.000	
	94.825		107.381

display a color specification. This graph is called the chromaticity diagram. It is recognized by the familiar horseshoe-shaped curve which defines the location of the spectrum colors. The ends of the spectrum colors are connected by a line

Table 2.2 10° 1964 CIE observer, illuminant A tristimulus weighting factors

Wavelength (nm)	16 Point data		
	$P\bar{x}$	$P\bar{y}$	$P\bar{z}$
400	0.034	0.003	0.139
420	0.792	0.081	3.780
440	1.896	0.305	9.734
460	1.978	0.859	11.522
480	0.718	2.135	6.770
500	0.037	4.886	2.299
520	1.523	9.652	0.747
540	5.674	14.463	0.200
560	12.437	17.484	0.005
580	20.545	17.580	−0.002
600	25.371	14.906	0.
620	21.591	10.080	0.
640	12.158	5.062	0.
660	4.635	1.819	0.
680	1.393	0.541	0.
700	0.374	0.145	0.
		100.000	
	111.155		35.194

Table 2.3 10° 1964 CIE observer, illuminant CWF tristimulus weighting factors

Wavelength (nm)	16 Point data		
	$P\bar{x}$	$P\bar{y}$	$P\bar{z}$
400	0.046	−0.001	0.119
420	2.171	0.256	10.576
440	6.233	0.948	31.726
460	2.697	1.234	15.834
480	0.906	2.601	8.339
500	0.021	4.501	2.141
520	0.946	7.000	0.546
540	7.182	17.339	0.202
560	15.709	22.445	0.022
580	26.090	22.284	−0.004
600	23.515	13.921	0.
620	12.236	5.685	0.
640	3.660	1.506	0.
660	0.668	0.257	0.
680	0.068	0.026	0.
700	−0.002	−0.001	0.
		100.000	
	102.144		69.501

called the purple line (non-spectrum colors) to form a defined space. All real colors fall inside the space. The chromaticity diagram is illustrated by Figure 2.6.

The chromaticity diagram defines the chromaticity of the color (hue and saturation) but the lightness of the color is defined by the Y tristimulus value. This is a third dimension of the chromaticity diagram and represents a point perpendicular to the x–y plane (Figure 2.7). This system is called the CIE Y, x, y system. A color can now be defined by its Y, x, and y values which serve as a useful specification of that color.

Since color can be specified or addressed by its position in the diagram, the difference between two colors can be described by the difference in their positions in the diagram. The chromaticity diagram has many practical uses in a variety of industrial color applications. Sometimes we overlook just how useful it can be. However, it is not the purpose of this chapter to discuss all the useful applications of the chromaticity diagrams. In fact, at this point, we will emphasize one of the major deficiencies of the chromaticity diagrams.

2.3.5 *Color difference calculations and the chromaticity diagram*

If we consider using the chromaticity diagrams as a possible means of setting up an objective procedure for color tolerance determinations, we run into some serious problems. The physical distance between two samples in one part of the diagram (e.g. red) will be different from the physical distance between two other samples in a different part of the chromaticity diagram (e.g. blue), although

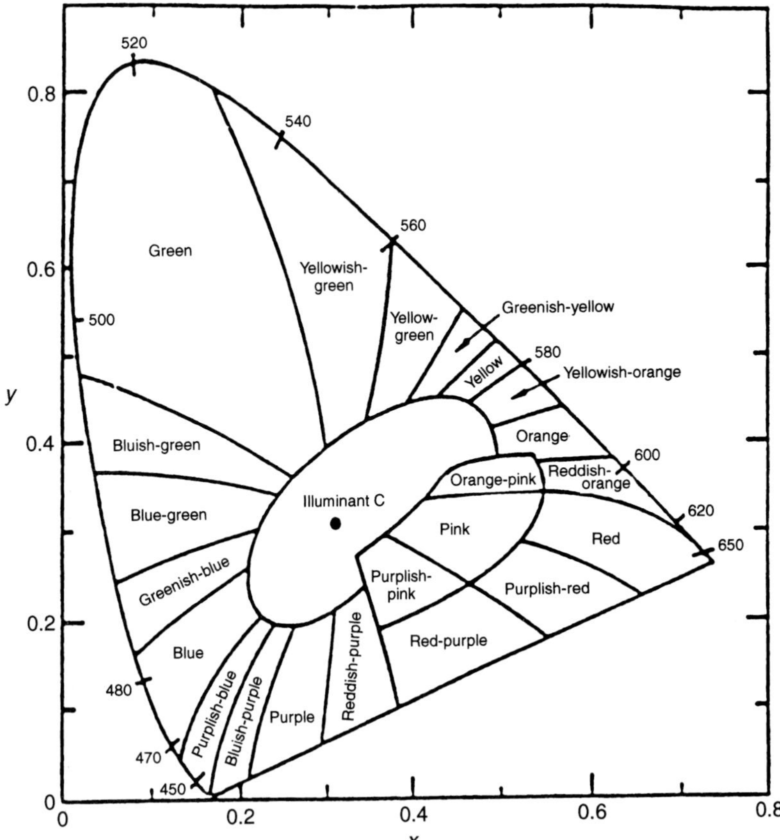

Figure 2.6 Chromaticity diagram.

visually both sets appear to have the same degree of color difference. The chromaticity diagram is not uniform from a perception standpoint.

These deficiencies were documented by David MacAdam in an experiment based on visual assessment of small color differences in CIE Y, x, y color space. MacAdam used colored lights just like the lights used in the original CIE experiments to produce a standard color on a background. He then used another set of lights to match that color. He then changed the settings on the lights to produce a just perceptible visual difference in all directions around the standard color. He did this for many different colors and plotted the results on a chromaticity diagram. If the chromaticity diagram was visually uniform or 'linear', the standard color should plot as a point with the differences forming circles of an equal radius around each of the standard colors. In fact, ellipses of various sizes around each of the standard colors (Figure 2.8) were obtained. This indicated that the use of CIE Y, x, y color space for color difference analysis would produce data which depended on what type (hue) of color was being compared.

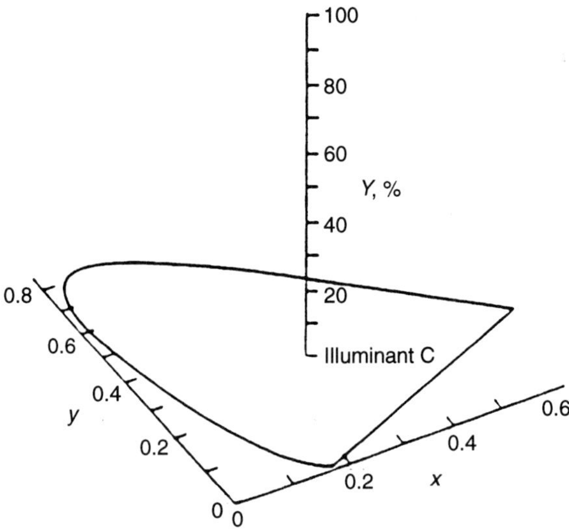

Figure 2.7 The third dimension of the chromaticity diagram.

Over the years a number of attempts have been made to develop linear and non-linear transformations of the chromaticity diagram to form a more visually uniform space. Perhaps it is analogous to plotting the chromaticity diagram on a rubber sheet, and then trying to pull the sheet so that the horseshoe-shaped diagram becomes a circle. In any event, to date, none of the transformations has produced a totally visually uniform color space, although some are better than others. In 1976, the CIE decided to recommend two different transformations. The one of interest for most industrial applications is the 1976 CIE $L^*a^*b^*$ (CIELAB) formula.

2.4 Applications of the 1976 CIE $L^*a^*b^*$ formula for the calculation of color tolerances

The CIELAB formula and diagram represents a non-linear transformation of the CIE *XYZ* values. All the equations are given in the Appendix.

2.4.1 *Color sensation*

There are three attributes which can be used to define the sensation of color. It is these attributes which give color its three dimensional characteristics. The attributes are hue, saturation (chroma) and lightness. They can be arranged in an orderly 'color space' like a cylinder (Figure 2.9).

Hue is defined as that attribute which determines whether a color is red, blue, yellow, green, etc. The geometric aspect of hue is a closed circle. Saturation (also referred to as 'chroma') is defined as the deviation from gray. The more a

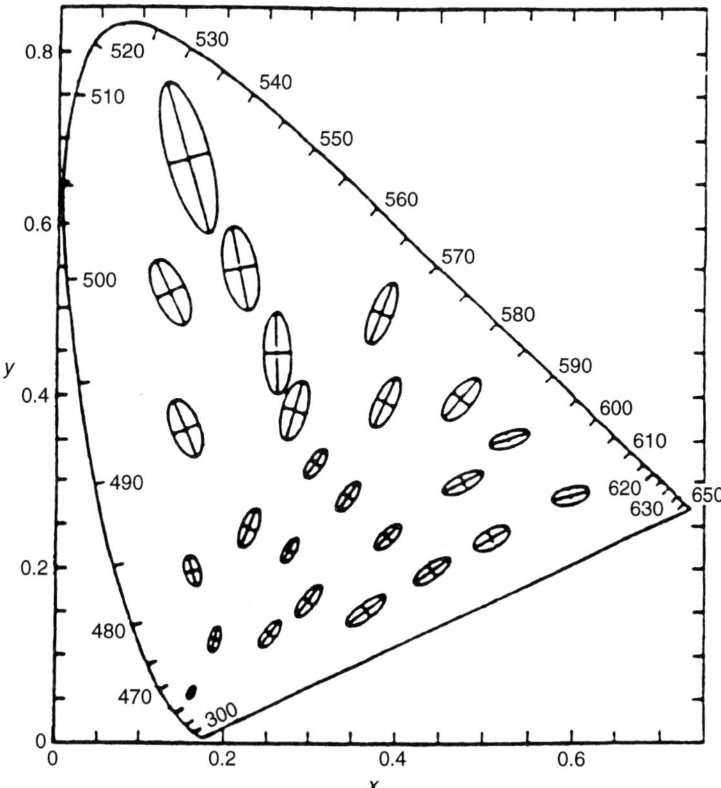

Figure 2.8 Ellipses obtained around each of the standard colors.

color deviates from gray, the more saturated it is. In terms of the cylinder (Figure 2.9), saturation increases with the length of the radius of the cylinder, starting with gray (zero saturation) and going out to the pure spectral color (highest saturation). Lightness relates the color to some continuous gray scale between white and black.

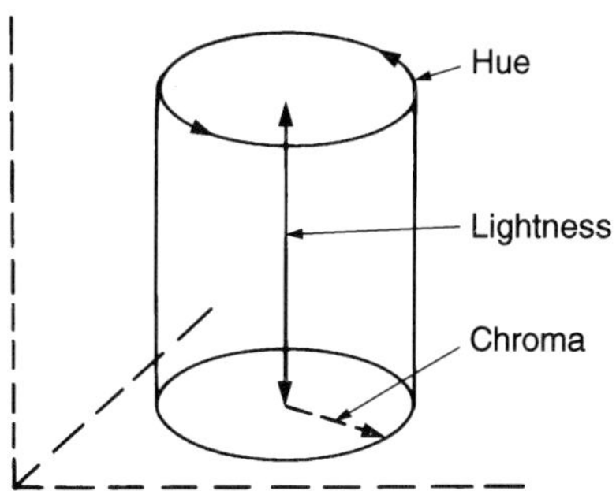

Figure 2.9 Arrangements of attributes in an orderly color space.

The space as illustrated by Figure 2.9 can be treated mathematically as a Euclidean space and any distances within the space can be calculated as the square root of the sum of the squares of the component differences. Therefore, when we properly scale the hue, saturation and lightness differences, the color differences between any two colors in this space can be calculated as the square root of the sum of the squares of the hue, saturation and lightness differences.

2.4.2 *CIE 1976 $L^*a^*b^*$ (CIELAB) color space*

The CIELAB color space is based on the concept described above in Figure 2.9, although the cylindrical nature of the space is not always recognized because of the manner that it is usually presented. The space is illustrated by Figure 2.10. The red/green axis is designated by the term a^* with positive values indicating a red hue, and negative values, a green hue. The yellow/blue axis is called b^* with positive values indicating a yellow hue and negative values; a blue hue. The lightness scale, which is L^*, is situated in the center of the a^*–b^* plane and is perpendicular to it.

Two colors (A and B) are plotted in the diagram. A and B were measured on a spectrophotometer to obtain the spectral reflectance curves. From these curves, the tristimulus values (X, Y, Z) were computed.

The tristimulus values (X, Y, Z) were input into the CIE $L^*a^*b^*$ equations to produce L^*, a^* and b^* coordinates. These coordinates place the color in a specific position in the three-dimensional CIELAB color space.

The two colors (A and B) of our example have been positioned in the color space based on their respective L^*, a^* and b^* values. In this example, we are

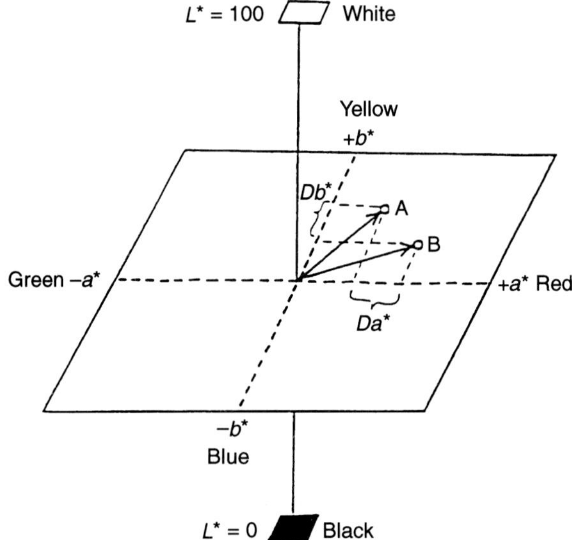

Figure 2.10 The CIELAB color space.

assuming that the L^* value of both A and B is the same. This means that both points are on the same plane. A and B are both between the $+a^*$ (red) and $+b^*$ (yellow) axis which indicates they are some type of yellowish–red (orange) hue. There is an arrow drawn from the center of the diagram to each of the points. This arrow is called a vector which indicates the hue direction and the saturation magnitude of the color. The vector starts at the center which is the gray, determined by the L^* value. The points A and B indicate that a certain amount of color of the respective hue has been added to the gray. The length of the vectors represent the CIELAB term C^* which is saturation.

Da^* is the difference between the two colors in the red/green direction while Db^* is the difference in the yellow/blue direction. Color B is closer to the $+a^*$ (red) than is A, so we can say that B is redder than A. The Da^* is calculated by taking the a^* of the standard and subtracting it from the a^* of the batch. A Da^* which has a positive value will indicate 'redder than' and a negative value 'greener than'. If we compare two reds and one is 'greener than' the other, we really mean to say that it is less saturated or less red. Color B is also closer to $-b^*$ (blue) than A so we would say that B is bluer than A. The difference in b^* values of the two colors make up the Db^* with the sign telling us what direction. In this example Db^* would be negative indicating that B was bluer than A.

The angle formed by the vectors A or B with the $+a^*$ axis is called the hue angle and denoted by the letter h. The angle between the two vectors is the delta hue angle and this angle indicates the hue difference between the two colors. If two colors fall on the same vector, this will indicate that the delta hue angle is zero and the colors will be the same hue. The hue angle concept is not used for most color difference analysis because it is hard to gauge what a hue difference of 2°, for example, means in visual terms. The term DH^* has been adopted to indicate hue difference in CIELAB color difference analysis. DH^* is defined as that part of the total color difference (DE^*) that is due to hue alone. The lightness difference (DL^*) and the saturation difference (DC^*) are subtracted from the total color difference (DE^*) to produce DH^*. This subtraction is a sum of the squares calculation according to the formula in Section 2.6. The sign associated with DH^* indicates which direction the vector is from the other. A positive value indicates it is counter-clockwise and a negative value indicates clockwise using the $+a^*$ axis as a reference. The sign of DH^* is not usually used in most color difference analysis. The numerical value is important in that it indicates the proportion of the color difference due to hue.

The difference in saturation in CIELAB is called DC^*. This is the difference in the length of the respective vectors. A positive DC^* indicates that the batch is more saturated or more vivid than the standard and a negative DC^* that the batch is less saturated or duller than the standard.

The difference in lightness in CIELAB is called DL^*. This is the difference in L^* or gray value of the color. A positive DL^* indicates that the color is lighter and a negative indicates that the color is darker. Appraisals of actual colors with the associated CIELAB color differences will help to understand these color difference terms.

Many users of color difference equations attempted to use only one color difference value, such as DE^*, to set up acceptability tolerances. This procedure will cause problems because only one term (DE^*) will not relate to our visual assessment. There are three aspects to our perception of color which are hue, saturation and lightness. These three do not have an equal effect on our visual perception of the acceptability of color. Often the concept of hue is more critical than either saturation or lightness. If a color has the correct hue, it can be off in saturation or lightness and still be visually acceptable. If only the total color difference DE^* was considered the color might have been rejected. A check of the DH^* of this color would indicate a very small difference in hue which is why the eye found the difference acceptable. This concept is illustrated by Figure 2.11 and Figure 2.12. Figure 2.11 shows two colors A and B which are of the same hue and differ only in saturation. The Da^* and Db^* values are both 2.0. Figure 2.12 shows two colors A and B which have the same saturation and differ only on hue. Notice that these two examples each calculate to the same DE^*. Note also the DH^* of the two colors. The colors in Figure 2.11 could be a good visual match while the colors of Figure 2.12 might not be acceptable. These examples also show the difficulty of using Da^* and Db^* values only. In both examples, the Da^* and Db^* values are 2.0. In Figure 2.11 they were a function of the saturation difference DC^* and in Figure 2.12 they were a function of the hue difference DH^*. Most real examples will fall somewhere between these examples with the hue difference and the saturation difference each becoming a part of the total. Note that in this example the DL^* was assumed to be zero.

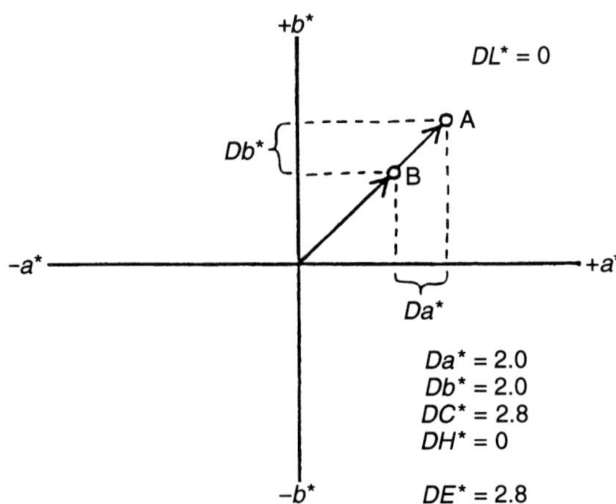

Figure 2.11 Two colors of the same hue but different saturation.

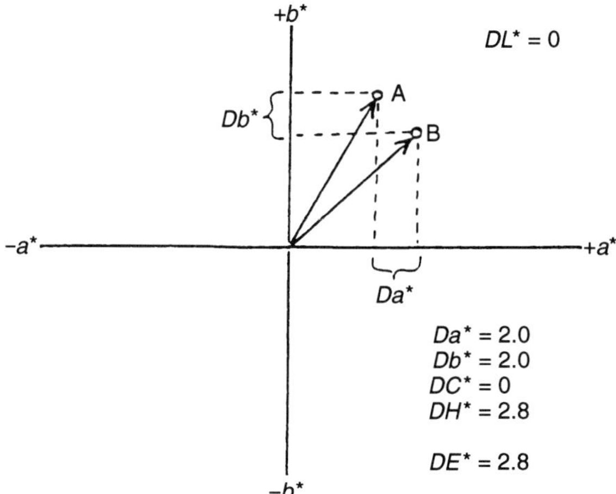

Figure 2.12 Two colors of the same saturation but different hue.

2.4.3 *Choice of color difference components*

We could see from the examples given above that is we use only the total color difference *DE* calculated from the equation

$$DE* = [(DL*)^2 + (Da*)^2 + (Db*)^2]^{1/2}$$

we could often have situations where the single calculated color difference (and tolerance) do not agree with visual assessments. We can find significant improvement by breaking the color difference into three components and having a different tolerance for each component. Some users develop a *DL**, *Da**, and *Db** for each standard color. This certainly can be done, but it takes considerable effort and a lot of real samples need to be generated. Before we decide what is best, let us look at the results of some visual experiments.

In Figure 2.13 (from Kuehni [1]) we see plotted in CIELAB color space a series of ellipses representing different color groups. The outer edge of each ellipse represents the position of colors which were *visually* judged to have equivalent color differences from the standard (dot in the center of the ellipse). These plots indicate that CIELAB space is not completely uniform with regard to color differences. We can see, from Figure 2.13, ellipses representing color equally different from a central color tend to align along a radial line from the center of the *a**, *b** diagram. They are aligned along lines of equal hue.

In Figure 2.14 (from Kuehni [1]) we see three ellipses representing colors equally different from the center color (dot in center of ellipse) and the corresponding tolerance boxes in terms of *Da** and *Db**. Even though the ellipses are of equal size and shape the resulting tolerances in *Da** and *Db** vary considerably.

On the other hand, if we consider the color difference components to be *DL**, *DC** and *DE** as in Figure 2.15 and look at the ellipses from Figure 2.14 plotted

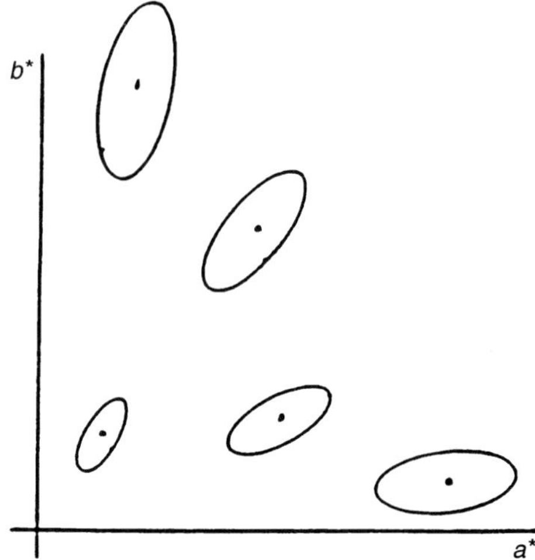

Figure 2.13 Cross section of CIELAB color space with colors on ellipses equally distant from central colors, based on industrial color difference evaluation.

in terms of DC^* and DH^* (Figure 2.16), the resultant tolerance boxes appear to make more sense.

This concept of hue, saturation and lightness should be used whenever we attempt to correlate numerical differences to visual assessments. The CIELAB terms of DH^*, DC^* and DL^* are very useful for this type of comparison. Remember, there is no perfect color space in existence which will correlate exactly to the eye but the use of these CIELAB attributes will usually allow good agreement of visual to computed differences. You will find that the

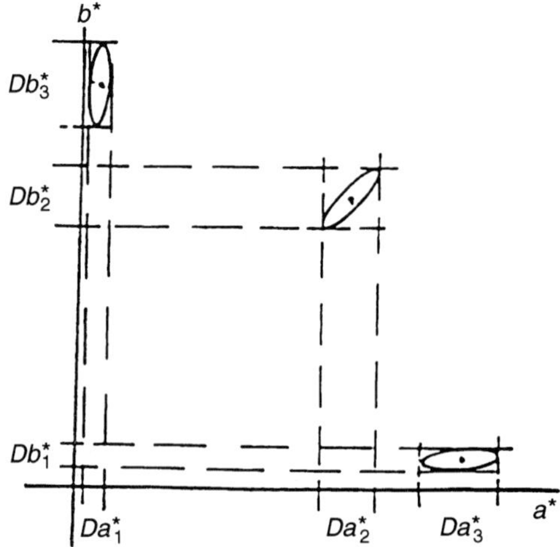

Figure 2.14 Ellipses representing colors equally different from the center color.

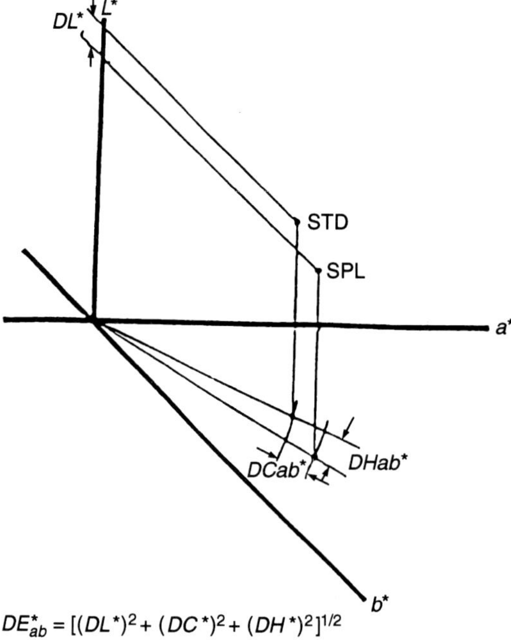

$$DE^*_{ab} = [(DL^*)^2 + (DC^*)^2 + (DH^*)^2]^{1/2}$$

Figure 2.15 Total color difference between standard (STD) and sample (SPL) and its components DL*, DC* and DH*.

parameter DH^* must be kept smaller than the DC^* or DL^*. An industrial operation can measure their colors which were accepted and rejected and determine the DH^*, DC^*, and DL^* of all the colors. A judgement can then be made on the best value to use for tolerance specifications.

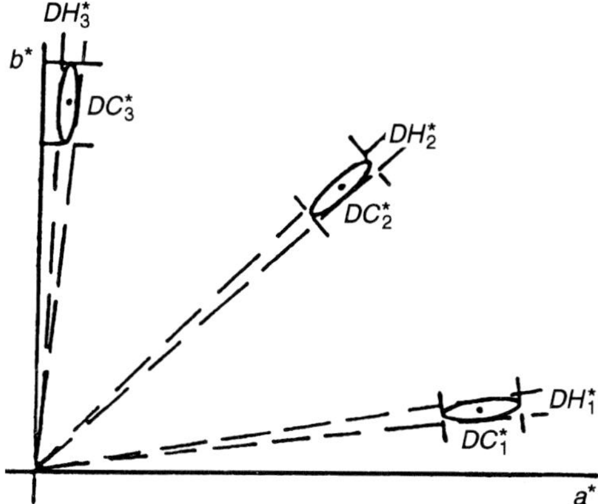

Figure 2.16 As Figure 2.14 but with tolerance boxes in terms of DL*, DC* and DH*.

2.4.4 *The CMC color tolerance formula*

For more than 25 years, British scientists have been involved in a considerable effort to develop a pass/fail color quality control method based on instrumental measurement. The CIE 1976 $L^*a^*b^*$ (CIELAB) system provided a common language or standard 'color map' for communicating color differences. Unfortunately, as we have learned in Section 2.4.3, the CIELAB single *DE* value does not always correlate with visual assessment.

Color scientists like McDonald [3] and his colleagues at J & P Coates Ltd, and members of the Color Measurement Committee (CMC) of The Society of Dyers and Colourists tried to develop a formula where a single value could be used to set up industrial color tolerances between a sample and standard. After a great deal of work and a large number of visual and instrumental assessments, they developed a modification of the CIELAB color difference equations which provides improved correlation between visual assessment and instrumentally measured color differences. In a sense they were able to determine how to pre-weight the values of hue, saturation and lightness for any color. This modification is known as the CMC formula. This modification of the CIELAB formula provides a single DE_{CMC} value. This development makes it possible to use a single number tolerance for judging the acceptability of a color match. The CMC modification of CIELAB provides a tolerance volume. This volume takes the shape of an ellipsoid whose semi-axes relates to the attributes of lightness (*SL*) saturation (*SC*) and hue (*SH*) (Figure 2.17). The volume may be modified based on mutually agreed upon acceptability limits. An ellipsoid whose surface boundary is 1.0 DE_{CMC} units from the standard becomes the unit of measurement for describing the acceptability of a colored batch. It is important to understand that the 1.0 unit CMC will represent a different size and shape ellipsoid for different regions of color space. Some appreciation for the changing size and shape of the tolerance ellipsoids in CIELAB space can be seen in Figure 2.18.

A commercial factor (*CF*) can be applied to the CMC tolerance. The *CF* can be used to modify the volume without changing the ratio of its semi-axes. For example a *CF* of 0.75 DE_{CMC} indicates the semi-axes of the acceptability ellipsoids are $\frac{3}{4}$ of the semi-axes of the normal CMC unit ellipsoid. A complete description of the components of the CMC equation are given in the Appendix.

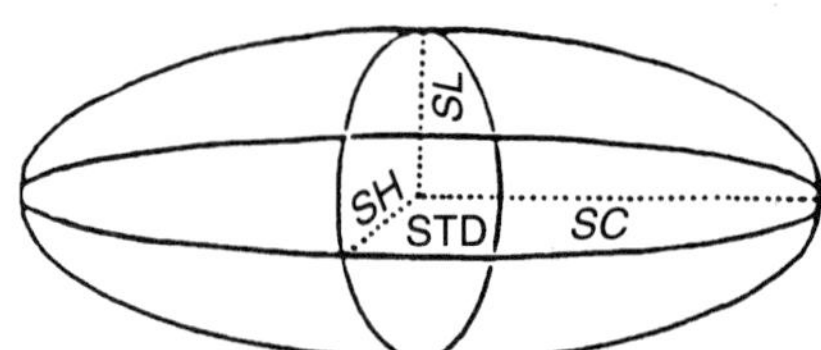

Figure 2.17 Acceptance volume whose surface is 1 CMC unit from the standard.

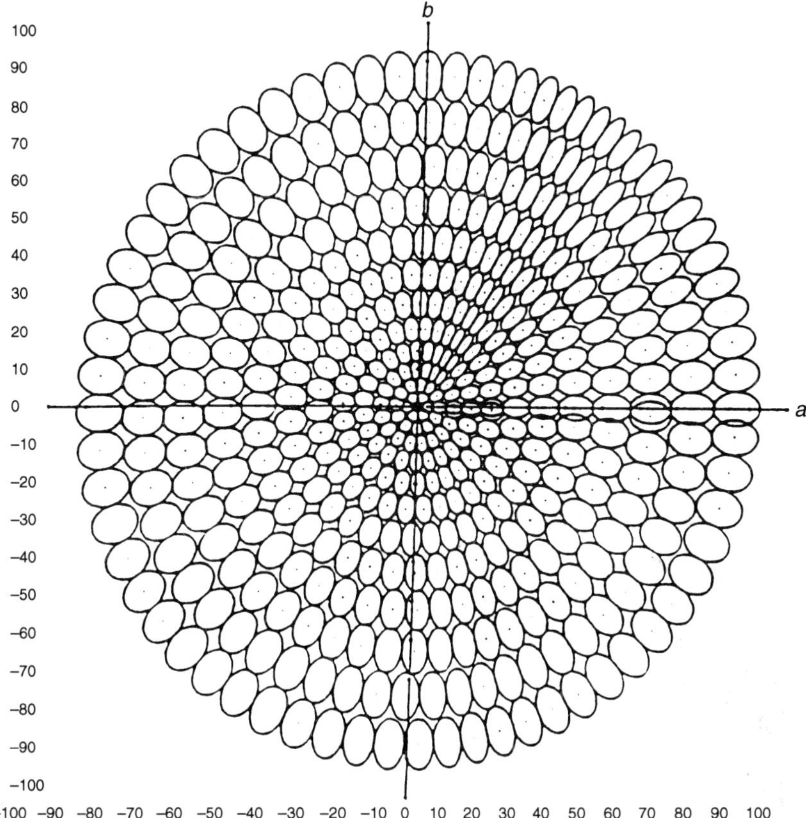

Figure 2.18 CMC two unit ellipsoids in CIELAB space.

In Figure 2.18, the area around *a**30, *b**30 represents the color area where tans or browns are located. We can see that the ellipsoids in this area are more narrow than the ellipsoids in other areas of color space. This means that the hue differences between two samples, in this area, in terms of their positions in CIELAB space, are more objectionable or less acceptable than similar distances in other parts of the CIELAB space.

Perhaps we can understand this better if we examine Figure 2.19. Figure 2.19 shows a blown up area of Figure 2.18. The ellipsoids shown are not actual size, but show the relative shapes of the ellipsoids. The large circle in the center of the figure represents all the colors, or points, that are 30 metric chroma units *C** from the neutral point. The small stars (*) and squares (■) are the positions of a standard and batch in each of the eight areas of the CIELAB color space. The samples plotted are all the same lightness and chroma. Therefore the *DL** and *DC** are zero for each pair of samples. The only difference between the standard and batch, in each case, are in the hue (*DH**). Each pair has a total *DE* of 2.0,

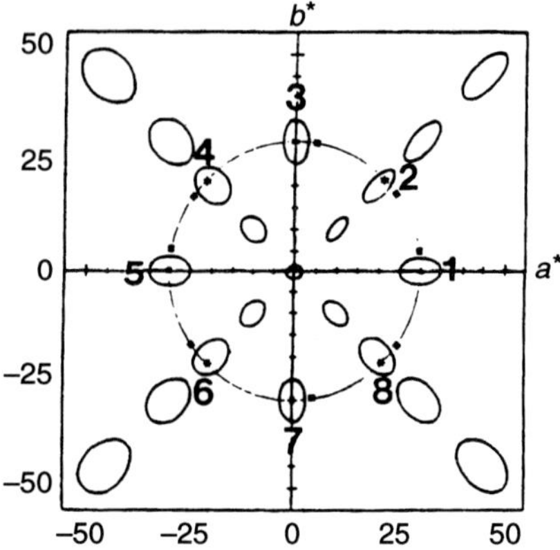

Figure 2.19 An enlarged area of Figure 2.18. For an explanation of this figure, refer to the text and Table 2.4.

and a DH^* of 2.0. It is interesting to look at the DE_{CMC} for each pair. These are shown in Table 2.4, which indicates, for example, that in terms of distances in CIELAB space, a difference in area 2 (tans) is more visually objectionable than the same distance in area 5 (greens).

2.5 Calculations: detailed example

In this section we will go through all the necessary calculations for the determination of a CIELAB color difference and CMC tolerance. The writer realizes that all these calculations are easily handled by a computer connected to a modern spectrophotometer. We go through the calculations step by step so that the reader will have a better understanding (at least once) of how the calculations are made. When we get to the CMC calculations we will take a short cut so that the reader will understand the principle of the procedure without getting bogged down in the mathematics.

Table 2.4 DE_{CMC} values for pairs of samples

Pair no.	*DE* CIELAB	*DE* CMC (2:1)*
1	2.00	1.44
2	2.00	2.35
3	2.00	1.68
4	2.00	1.34
5	2.00	1.31
6	2.00	1.39
7	2.00	1.65
8	2.00	1.51

For all samples $L^* = 65$ and $C^* = 30$
* See Section 2.6 for explanation of (2:1)

2.5.1 *Calculation of tristimulus values X, Y, Z*

Before we can calculate CIELAB values, we must first calculate tristimulus values X, Y, Z. We start out by obtaining reflectance values for a colored sample using a suitable spectrophotometer. The reflectance values are taken at equally spaced wavelengths across the spectrum. For this example, we will use the data given in Table 2.6 for Green C5.1. In all cases we will use 16 wavelengths from 500 to 700 nm at 20 nm intervals. (Other intervals could also be used, e.g. 10 nm, 5 nm or 1 nm.) The values of R are multiplied together with the values of P (Figure 2.2) and $\bar{x}$, $\bar{y}$ and $\bar{z}$ (Figure 2.4) to give products at each wavelength $RP\bar{x}$, $RP\bar{y}$ and $RP\bar{z}$. These values are summed to give the tristimulus values of X, Y, Z. By convention, the value of $Y = 100$ is given to an ideal white reflecting 100% at all wavelengths. Normally, tables are available which contain tristimulus weights which are the product of $P\bar{x}$, $P\bar{y}$ and $P\bar{z}$ at each wavelength. (See Table 2.1.) The P values describe a particular illuminant (D65 or A) and the $\bar{x}, \bar{y}, \bar{z}$ values of a particular observer (10° or 2°). The values in the tables are normalized so that the sum of all the values of $P\bar{y}$ is equal to 100 or 1.0. If the sum is 100 then the reflectance values are used as a decimal from 0 to 1. If the sum is 1.0, then the reflectance values are used as percentages from 0 to 100. The values in Table 2.1 are normalized so that the sum of all the values of $P\bar{y}$ is equal to 100.

Table 2.5 illustrates how the computations are made. Column 1 lists the wavelength intervals from 400 to 700 nm. Column 2 is the reflectance values (as decimal) from the Green C5.1 standard (Figure 2.20A). Column 3 contains the tristimulus weights for $P\bar{x}$ (illuminant D65, 10° observer from Table 2.1).

Table 2.5 Data required to calculate tristimulus values

1 nm	2 %R	3 $P\bar{x}$	4 $RP\bar{x}$	5 $P\bar{y}$	6 $RP\bar{y}$	7 $P\bar{z}$	8 $RP\bar{z}$
400	0.1359	0.251	0.034	0.023	0.003	1.090	0.148
420	0.1531	3.232	0.495	0.330	0.050	15.383	2.355
440	0.1627	6.679	1.087	1.106	0.180	34.376	5.593
460	0.1842	6.096	1.123	2.620	0.483	35.355	6.512
480	0.2283	1.721	0.393	4.938	1.127	15.897	3.629
500	0.2558	0.059	0.015	8.668	2.217	3.997	1.022
520	0.2544	2.184	0.556	13.846	3.522	1.046	0.266
540	0.2385	6.810	1.624	17.355	4.139	0.237	0.056
560	0.2085	12.165	2.536	17.157	3.577	0.002	0
580	0.1698	16.467	2.796	14.148	2.402	−0.002	0
600	0.1319	17.233	2.273	10.105	1.333	0	0
620	0.1130	12.894	1.457	6.020	0.680	0	0
640	0.1077	6.226	0.670	2.587	0.279	0	0
660	0.1087	2.111	0.229	0.827	0.090	0	0
680	0.1179	0.573	0.068	0.222	0.026	0	0
700	0.1245	0.120	0.015	0.047	0.006	0	0

Sum: $X = 15.37$, $Y = 20.12$, $Z = 19.58$

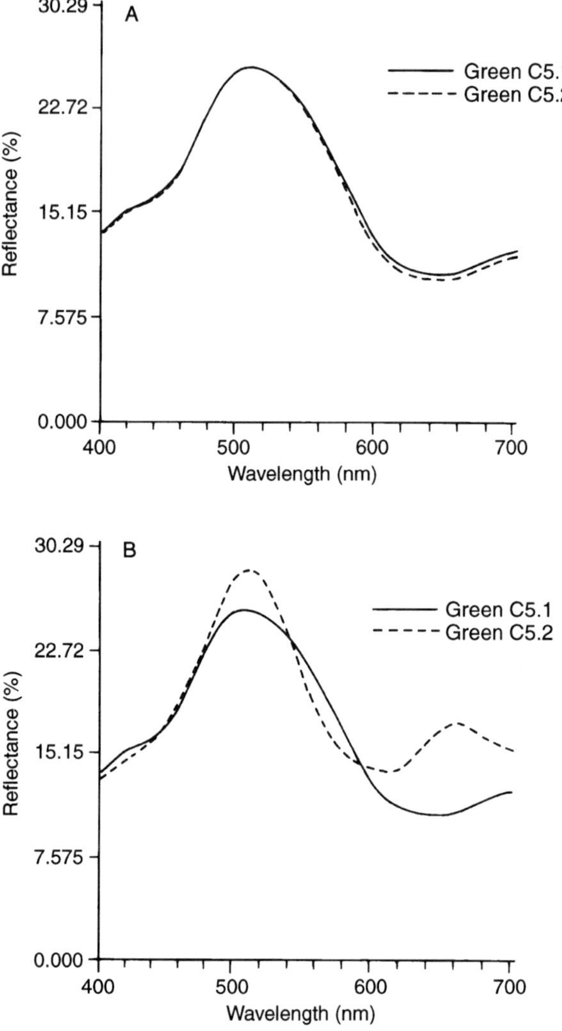

Figure 2.20 The spectrophotometric curves of the green standard (C5.1) (A); and two different green batches (C51 and 5.3) (B).

Column 4 contains the product of Column 2 and Column 3. The sum of all of the values in Column 4 is the *X* tristimulus value. Column 5 contains the weight for $P\bar{y}$. Column 6 contains the product of Column 2 and Column 5. The sum of the values in Column 6 is the *Y* tristimulus value. Column 7 contains the weights for $P\bar{z}$. Column 8 contains the product of Column 2 and Column 7. The sum of the values of Column 8 is the *Z* tristimulus value. Now we have the tristimulus values for the green standard C5.1 calculated for illuminant D65, and the 10° observer:

$$X = 15.37 \qquad Y = 20.12 \qquad Z = 19.58$$

You can see that the calculations were not difficult, but it would be very time consuming to do if we needed the data for a lot of samples. Fortunately, computers associated with the spectrophotometers carry out these calculations very quickly. All the other tristimulus values given in this chapter (Tables 2.6 and 2.7) were calculated by computer. Figure 20A and B shows the spectrophotometric curve of the green standard (C5.1) and two different green batches, C5.2 and C5.3.

Table 2.6 Tristimulus values calculated by computer (part 1)

(a) 20 nm spectral listing

Wavelength (nm)	Std %*R*	Bat (1) %*R*	Bat (2) %*R*
400	13.59	13.50	13.20
420	15.31	15.19	14.67
440	16.27	16.18	16.06
460	18.42	18.40	18.88
480	22.83	22.87	23.23
500	25.58	25.62	27.86
520	25.44	25.43	28.27
540	23.85	23.70	24.12
560	20.85	20.52	18.48
580	16.98	16.54	15.30
600	13.19	12.75	14.16
620	11.30	10.91	14.12
640	10.77	10.40	16.16
660	10.87	10.50	17.58
680	11.79	11.40	16.47
700	12.45	12.08	15.47

Large area/spec incl. d/O
Std: green C5.1
(1): green C5.2
(2): green C5.3

(b) CIELAB difference

	Ill/Obs	*DE**	*DL**	*Da**	*Db**	*DC**	*DH**
Green C5.2	1	0.78	−0.25	−0.65	−0.34	0.58	0.45
	2	0.94	−0.35	−0.70	−0.53	0.73	0.48
	3	0.75	−0.34	−0.50	−0.45	0.41	0.53
Green C5.3	1	1.19	0.55	0.86	0.62	−0.72	−0.78
	2	4.00	0.68	3.90	0.56	−3.93	−0.37
	3	1.09	−0.18	0.99	−0.41	−1.05	0.22

Large area/spec incl. d/O
Std: green C5.1
1: D65/10°
2: A/10°
3: CWF/10°

Table 2.7 Tristimulus values calculated by computer (part 2)

(a) Tristimulus values	Ill/Obs	X	Y	Z	x	y	
Green C5.1	1	15.37	20.12	19.58	2791	3653	
Green C5.2	1	15.08	19.89	19.53	2767	3650	
(b) CIELAB difference	Ill/Obs	L^*	a^*	b^*	C^*	h^*	
Green C5.1	1	51.97	−20.34	3.77	20.69	169.50	
Green C5.2	1	51.71	−20.99	3.43	21.27	170.72	
		DE^*	DL^*	Da^*	Db^*	DC^*	DH^*
	1	0.78	−0.26	−0.65	−0.34	0.58	0.45
(c) CMC (2.00 : 1.00) difference	Ill/Obs	L^*	a^*	b^*	C^*	h^*	
Green C5.1	1	51.97	−20.34	3.77	20.69	169.50	
Green C5.2	1	51.71	−20.99	3.43	21.27	170.72	
		DE	DL^*/SL	Da^*	Db^*	DC^*/SC	DH^*/SH
	1	0.51	−0.12	−0.65	−0.34	0.35	0.36

Large area/spec incl. d/O
1: D65/10°

2.5.2 *Calculation of L*, a*, b*, C**

(See Section 2.6 for complete definition of symbols.)

2.5.2.1 *Green standard C5.1.* $X = 15.37$, $Y = 20.12$, $Z = 19.58$. (From Table 2.5.) X_n, Y_n, Z_n are the tristimulus values for the standard illuminant (see Table 2.1)

$$L^* = 116\,(Y/Y_n)^{1/3} - 16$$

$$= 116(20.12/100)^{1/3} - 16$$

$$= 116(0.2012)^{1/3} - 16$$

$$= 116(0.5860) - 16$$

$$= 67.976 - 16$$

$$= 51.98$$

$$
\begin{aligned}
a^* &= 500[(X/X_n)^{1/3} - (Y/Y_n)^{1/3}] \\
&= 500[(15.37/94.825)^{1/3} - (0.2012)^{1/3}] \\
&= 500[(0.1621)^{1/3} - (0.2012)^{1/3}] \\
&= 500[0.5453 - 0.5860] \\
&= 500[-0.0407] \\
&= -20.35 \\
b^* &= 200[(Y/Y_n)^{1/3} - (Z/Z_n)^{1/3}] \\
&= 200[(0.5860) - (19.58/107.381)^{1/3}] \\
&= 200[(0.5860) - (0.1823)^{1/3}] \\
&= 200[(0.5860) - (0.1823)^{1/3}] \\
&= 200[(0.5860) - (0.5670)] \\
&= 3.80 \\
C^* &= (a^{*2} + b^{*2})^{1/2} \\
&= [(-20.35)^2 + (3.80)^2]^{1/2} \\
&= (414.12 + 14.44)^{1/2} \\
&= 20.70
\end{aligned}
$$

2.5.2.2 *Comparison of standard and batch*

	X	Y	Z	L^*	a^*	b^*	C^*
Standard (C5.1)	15.27	20.12	19.58	51.98	−20.35	3.80	20.70
Batch (C5.2) (from Table 2.7)	15.08	19.89	19.53	51.72	−20.99	3.43	21.27
Delta (D) Batch – Std.				−0.26	−0.64	−0.37	0.57

$$
\begin{aligned}
DE &= [(DL^*)^2 + (Da^*)^2 + (Db^*)^2]^{1/2} \\
&= [(-0.26)^2 + (-0.64)^2 + (-0.37)^2]^{1/2} \\
&= (0.6141)^{1/2} \\
&= 0.78 \\
DH &= [(DE)^2 - (DL^*)^2 - (DC^*)^2]^{1/2} \\
&= [(0.78)^2 - (-0.26)^2 - (0.57)^2]^{1/2} \\
&= (0.2159)^{1/2} \\
&= 0.46
\end{aligned}
$$

Note: All of the values computed in this exercise compare closely to those values generated by a computer (Table 2.6). There are only some small round off differences.

2.5.2.3 *Calculation of the color difference (between C5.1 and C5.2) by the CMC modification of CIELAB*

$$DE_{\mathrm{CMC}} = \left[\left(\frac{DL^*}{l(sL)}\right)^2 + \left(\frac{DC^*}{c(sc)}\right)^2 + \left(\frac{DH^*}{sH}\right)^2\right]^{1/2}$$

where $l = 2$ and $c = 1$

$$DE_{\mathrm{CMC}} = \left[\left(\frac{-0.26}{2(1.11)}\right)^2 + \left(\frac{0.57}{1.68}\right)^2 + \left(\frac{0.46}{1.25}\right)^2\right]^{1/2}$$

$$DE_{\mathrm{CMC}} = [(0.014) + (0.115) + (0.135)]^{1/2}$$

$$DE_{\mathrm{CMC}} = 0.51$$

Note: The method for determining *SL*, *Sc* and *SH* are explained in Section 2.6.

The computer outputs for the *X Y Z*, CIELAB and CMC calculations for these two samples are given in Table 2.7.

2.5.3 *Metamerism*

Sometimes it is useful to plot the delta a^*b^* values in a Da^*, Db^* diagram. The standard plots in the center of the diagram. Each batch is then plotted relative to the standard. In Figure 2.21, we have two green batches C5.2 and C5.3 plotted relative to our standard green C5.1.

The two batches are plotted for daylight (D-65) and tungsten light (A). The data for green C5.2 and C5.3 are given in Table 2.6. From the plots we can see

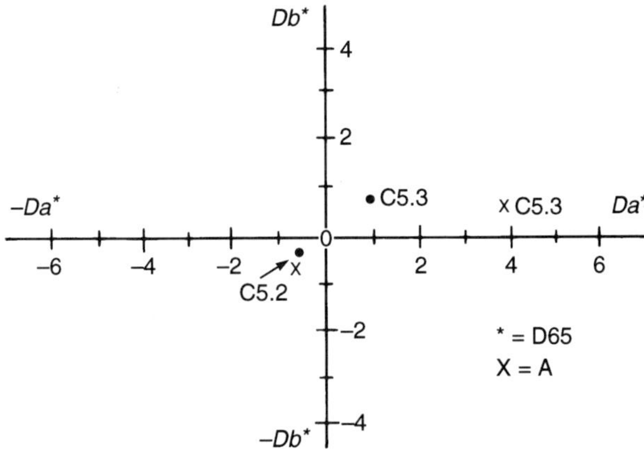

Figure 2.21 Two green batches, C5.2 and C5.3, plotted relative to the standard green, C5.1.

that both are green C5.2 and green C5.3 are within 1 Da^* and 1 Db^* value in daylight. Green C5.2 is also within 1 Da^* and 1 Db^* value in tungsten light, but green C5.3 is considerably further from the standard when its tungsten values are compared to the tungsten values of green C5.1. Green C5.3 is a metameric match to the standard. It can be seen from the curves of C5.1 and C5.3 (Figure 2.20B) that the standard was made with a phthalocyanine green pigment as a major component, while green C5.3 contained phthalocyanine blue. Whenever we use color difference calculations as a quality control tool, serious limitations exist if significant metamerism is present. If an original sample from a customer cannot be matched with essentially the same colorants, then the original standard should not be used as the quality control standard. A match which was judged to be acceptable by the customer should then become the quality control standard. Each successive batch should then be compared to the standard which is made with the same colorants that are used in the batches.

It is hoped that you will closely examine the data for C5.2 and C5.3. Try to calculate the tristimulus values, the CIELAB values and the various color difference values. This exercise will help the reader get a 'feel' for the whole concept of color difference calculations.

2.6 CIELAB color difference equations

$$L^* = 116(Y/Y_n)^{1/3} - 16$$

$$a^* = 500[(X/X_n)^{1/3} - (Y/Y_n)^{1/3}]$$

$$b^* = 200[(Y/Y_n)^{1/3} - (Z/Z_n)^{1/3}]$$

X, Y, Z are the tristimulus values of the colored object. X_n, Y_n, Z_n are usually the tristimulus values of the sandard illuminant.

The above equations for L^*, a^* and b^* apply if the ratios X/X_n, Y/Y_n and Z/Z_n are larger than 0.008856. If they are equal to or smaller than this value, the following equations are to be used:

$$L^* = 116f(Y/Y_n) - 16$$

$$a^* = 500[f(X/X_n) - f(Y/Y_n)]$$

$$b^* = 200[f(Y/Y_n) - f(Z/Z_n)]$$

where:

$$f(X/X_n) = 7.787(X/X_n) + 16/116$$

$$f(Y/Y_n) = 7.787(Y/Y_n) + 16/116$$

$$f(Z/Z_n) = 7.787(Z/Z_n) + 16/116$$

$h = \arctan(b^*/a^*)$, angle formed by sample vector to reference line

$$C^* = (a^{*2} + b^{*2})^{1/2}$$

$$DL^* = L^*_{\text{sam}} - L^*_{\text{std}}$$

$$Da^* = a^*_{\text{sam}} - a^*_{\text{std}}$$

$$Db^* = b^*_{\text{sam}} - b^*_{\text{std}}$$

$$DC^* = C^*_{\text{sam}} - C^*_{\text{std}}$$

$$DH^* = [(DE^*)^2 - (DL^*)^2 - (DC^*)^2]^{1/2}$$

(DH^* is always at right angles to DC^*)

$$DE^* = [(\text{DL}^*)^2 + (Da^*)^2 + (Db^*)^2]^{1/2}$$

$$Dh^* = h_{\text{sam}} + h_{\text{std}}$$

2.6.1 *CIE L*a*b**

DL^* – Lightness/darkness difference. The shade of gray (white/black) difference.
+ = Lighter than
− = Darker than

Da^* – Red/green color difference.
+ = color is redder than (or less green than)
− = color is greener than (or less red than)

Db^* – Yellow/blue color difference.
+ = color is yellower than (or less blue than)
− = color is bluer than (or less yellow than)

DC^* – Difference in chroma or saturation of a color.
+ = more saturated than (more color of that hue)
− = less saturated than (less color of that hue)

DH^* – Difference due to hue only.

DE^* – Total color difference.
(includes hue, chroma and lightness)

2.6.2 *CMC*

The calculation of a color tolerance in CMC (2 : 1) units uses the following equation:

where $l = 2$ and $c = 1$:

$$DE_{\text{CMC}} = \left[\left(\frac{DL^*}{lSl}\right)^2 + \left(\frac{DC^*_{ab^*}}{cSc}\right)^2 + \left(\frac{DH^*_{ab^*}}{SH}\right)^2\right]^{1/2}$$

where for the standard:

$$\text{for } L^* > 16, \qquad SL = \frac{0.040975L^*}{1 + 0.01765L^*}$$

$$\text{for } L^* \leq 16, \qquad SL = 0.511$$

$$SC = \frac{0.0638C_{ab^*}}{1 + 0.0131C_{ab^*}} + 0.638$$

$$SH = (FT + 1 - F)SC$$

where
$$F = \left[\frac{(C_{ab^*})^4}{(C_{ab^*})^4 + 1900}\right]^{1/2}$$

and
$$T = 0.36 + \text{abs}[0.4 \cos(35 + h_{ab})]$$

unless h is between 164° and 345° then:

$$T = 0.56 + \text{abs}[0.2 \cos(168 + h_{ab})]$$

and for the last two equations 'abs' indicates the absolute, i.e. positive value, of the term inside the square brackets.

Note: When $l = 2.0$ this equation fixes the ratio of the three terms ($SL : SC : SH$) to correlate with visual assessment to typical textile samples. Other values of l may be required in cases where the surface characteristics dramatically differ. Alternate values of l may sometimes be required when measuring very dark samples, but the user should assume an l value of 2.0 until actual results indicate a need for adjustment. The value of c is always left at 1.0, and may be excluded from the formula altogether. It is left here to maintain agreement with the British Standard.

2.6.3 *Concept of unit volume/tolerance*

The equation of $DE_{CMC} \leq 1.0$ describes an ellipsoidal volume, with axes in the direction of lightness, chroma and hue centered about a standard. Ellipsoid semi-axes lengths of lSL, cSc and SH calculated for a given standard describe a unit volume of acceptance within which all samples are less than 1.0 DE_{CMC} unit from the standard.

This volume, and the size and ratio if its semi-axes, become the basis for the establishment of an appropriately sized volume of acceptability for a given commercial situation by the application of a commercial factor (cf) to all dimensions.

The volume of acceptance is defined by $DE_{CMC} \leq cf$.

Note: The size and orientation of the diagram shown in Figure 2.17 varies considerably depending on the location of standard in color space and is used here as a means of conceptualizing the semi-axes.

References

1. Kuehni, R.G. (1985) Detroit Color Council Technical Bulletin No. 1, *CIELAB color differences and lightness, hue and chroma components for objective color control.*
   (Figures 2.9, 2.13, 2.14, 2.15 and 2.16 were taken from this paper.)
2. Billmeyer, F.W., Jr. and Saltzman, M. (1981) *Principles of Color Technology*, 2nd Edition.
   (Figures 2.2, 2.3, 2.4, 2.5 and 2.7 were copied from this book.)
3. McDonald, R. (1988) Acceptability and perceptibility decisions using the CMC color difference formula. *Textile Chemist and Colorist* June issue.
   (Figure 18 was taken from this paper.)
4. CMC Calculation of Small Color Differences for Acceptability. AATCC Test Method 173-1989. *Textile Chemist and Colorist* November 1989.
   (The CMC equations came from this publication.)
5. The set of green paint chips was provided by Collaborative Testing Services, Inc.

3 Flexographic printing presses

D. TUTTLE

3.1 Introduction

When the original name of the printing process we now know as flexography was changed from aniline printing, many printers believed the name was adopted primarily because the process was so widely used for printing flexible packaging materials. It is, of course, true that flexography is the major printing process used in the flexible packaging industry, but at the time several thousand members of the printing and converting industries chose the new name for the flexographic printing process, a number of reasons accounted for their choice, in addition to the fact that the new name was deemed appropriate for the name of the most widely used printing process in the flexible packaging printing and converting industries.

Among the reasons commonly given by industry members for their choice of 'flexography' as the most fitting name for this printing process were the following:

1. The process used flexible rubber or elastomeric printing plates
2. It was admittedly the most widely used process for printing flexible materials of all kinds, including papers, films, foils, wallcovering, floor covering and other decorative materials, as well as flexible packaging and wrapping materials
3. The printing process was itself inherently 'flexible', being adaptable to the printing of an unlimited variety of non-flexible, rigid materials, including sheets of wood, plastics, glass and metals
4. As many printers and converters had learned prior to 1952, this 'flexible' process was readily adaptable to operations other than printing, including coating, sizing, laminating and related functional and decorative operations

Anyone familiar with the extreme versatility of the flexographic press and process, knows that flexo presses may be found 'trademarking', 'tinting', 'stripping', 'ruling' and 'water-marking', as though these jobs were just what flexo presses were originally designed for. Thus it is little wonder that many of those who voted in favor of the new name 'flexography' in 1952 did so with the versatility and flexibility of the process in mind.

3.1.1 *Versatility of flexo presses*

Many years ago, when the flexible packaging industry was still struggling to break out of its shell, those converters who had flexo presses became familiar with many of the jobs other than printing which they could do on these machines, partly because at that time much of today's specialized equipment was not yet available.

Today, a rather surprising number of flexographers never think of their flexo presses as anything but printing machines, and they are therefore often not getting all the use, versatility and profits which their flexo presses are potentially capable of providing. It is to those younger members of the flexo fraternity, therefore, to whom these thoughts are primarily aimed, in the hope they may appreciate what else may have been included, free of charge, when they bought their flexo presses, and in the hope they may be stimulated to think about how they can take greater advantage of their flexo presses.

3.1.2 *The flexo press as an excellent coating machine*

For example, any flexo press is an excellent coating machine, without any alterations or modifications whatever. For 'pattern coating', where a pattern application of varnish or overlacquer is desired, the varnish or lacquer is printed in the desired pattern from whatever type of conventional flexo printing plate will best withstand the solvents involved. Where a continuous solid coating is desired, whether it is to cover the full width of the web, leave a 'dry' edge, or consist of one or more broken bands of varnish or lacquer running with the web, a coating roller of the desired width is used in place of the 'plate cylinder' in the printing unit or station where the lacquering is to be done. No other changes are necessary, and for this kind of work the coating roller need not even have the same circumference or 'repeat length' as the printing cylinders, so that a single coating roller may, for example, be used for many different jobs, so long as the width is suitable.

Solid, band or pattern varnishing or lacquering is frequently used to enhance gloss, rub-proofness, abrasion resistance, to improve machinability and/or to provide protection for the contents of a package from adverse environmental factors.

Where functional, protective, or heat-sealing varnishes, lacquers or coatings must be applied to the side opposite from that which is printed and which must be applied in the same pass through the press, it is necessary that either the press be designed to permit one or more printing stations to be reversed or that the web itself must be reversed between the printing and coating or lacquering operations, as with turning bars. Such an arrangement, however, not only makes for an economical operation by permitting printing and/or varnishing, lacquering or coating on both sides of a web in a single pass through the press, but it provides a simple means for maintaining register between operations on opposite sides of the web.

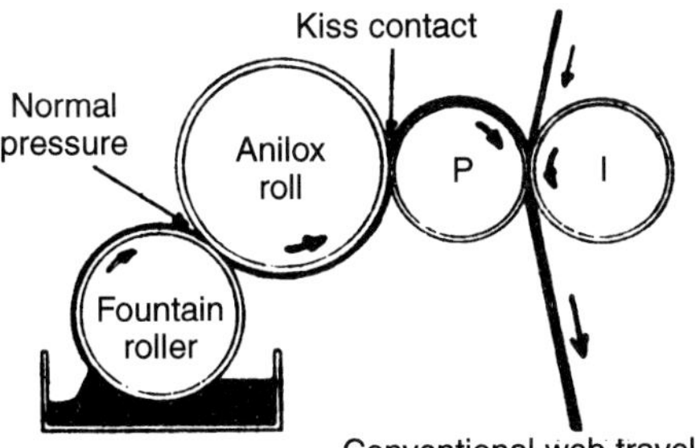

Figure 3.1 Conventional web travel.

There are times when it is advantageous to be able to apply a clear or colored coating to a web on a flexo press without having to buy a special coating roller for each width of stock to be run, especially where short runs are involved. Depending on the requirements of the job, there are several ways in which this sort of thing can be done. Some methods involve mechanical modifications to a conventional flexo press, and we will discuss those later. One method quite commonly used for certain types of work does not involve any change other than the way the press is webbed, and it provides the additional bonus of coating all the way to the edge of the stock, on both edges.

3.1.3 *A change in webbing*

In this method, the web, instead of passing between the plate cylinder and the impression cylinder and on to the next station, as in Figure 3.1, is passed between the plate cylinder and impression cylinder, then wrapped halfway around the plate cylinder, halfway around the anilox roll, and thence on to the next station, as in Figure 3.2, or, in the case of the last printing station on the press, the web is passed directly into the dryer. The fountain roller and anilox roll are adjusted with approximately normal pressure, depending on job requirements, but no pressure is maintained between either the anilox roll and plate cylinder or between the plate cylinder and impression cylinder (Figure 3.1). In

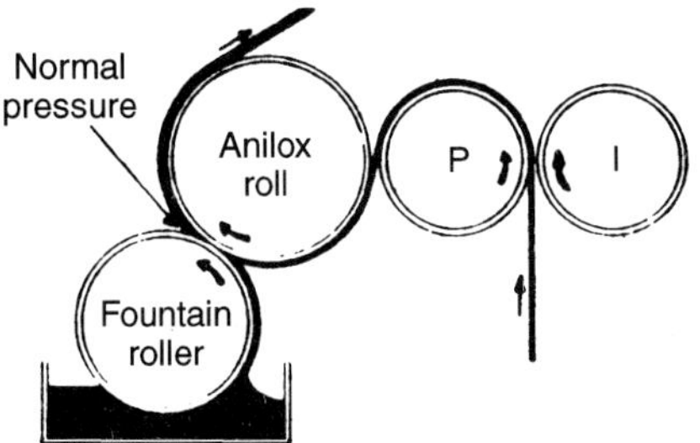

Figure 3.2 Webbing for nip coating operation.

fact, some operators use a bare plate cylinder when running this way, with a resulting gap of at least $\frac{1}{8}''$ between the plate cylinder and both the impression cylinder and form roller. Other operators prefer to omit the drive gear from the plate cylinder, letting it rotate freely with the web shaft on presses with independent drive systems for the inking units.

None of the arrangements mentioned here involve any modifications or changes to commonly used production flexo presses. They are typical of the sort of things that are being done in production on many existing flexo presses and which can be readily duplicated by any flexographer.

3.1.4 *Expanding the usefulness of a flexo press*

We will now consider some other ways in which the usefulness of a flexo press can be expanded and broadened, including some operations which involve modifications and additions to conventional standard flexo presses.

We have mentioned an unconventional 'webbing' which permits various coating, varnishing or lacquering operations in any range of web widths up to the maximum width the press will accommodate, without the necessity for any special coating roller, let alone a separate one for each web width. A conventional printing operation may be desirable for many coating, lacquering or varnishing operations. However, for reasons such as better control of thinner coating films, higher solids coatings, smoother coating surfaces or more efficient use of material, and where the volume or the frequency of such work justifies the modification, any width web the press will handle can be coated with a clear or colored material.

This can be accomplished by using a conventional coating roller in the maximum width the press will accommodate and by using an impression bar in place of the usual impression cylinder in the printing station involved. The bar in such an installation is essentially a piece of steel drill rod about $\frac{1}{4}''$ in diameter, held securely in a sturdy clamp (see Figure 3.3). This drill rod can be quickly, easily and cheaply cut to a length a trifle less than the web width to be run so that, in operation, the coating roller surface does not contact any surface beyond the edge of the web to be coated, at the point of impression (see insert, Figure 3.3). Obviously, this arrangement can be used for printing continuous design patterns for trademarking, decorating, etc., as well as for overall solid or band coating.

(a) *Converting a coating station.* Occasionally a flexographer is confronted with a situation which completely stymies a conventional flexo operation. An example is where a coating must be applied from solution in an aromatic hydrocarbon solvent such as toluol. There are a few such materials which not only contain 'impossible' flexo solvents but which must be applied at elevated temperatures. These materials can be applied on a gravure press where neither the engraved metal cylinder nor the metal doctor blade, which are in contact

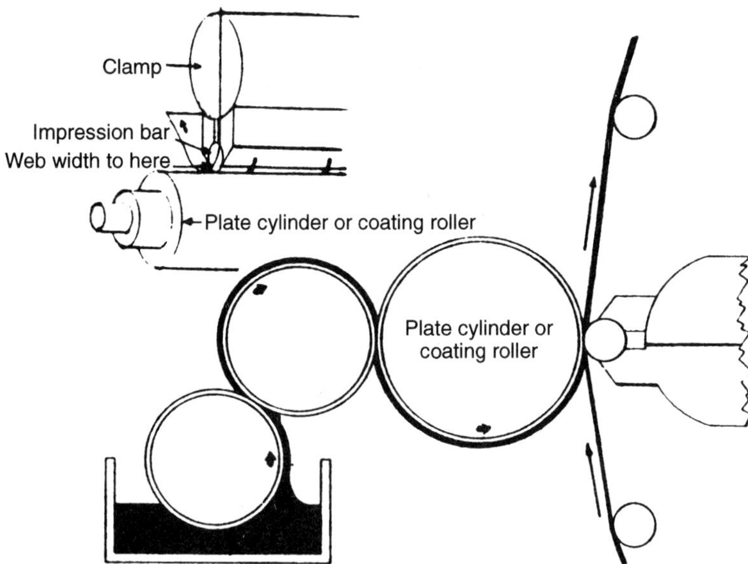

Figure 3.3 One modification of a flexo press to expand its usefulness.

with the solution, are affected by either the solvent or the elevated temperature. So, the resourceful flexographer who is faced with this problem may convert one of his printing units to a sort of gravure coating station, as shown in Figure 3.4.

In this example, we have assumed that the anilox roll in the press was interchangeable with the fountain roller and that the anilox roll was engraved with a screen which applied the desired amount of material, so that the only addition required was a doctor blade and holder plus the necessary heating unit for the fountain. If the anilox roll was not interchangeable with the fountain roller, or if the screen did not correspond to that required for the desired

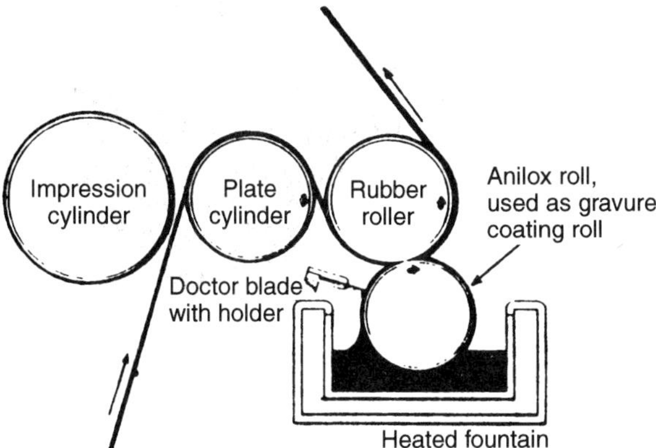

Figure 3.4 Conversion of a flexo press to a gravure coating unit.

application, a new engraved roll would be necessary for such an operation. It is, of course, also true that the width of the rubber roller being used as the impression roller in Figure 3.4 must correspond to the width of the stock being coated (actually a trifle less), to avoid contacting the rubber compound at the ends of the roller with the harmful solvent in the coating carried by the engraved roller. Such a relatively simple modification is not equivalent to a sophisticated rotogravure unit, but it has served adequately for specific uses similar to those outlined.

(b) *Offset-gravure arrangement.* The gravure modification to which the standard flexo roller arrangement lends itself admirably is the printing or coating method known as gravure-offset, or offset-gravure. In this modification, the gravure coating or design roller is put in the position of the anilox roll. A doctor blade in a suitable holder is mounted so as to wipe the engraved design roller, and a rubber-covered transfer or offset roller, identical in circumference to the engraved design roller, is mounted in the plate cylinder position (as shown in Figure 3.5). The most common kinds of work where this sort of approach is of value are those involving intricate designs with fine tonal gradations on coarse, irregular surfaces not suitable for either conventional flexo or gravure processes.

(c) *Applying heavy coatings using a flexo press.* When using a flexo press for coating operations, one frequent problem is that it is difficult to apply heavy coatings as desired, and when adequate coating weights are achieved, these tend to exhibit irregular patterns such as ‘crow’s feet’, striations and so forth. If the coating is being applied to provide barrier properties, it may be very important that the application be as smooth as possible. A modification

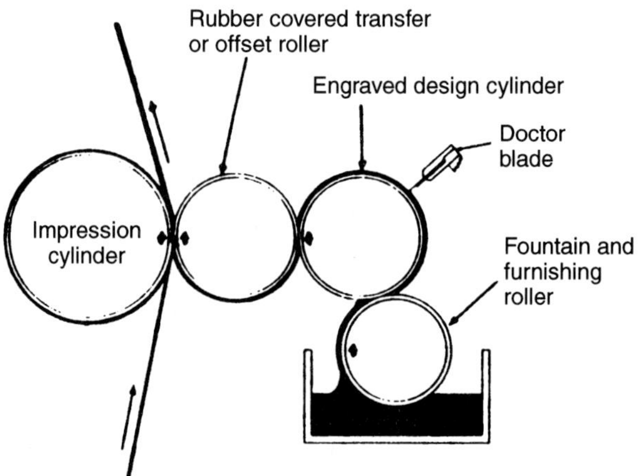

Figure 3.5 Conversion of a flexo press to a gravure-offset unit.

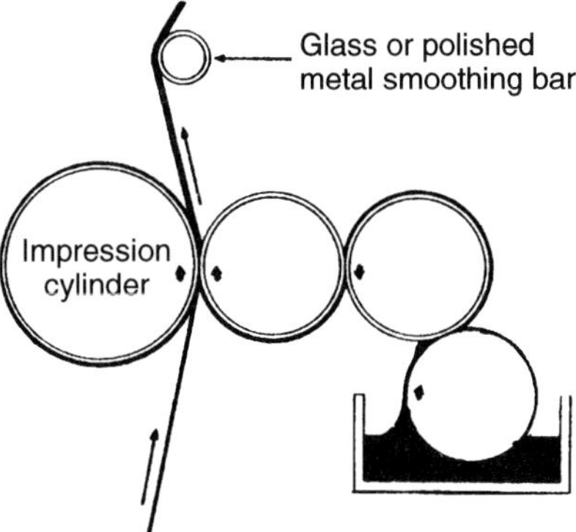

Figure 3.6 A modification of a flexo press for applying heavy coatings.

which is frequently used to smooth heavy coatings following their application is shown in Figure 3.6.

This modification consists simply of a smoothing bar or rod mounted after the application point, in such a way as to lightly wipe the surface sufficiently to smooth or level the coating. The amount of pressure used between the rod and the web and the distance between the rod and the application point may vary for maximum effectiveness, depending on the drying rate and the application weight, so the bar should be mounted so as to provide a range of adjustments on both counts.

These modifications we have discussed do not of course, include all the changes that have been effected in adapting flexo presses to kinds of work other than the printing jobs for which they were originally and primarily designed. But the things mentioned are typical of the ways in which ingenious flexo printers have expanded the range and scope of the flexo presses to better serve their customers.

Perhaps the next time an impossible job comes your way, these examples may cause you to take a second look at your flexo press, and see it with added respect for its hidden versatility.

3.2 Evolution of flexo-type litho

In Section 3.1, we discussed the evolution and growing application of flexo-type ink distribution systems in letterpress printing markets. Since there is considerable similarity between traditional letterpress ink distribution systems and the ink distribution systems used for lithographic inks and presses, the question has frequently been asked whether flexo-type letterpress ink distribution systems would not work just as well on lithographic presses. In this section, we will discuss this question.

First, there are significant differences between the letterpress and lithographic printing systems, in that letterpress printing is a typographic method in which the printing is done from the surface of raised type. Lithographic printing is a planographic printing method in which the printing is done from a smooth surface. Lithography is based on the fact that oil and water will not mix. The printing areas of the lithographic printing plate are treated to accept oil and repel water while the non-printing areas of litho printing plates are treated to accept water and repel oil.

The lithographic ink distribution system therefore consists of not only an ink distribution system similar to that of the paste-bodied letterpress ink distribution system to distribute the oil-base, paste-bodied lithographic inks and apply a uniformly thin film of these litho inks to the printing area of the lithographic printing plates, but also a 'dampening system' to apply a continuous thin film of water to the non-printing surface of the same lithographic plate.

Because the film of oil ink and the film of water are applied to the same flat surface of the lithographic printing plate, these two films, i.e. of oil-base ink and of water, are in essentially constant contact with each other, under conditions of considerable agitation. Thus, there is a tendency for the oil and water films to emulsify, forming a colloidal suspension of water from the dampening solution in the oil-base litho ink film. The proportion of water in this emulsion or colloidal suspension, and its behavior versus the oil-based litho ink in the composition, are factors which have a great deal to do with the successful performance of a lithographic printing system.

Similarly, the surface energy of the ink rollers in a lithographic ink distribution system are an important factor in determining the composition and behavior of the ink/water emulsion which evolves and stabilizes during the operation of a lithographic press. This, in turn, helps to explain why some of the early experiments with chrome-plated anilox rolls in flexo-type ink distribution systems for litho presses met with variable and limited success.

3.2.1 *Flexo-type ink systems*

Following the early trials of flexo-type ink distribution systems for lithographic presses, the original chrome-plated anilox rolls were replaced with anilox rolls surfaced with materials other than chrome, and the best litho versions of such anilox rolls have given results equal in printing quality to the best conventional litho press ink distribution systems. In addition, the flexo-type ink distribution systems proved significantly more energy-efficient, more labor-efficient and resulting in less makeready time and less waste.

The flexo-type ink distribution systems for lithographic presses reduce the total number of ink distribution rollers from a dozen or more, in the case of the traditional litho system, to as few as three. The entire flexo-type litho ink distribution system consists of the anilox metering roll, a reverse-angle doctor blade, and two rubber-covered form rollers. Conventional oil-based (water immiscible) lithographic inks are used, along with a conventional dampening system and fountain solution. Gone are the ink fountain adjusting keys, doctor roller, vibrator rolls, oscillating rollers and the rest of the ink roller train, plus most of the time and trouble of adjusting all these components of the traditional lithographic ink distribution system.

As one lithographer remarked on the occasion of his first introduction to one of the new flexo-type litho ink distribution systems;

> 'a lithographer used to have to adjust the whole ink distribution train, from the fountain keys to the form rollers when setting up a job, and then adjust the dampening system, alternately working from one to the other to complete the makeready and get the job running. With the flexo-type inking system and anilox rolls, the ink distribution system is virtually automatic and all the pressman has to do is to adjust the dampening system and he is off and running!'

(a) *Reduction in the number of rollers.* With only about 20% as many rollers as in the old, traditional litho inking system, significantly less power is required for the flexo-type anilox inking system, and there are obviously a lot fewer rolls to maintain and replace as well as to adjust. Since all the rolls in the flexo-type inking system rotate in the same direction, this makes for a smoother, quieter operation. The flexo-type, anilox roll system minimizes or eliminates ink misting in the pressroom and reduces both the weight of the press and the space required for a new press of this design, in addition to which it improves accessibility.

To date, the only lithographic presses we know of that are equipped with flexo-type, anilox ink distribution systems are those equipped with retrofit units, but the results being obtained with these units, and the enthusiasm of the pressmen as well as the users and suppliers of the retrofit units all indicate that we can expect to see new offset lithographic presses with similar designs of simplified ink distribution systems being marketed and used with increasing frequency.

(b) *Reduction in waste.* One of the most obvious reasons for the growth of such simplified ink distribution systems for offset litho presses in that this development gets directly to the heart of one of the major problems of the lithographic process, which is waste. Part of the relatively high waste factor associated with offset lithography is related to press makeready, downtime and setup time connected with adjustments of the relatively antiquated litho ink distribution system; and part of the high litho waste factor refers to the printed paper stock which is wasted during the same operations.

Thus, by simultaneously reducing all these excess waste elements, the adoption of the automatically controlled, accurately metered, simplified flexo-type ink distribution system for lithographic presses promises to quickly and significantly improve the economic status of the lithographic printing process and industry more, in a few short years, than any other development since that of the sheet metal lithographic printing plate, with the compliments of flexography!

(c) *The possibility of faster drying inks.* An additional advantage of the new, simplified ink distribution system(s) for lithographic presses, and a plus which does not appear to have been generally recognized to date, is the fact that the combination of the greatly reduced area of the total surface of the ink distribution rollers and the predetermined ink film metering effect of the anilox roll surface make it possible to formulate and print much faster drying litho inks than ever before. Many offset-litho pressmen may well question the truth of this statement, but most litho ink chemists and formulators will appreciate the new vistas in litho ink formulation and performance which will now be opened to them.

Because of the multiplicity of ink distribution rollers in a traditional lithographic ink distribution 'train', and because of the ink circulation pattern followed by the ink, from the time it first leaves the ink 'fountain' until that 'portion' of ink is completely consumed, after numerous circuits of the complete train of ink distribution rollers, an average of about 4 h time has elapsed. This means that litho ink chemists and formulators have long been accustomed to observing the 'rule of thumb' that the average litho ink must not dry, in a film of printing thickness, on a metal, glass or plastic slab, in less than 4 h.

The flexo-type (anilox roll plus two form rollers) litho ink distribution system reduces the total surface area of the ink distribution rollers to an average exposure area equal to perhaps one quarter that of the traditional system. This, in itself, and assuming the same ink film thickness and circulation pattern as on the old familiar litho ink distribution system, would indicate that the drying rate of an ink film or printing thickness on the flexotype system could logically be reduced to about 1 h. In fact, however, we gain even more than that because of the metering characteristics and behavior of the anilox roll.

The anilox roll surface, in the case of the flexo-type system, is in contact with the ink form-roll surfaces continually. Therefore, ink which is not transferred to non-printing areas of the litho plate in any revolution of the form rollers tends to be transferred to the surface of the anilox roll and is redistributed evenly to form roller surfaces with each revolution. This may sound confusing, but the net result is that litho inks to be run exclusively on presses with the new flexo-type ink distribution systems can now be formulated with a drying rate at least ten times as fast as the old established 4-h rule-of-thumb standard, with satisfactory printing quality results and much faster drying of the printed copy.

The changes which lithographers will experience in the near future, due to the evolution of and conversion to new flexo-type anilox ink distribution systems, will be both interesting and exciting to every phase of this industry.

3.2.2 *Advances in flexography and related technology*

Flexography's early history, which dates back to the early 1900s, was related to the use of coal-tar dyestuffs dissolved in alcohol, to print roll-to-roll wrapping papers, paper bags, and so on. By the early 1930s, these paper bags and wrapping materials evolved into transparent film for bags and other packaging materials, including cellophane. These various transparent plastic films and packaging materials were gradually modified, making them successful in a wide range of printed film combinations. Temperature, humidity, pressure and an infinite range of variables in the materials to be packaged are mong the other factors also affecting the finished packages and their contents.

Flexography's total growing printing process, including inks, presses and equipment has become increasingly complicated. As the growth of plastic films increased, the use of flexo inks also increased. Anilox ink rolls were developed in the period between 1938 and 1939, and 100% pigment inks became common during World War II. Higher press operating speeds and faster ink drying rates developed rapidly by the end of World War II, and 'water-reducible' flexo inks grew rapidly for use first with paper stocks and then gradually with plastics, films, foils and increasingly complex structures.

As flexo presses, inks and equipment increased in complexity and numbers, operating personnel per flexo press steadily increased. Today's higher press speeds, longer running times and higher quality printing standards have resulted in more cooling rollers per press. Some flexo printers experimented with various types and grades of paper printing stocks to test press speeds of 500 fpm and more in early flexo press trials.

In metal foil printing, many jobs were run at speeds from 100 fpm to 200 fpm or more; and in cellophane printing, early flexo printing speeds of 500 fpm were not uncommon. In some categories, including boxboard and paperboard, such materials became important to many growing segments of the paper and paperboard industries in recent years. An example is the growing corrugated container printing and converting industry in which flexographic printing is steadily increasing in importance. There is a growing number of in-line flexo presses today, which is attributable to their use in folding box and corrugated container operations.

Average operating speeds for flexo printing in tandem on non-absorbing paper stocks during the initial strong growth period for flexography ranged from 300 fpm upward to about 600 fpm. Most flexo printers felt that printing was not the limiting factor on press speeds when running other operations in tandem, for the following reasons:

1. Flexo press drying equipment at that time was being greatly improved
2. With respect to printing on paper, particularly with bag machines, the widespread adoption of water-reducible flexo inks resulted in faster drying (by absorption), cleaner printing, and less tack (with consequently less tendency for ink film to 'pick' off on formers, draw-rollers, etc.)
3. A substantial number of coupled operations involve reciprocating cutters or creasers. These had an initial maximum speed of about 300 fpm. Naturally, this resulted in a speed-limiting factor which was not present in earlier years

The net result of these combined developments was that, although the printing operation was no longer a prime factor in limiting the speed of coupled operations, the converting operation had moved into its place as a prime 'bottle-neck'. Thus, tandem operating speed increased after 1945, although it did not show continuing sharp increases over prior experience.

An early industry survey asked questions about flexo printing of all types. Of all those flexo printers reporting, 11% said they currently had presses that combined flexo and gravure printing. At the time of the previous survey, this figure would have been closer to 2%. Three possible causes of this development are cited:

1. An increasing demand for tone work, shading, and so on had arisen
2. Gravure cylinders and other equipment were more readily available than in the past
3. Better liaison and greater cooperation between gravure cylinder suppliers and flexo printing plate and design-roll manufacturers had been developed

At the time of the earlier flexo survey, only a handful of printers would have said that they encouraged the use of halftones or benday screens for shading. The later survey reported that some 46% do just that. Some 34% of the flexo printers surveyed reported that they produced commercial color-process jobs. At the earlier date there were no such jobs reported in actual production, although there was extensive experimentation.

Both of these comparisons point to the marked progress in terms of quality flexo work and to significant improvements in presses, plates, inks and technology during the period of time between the two flexo surveys.

One survey question read, 'Do you proof plates for registration of equipment other than the printing press?' Some 31% of reporting printers said, 'Yes, on all jobs'. In the prior survey such proofing was rarely done except for occasional jobs in one or two plants. There are two probable reasons for this development: (a) mounting and proofing equipment had become more readily available since the prior survey; and (b) pre-makeready equipment and procedures had grown in importance, due to higher press speeds and keener competition.

As the question and answer period ended at the Packaging Institute's 1954 Flexo Survey Report to the Rochester, N.Y., Club of Printing House Craftsmen, a member of the Rochester group of pressmen remained after the rest of the audience had left. He introduced himself, as a long-time employee of the Rochester Folding Box Company, to the head speaker from the Packaging Institute group.

He gave his name as Frank Spies, inventor of the 'Quad-Lok Box', the patent of which was owned by Rochester Folding Box. Spies explained that the Kiekheifer Corrugated Container Company had acquired Rochester Folding Box and its rights to the 'Quad-Lok Box' patents.

Spies expressed his personal and corporate interest in the growth and progress of the flexo industry as well as that of the packaging industry—especially in the area of corrugated containers. He expressed particular interest in the new flexo pigment inks with emphasis on water-reducible materials for corrugated paperboard containers, such as his own 'Quad-Lok Box'.

He said that he felt sure Kiekheifer would have a strong interest in furthering the growth and the improvement of flexo printing on the 'Quad-Lok Box'. The speaker assured Spies that the widely used water-reducible flexo inks described would indeed perform as explained in the presentation and he would, in fact, be happy to prove this to Spies' satisfaction.

Spies told the speaker that if he would send him a sample of the new water-reducible flexo ink, together with instructions for its use, and if the ink worked as promised, his company would open the plant premises to him from that time on. The speaker did just that, and within a week or so, Spies telephoned him, to confirm that the initial sample had performed as promised, and invited him to the plant as his company's guest during subsequent trials.

After several visits to the plant in Rochester, he was informed by Spies that the plant executives had decided to change their original plans for making a number of complete manufactured and printed 'Quad-Lok' test units at each plant location. The new plan called for Rochester Folding Box Company, Keikheifer and the Samuel Langston Company, working together, to construct separate production facilities for the Quad-Lok box, with associated flexo printing, at each plant location.

Before this plan could be implemented, however, Frank Spies passed away. Samuel Langston II was then called upon to carry out the project but was soon after killed in an airplane crash. The project was never carried through, and it was not until the early 1960s that flexography became a formidable process in the corrugated industry.

3.3 Testing incoming inks

Every flexographer is aware of the fact that according to the official definition provided by the Flexographic Technical Association flexography is a typographic or 'letterpress' printing process. We are also aware that flexography uses fluid, fast-drying inks while the traditional letterpress printing process uses viscous, paste-bodied, slow-drying inks. Because of these very important differences between flexo inks and letterpress inks, the letterpress pressman does not normally measure or alter the physical properties of his letterpress inks, using them as they are received from his ink supplier.

The flexo pressman, on the other hand, knows that his fast-drying flexo ink usually, if not always, requires an initial press-side reduction with an addition of solvent, such as alcohol or water, to arrive at the optimum running viscosity and to maintain this ideal ink viscosity through any given press run.

In other words, the letterpress pressman is primarily and typically concerned with ink fountain adjustments and control while the flexo pressman is primarily and typically concerned with ink composition and control.

It is important for a flexographer to understand the basic physical and chemical properties which define and constitute a quality flexo ink for the press he is using, the substrate on which he is printing and the end results which are required. When we refer to the basic ink properties with which a flexographer should be familiar, we do not mean to imply that the pressman himself must personally evaluate ink properties for specific jobs, but in any flexographic printing operation it is desirable to have a qualified technical person in the company who can make at least a half dozen simple comparison tests of incoming ink quality to ensure uniformity of performance and printing results from run to run and throughout any particular run.

3.3.1 *Six simple tests*

For this purpose we suggest the following six simple tests with which every flexo printer should be familiar, in order to test and check any batch or shipment of ink he/she receives from a supplier. Such tests often prevent costly mistakes in printed results or loss of press time.

In making and in interpreting these six tests it is important to remember that in flexo inks, particularly alcohol-reducible formulations, the solvent component is very volatile and may be lost to a greater or lesser extent while handling the ink and conducting the test. Solvent loss may also take place in stored sample containers if these are not properly closed each time a standard is used for comparison. Handling these standards carefully will minimize loss of solvent. It is also wise to renew each standard with a fresh supply, in a new container, at frequent intervals.

(a) *Ink color and strength (or hue and print density).* A drawdown will show both the color and strength of the sample, compared with the retained standard. This drawdown should be made on a hard-finished fine-grain bleached paper stock. After carefully stirring the sample and the standard, pull down a spot of each ink side by side with a wide-blade drawdown knife that is pulled across the two spots with relatively light pressure for a distance of about an inch. Heavy pressure is then applied for the remainder of the drawdown.

The first part of the drawdown is probably the most important. This simulates the solid print that will be obtained on the press. The light portion of the drawdown gives the undertone, and brings out any finer shade differences which may exist.

A heavy viscosity ink will tend to give a heavier film on the drawdown, and this often gives the impression of greater strength and depth of color. For this reason, great care and judgement are necessary in making this test. If the sample is at a higher viscosity than the standard, it may appear to be darker and stronger than the standard. If, on the other hand, the standard has lost solvent and increased in viscosity, the sample may appear light and therefore weak in color. Since flexo inks lose solvent quickly, due to both evaporation and penetration, the two spots of ink to be compared should be placed on the drawdown paper and drawn down as quickly as possible so that there will be a minimum of evaporation and change in viscosity, with resulting apparent changes in strength, gloss and penetration.

In determining color strength, some flexo printers try to use a 'bleaching' technique, such as is often used for offset and letterpress inks. This bleach test involves mixing equal amounts, by weight, of the standard and the sample ink with equal amounts of a similar opaque white ink formulation and then making a comparison drawdown of the two 'bleached' or 'extended' mixtures. This procedure, however, is difficult, time-consuming, and of doubtful value because of the loss in solvent and change in weight during the weighing-out operations of the two samples.

(b) *Drying time.* For alcohol-reducible inks, determination of drying time is usually done by making a heavy rollout of the standard and sample side by side on a non-absorptive surface. The drying speed is then determined by simultaneously tapping the two films with the index and adjacent finger tips until no tackiness is apparent. Another comparative drying-time test can be made by dipping a fingertip into the batch of ink and another fingertip into the standard sample, tapping or rubbing out the inks side by side on a clean, non-absorptive surface, and timing the relative speed at which a tack-free state is achieved.

Water-reducible flexo inks dry more by penetration than by evaporation, and the simplest method for testing the drying rate of water-borne flexo inks is to make a drawdown of the batch versus the standard on the stock to be printed. Here again, tapping the two ink films side by side with two fingers will determine whether the drying rate of the two samples are equal, and if not, which dries first.

(c) *Solids content.* Weigh by difference, into a flat-bottom metal dish, as rapidly as possible to avoid loss of solvent, a 5 to 10 g sample of the batch of ink to be tested. Friction-type can covers are suitable for use as dishes. Carefully evaporate the solution to dryness, avoiding loss of material by spattering, and heat the residual material to constant weight in an oven at 100° to 105°F. After cooling, determine the weight and calculate the solids content.

(d) *Fineness of grind.* This test is particularly important in the case of flexo inks to be used in printing critical quality color process work on high-grade non-porous substrates or where a smooth, glossy, printed ink surface is essential. The most accurate fineness of grind determination can be made by using a fineness of grind gauge, such as the Heyman gauge, the NPIRI grindometer, etc. A rough estimate of the fineness of grind of a flexo ink can be made by smearing a sample of the ink on the surface of a smooth glass plate with a spatula or a glass rod.

(e) *Weight per gallon.* Standard weight-per-gallon cups are available from laboratory equipment supply companies. Directions for use of any particular cup are funished by the supplier. In most cases, the cup is weighed before and after filling and the weight in grams of ink contained therein is divided by 10 to get the weight per gallon in pounds. The weight per gallon in pounds multiplied by 0.12 is the specific gravity.

(f) *Viscosity.* A viscosity test is one of the most important tests a flexographer can make. The most accurate measurements of viscosity that can be conveniently made require the use of rotational type laboratory instruments, but the simplest and most commonly used method uses an efflux cup. The efflux cups most familiar to the flexographer are the Zahn, Shell and Ford cups. Viscosity determination by these cups is made by measuring, with a stop watch, the number of seconds required for a cupful of ink to empty through a standard orifice in the bottom of the cup. These cups come in various orifice sizes, designated by numbers.

In recording viscosity, care should be taken to designate the type of cup used and the temperature of the determination. For example, a reading might be given as: 22 s, No. 2 Zahn cup, at 75°F. An increase in temperature causes a decrease in viscosity, but the degree of change varies from ink to ink. Therefore, it is not possible to apply an arbitrary correction factor which could be used to give the viscosity at a temperature different from that at which a reported measurement was made.

It is very common for flexographers to record the viscosity at which each flexo ink used on each job was previously run, and it is a normal production practice for each pressman to also record the supplier and identifying ink number used for each job. Very often these factors constitute the limit of the information provided for the guidance of the pressman who next sets up and runs the repeat run of the same job.

With increasingly common requirements for additional flexo job information, due to more frequent color-process print jobs, more complex substrate compositions and more demanding end-use package requirements, more of the kinds of information provided by tests such as those listed can be expected to become a part of more job production tickets in flexo printing plants.

3.4 Viscosity as a measurement to make production flow

Viscosity of a flexo ink is an extremely important characteristic which deserves careful consideration and constant attention on the part of every flexographer. This is a fact that every flexographer knows, but which, at one time or another, probably every flexo pressman has forgotten.

Viscosity is the property of an ink which determines how it will flow and transfer from the ink fountain to the surface of the substrate being printed. For practical purposes, ink viscosity is often referred to as ink consistency. It is obvious that the understanding, measurement and control of viscosity are necessary elements in the production of high quality flexo printing.

In the operation of a flexo press, ink distribution involves a number of factors. For instance, the volume and density of the ink film furnished to the printing plate and transferred to the surface of the substrate being printed will vary according to variables including:

1. The count and configuration of the engraved surface of the anilox roll and the extent to which it has been worn
2. The hardness of the rubber-covered fountain roller, in a two-roll ink distribution system, and the pressure setting between this fountain roller and the anilox roll
3. In the case of a reverse-angle doctor blade system, the contact angle and pressure of the blade assembly and adjustment
4. The speed at which the press is run
5. The viscosity of the ink

In the case of this example, we can assume that the first three factors are known and will remain constant throughout the run, and if the speed at which the press is running is also maintained at a constant figure, any significant variation in the ink film thickness or density will be due to variations in the viscosity of the ink.

If an ink is too viscous, the ink film thickness of the resulting print will be too heavy and the color will be off standard. A heavy ink film, in addition to causing an accumulation of ink on the edges of the plates, will also cause dirty printing and filling in of the type. Conversely, if the ink is too thin, the finished print, having a washed-out appearance, will be equally unacceptable. If throughout the run the viscosity is allowed to vary between extremes of too thick and too thin, the resulting prints will lack uniformity, will continually vary from standard and be seen as intermittently too light and too dark.

In establishing a criterion for determining the optimum ink viscosity for any particular job, it is common to assume that the optimum viscosity is that which provides the minimum ink film thickness while retaining adequate color density.

3.4.1 *Yield value of a liquid*

We mentioned, at the outset, that ink viscosity is often referred to as ink consistency. This is adequate and well understood among pressmen but it is not sufficiently accurate to be acceptable among physicists. The physicist tells us that a true liquid, such as water or honey, will flow when subjected to any force, no matter how small. On the other hand, certain plastic liquids, such as mayonnaise, possess an inner structure which prevents them from flowing until the force to which the 'plastic liquid' is subjected exceeds the minimum value, known as the yield value, required to break up the inner structure.

A 'kitchen experiment' which illustrates this phenomenon is interesting and easy to perform. Two similar funnels should be filled, one with honey and one with mayonnaise. The honey, yielding to the force of gravity, will flow down through the funnel. The mayonnaise, however, since it possesses an inner structure, will not flow for the obvious reason that the force of gravity is less than the yield value of mayonnaise.

Now, then, if a tight-fitting, inverted rubber funnel and tube is attached to each of the funnels containing honey and mayonnaise, and adequate pressure is applied, the mayonnaise will flow down its funnel very readily. The same pressure applied to the funnel containing honey, however, will cause only a relatively slight increase in velocity. The physicist explains this by saying that the honey (although having no yield value), has a higher viscosity than the mayonnaise. After seeing this experiment, it is easy to understand the more technical definition of viscosity as the resistance to flow which a fluid system exhibits upon application of a displacing force; or, in more mathematical terms, viscosity is the measurement of that resistance.

Where yield value is determined to be a property of any fluid system, it is usually due to some element in the structure of the system stemming from the interlocking relationship of the vehicle with some dispersed solid. The inner structure may differ widely from fluid to fluid. Substances such as mayonnaise and most lithographic inks have high yield values, while other fluids, such as many flexo inks, have yield values so low that their presence can often be detected only by a viscometer.

Some fluid systems have an inner structure which may be broken down by agitation, making the fluid flow more freely. This quality is technically known as thixotrophy. Like yield value, thixotrophy is a significant flow property if ink, more significant with letterpress and lithographic inks, but not at all unusual with flexographic and rotagravure inks.

3.4.2 *Confused by viscosity?*

Obviously, confusion has often been caused by the use of the word viscosity to include all the flow properties of an ink. We, too, were guilty of this in our first paragraph of this discussion, which should technically have been changed to read, 'The many flow properties of a flexo ink, i.e. viscosity, yield value and thixotrophy, can be defined as those physical properties which enable it to be transferred from the ink fountain to the surface of the substrate being printed.'

The importance of continuous and accurate measurement and control of flexo ink viscosity was recognized early in the history of flexographic printing, and, in fact, the main feature of the first annual technical forum held by the Flexographic Technical Association in 1959 was devoted to automatic viscosity control systems for flexography. At that time there were five different automatic viscosity control systems available for flexo printers, and manufacturers of each of these five systems were given equal time to describe and demonstrate equipment.

3.4.3 *Measuring the value of controls*

In these early stages of automatic viscosity control systems for flexo, the equipment was considered a relatively expensive investment of questionable value by many flexographers. One understandable reason for this opinion at that time was that flexography was then in the early stages of quality color process printing, and flexo press operating speeds rarely exceeded 300 feet per minute on any substrates except paper.

As flexo-printed films, laminated structures and flexible packaging structures became increasingly popular, however, the value of automatic viscosity controls became increasingly obvious. A popular demonstration of the dollar value of automatic flexo ink viscosity controls made by the developers of such equipment was often impressive. They would ask a doubting flexo printer to divide a long run into two equal parts. One half of this run would have the viscosity of the major color (usually white) manually controlled by the pressman. The ink and solvent used would be accurately weighed at the beginning and end of the first half of the run.

At the beginning and end of the second half of the run, identical records were kept on ink and solvent consumption, the only difference being that the viscosity for the second half of the run was automatically controlled, instead of depending on the pressman's judgement for the solvent additions necessary to maintain the initial ink viscosity and printed density uniformly throughout that portion of the run. The results were almost invariably a revelation to both the pressmen and management; in almost all instances the ink consumption of the manually controlled portion of the run was approximately twice that for the automatically controlled portion.

3.4.4 *The obvious conclusions*

An obvious conclusion was that since the ink consumption for the manually controlled portion of the run was twice that for the automatically controlled portion, the press operating speed was lower and the waste higher for the manually controlled part. Thus, it became apparent that the cost of the automatic viscosity control equipment invariably more than paid for itself, and became an important factor in steadily increasing flexo press operating speeds as well as improving flexo quality.

The initial phase of the industry's adoption of automatic viscosity controls for flexography, therefore, was largely dependent on economic factors, but as flexography in the ensuing years steadily improved in print quality, the industry has increasingly appreciated the fact that in order to consistently and uniformly produce high quality full-color process printing, the use of automatic ink viscosity controls has become virtually an absolute necessity.

Today, there are a variety of automatic viscosity controls available to meet the needs of every flexographer, from narrow-web label printers to flexible package printers, and corrugated and publication printing industries. Today's flexographer can choose the equipment to suit his needs, based on cost and performance, from a wide range of proven designs. Automatic ink viscosity controls are now one of the flexo industry's familiar accessories.

3.5 A unit that enables variable control of ink density and thickness

There has been a growing trend among letterpress and lithographic printers to convert their ink distribution systems to more simplified 'flexo-type' anilox roll systems. Recently a further and related development has occurred, which will be described here.

As every flexographer and an increasing number of letterpress and lithographic printers know, one of the greatest virtues of the anilox ink distribution roll is its ability to accurately meter a uniform film of ink to the surface of the printing plate, directly as in a flexo press, or indirectly, via form rollers, as in the case of retrofitted letterpress or litho presses. This is, of course, in many if not most cases, an obvious advantage. However, it is also a limiting factor when it is desired to vary the thickness or density of the ink film applied to the printing plate or substrate surface. In such a case, the property of an anilox roll to consistently and uniformly meter exactly the ink film thickness determined by the machine-engraved surface of that particular anilox roll becomes a limiting factor because the ink film can normally be varied only by changing the anilox roll to one with a different count and/or configuration.

Changing an anilox roll to vary its application volume may require from 10 min to as much as several hours of press downtime, depending on press width, so that where such changes are required frequently, the positive but limited ink volume metering characteristics of any one particular anilox roll may become a disadvantage.

3.5.1 *Various density control*

With these factors in mind, an increasing number of letterpress and lithographic printers interested in anilox of flexo-type retrofit projects are keenly interested in reports of a new development which promises to provide, on anilox ink distribution systems, continuously variable ink film density over a range of as much as ±50% during a press run, without changing the anilox roll and without altering press speed.

This really represents dramatic progress in the advances of the 'anilox' letterpress and litho ink distribution systems, in less than 5 years since the date of the first commercial installation of such a retrofit unit on a production letterpress newspaper press in June, 1978. When that installation was made, the production personnel were impressed by the fact that the second impression printed was salable, and by the fact that there were not even any fountain keys for the pressmen to adjust. In fact, at the end of the first day's run, the pressmen remarked that if all the units were converted to the same systems, the pressmen would have much less to do!

As additional installations of the flexo-type 'anilox' letterpress ink distribution systems were made in more newspaper plants, however, it was found that some newspapers preferred ink densities which were either a little higher or a little lower than that provided by the standard ANPA/RI anilox roll engraving. Such requirements can be met by modified anilox roll engravings, but all anilox roll surfaces wear with time, so that the printed ink density is subject to a corresponding gradual variation. When the anilox roll surface wear results in the maximum allowable printed ink density variation, the anilox roll is then replaced with a new or reconditioned one.

Obviously, if the new 'variable density' control development meets all the claims of its inventor, and in fact really makes it possible for the pressman to accurately vary the density of the printed ink film as much as plus or minus even 25% by simply rotating an adjusting wheel or screw, this indeed represents a significant advance. But in the 'preview' demonstration of the 'variable density' unit installed on a new flexo-type letterpress ink distribution system which we witnessed, the total density variation was indeed 100% from minimum to maximum densitometer readings and the readings at four different density settings remained constant until changed by the pressman.

3.5.2 *Two rollers and doctor blade*

The ink distribution system on which this demonstration was made consisted of an engraved, chrome-plated anilox roll, a reverse-angle Swedish blue steel doctor blade and two rubber-covered form rollers, as shown in Figure 3.7. The

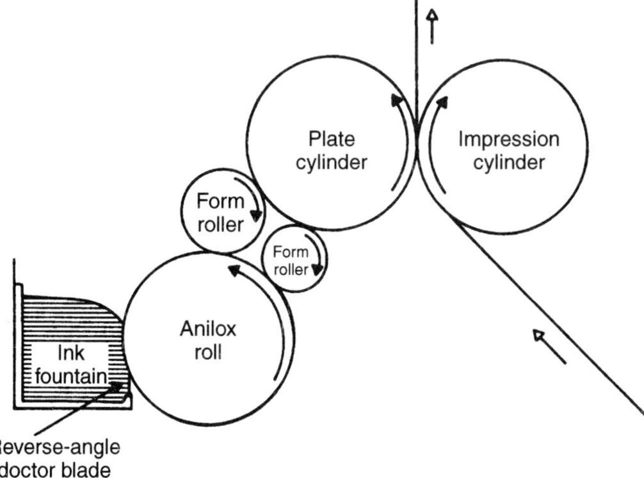

Figure 3.7 Flexo-type letterpress ink distribution system with anilox roll, reverse-angle doctor blade and two form rollers, similar to unit on which new variable-density control was demonstrated.

ink used for the test was identified as a conventional, un-modified letterpress newspaper printing ink.

We asked whether this same variable-density control is adaptable to currently available flexo inks and presses, and were told that both litho and flexo adaptations are included in the pending patent claims covering this variable-density concept and design, but neither a litho nor a flexo press with the variable-density installation is yet available for demonstration. However, an ink distribution system and variable density control very similar to the one we saw has been applied to a different type of letterpress unit using heavier viscosity inks than news inks, with results similar to those we witnessed.

During the preview of this operating, variable-density control on a flexo retrofit newspaper unit, we were able to see only the small exposed rotatable control wheel of the unit, plus portions of the basic ink distribution system and generous samples of newspaper pages printed at different ink densities, on which we were able to take densitometer readings. The only ink used throughout the demonstration was commercial letterpress newspaper, printing black ink. The extreme density readings ranged from about 0.75 to 1.15, or the density normally covered by anilox roll engravings from 360 TPB to 200 Q, but most of the samples were run at four density readings of 0.85, 0.95, 1.05 and 1.10.

The density-control unit itself on the demonstration press was totally enclosed, but it is expected that one or more production letterpress and/or litho presses will soon be in operation, when more details of this exciting development's design, construction, operation and cost will become available.

In the meantime, interested segments of the letterpress, litho and flexo printing industries will be anxiously awaiting that day.

3.6 Drying of inks

An ink which consists of binders or pigments dispersed or dissolved in a volatile fluid does not serve its purpose until the fluid is removed from the ink so that only the non-volatile portion is left on the substrate to which the ink had been applied. This process by which the volatile part of the ink is removed is called 'drying'. Unfortunately, drying is not as simple a matter as we would like it to be.

Drying of a fluid ink involves three different kinds of technology: the technology of flow, the technology of heat transfer and the technology of ink formulation. The first two technologies are equipment problems. The last technology is, of course, a problem for the ink manufacturer.

In flexography, the drying of ink is generally accomplished by using heated air as the drying medium. This process is further complicated by the need to direct and control both the flow of air and its temperature.

For the inkmaker, the problem of ink drying represents a compromise between drying ability; that is, speed of drying and flow-out of ink. If the inkmaker uses a fluid dispersant of too great a volatility (a 'fast' solvent for quick drying), there is the danger of ink drying on the printing plate or in the cells of the anilox cylinder. Conversely, if the inkmaker uses solvents of low volatility (for 'slow' drying), there is the danger of smearing of one color by another, of offset of wet ink on the imprinted side of the web, and of residual odor and blocking.

Since the ink manufacturer has customers whose equipment may be either very old, very new, or somewhere in between, or with very efficient dryers or virtually no dryers at all, it is difficult for him to standardize on formulations. What is good for one customer whose press has a very efficient drying system may be no good at all for the customer whose press uses a simple air drying system.

Fortunately, the general trend in recent years has been one of considerable improvement in drying techniques and equipment. During the past several years, flexographic presses have increasingly been equipped with between-color dryers to aid in ink trapping. Several companies have developed high-velocity heated air dryers which can be fitted to existing presses. The more advanced flexographic presses include tunnel dryers, using high velocity air streams with high capacity heaters as standard equipment. Many of these developments have come about from a more thorough engineering analysis of the problems of drying by using heated air.

3.6.1 *Concept of boundary layer*

Current theories of the nature of liquid or gas flow over a surface involve the concept of a boundary layer of fluid (which can be either gas or liquid) adjacent to the surface over which the fluid is flowing. The velocity of the fluid stream is considerably less than the velocity in the main stream. This 'boundary layer' acts as a sort of porous barrier to the transfer of heat from the fluid to the surface and to the transfer of mass, in this case, solvent vapor from the surface to the heating fluid. The effect of forced flow is three-fold:

1. It improves transfer of heat from the heated air stream to the web being dried
2. It removes residual solvent from the surface of the boundary layer, thus permitting further diffusion of solvent from the web through the boundary layer
3. It reduces the thickness of the boundary layer

However, as with most natural phenomena, the relationship between air velocity and boundary layer thickness is not a linear one; i.e. if you double the air velocity you do not halve the boundary layer thickness. Figure 3.8 shows, in a semi-quantitative way, what happens to boundary layer thickness. As the velocity gets greater, the effect of a further increase in boundary layer thickness becomes considerably less.

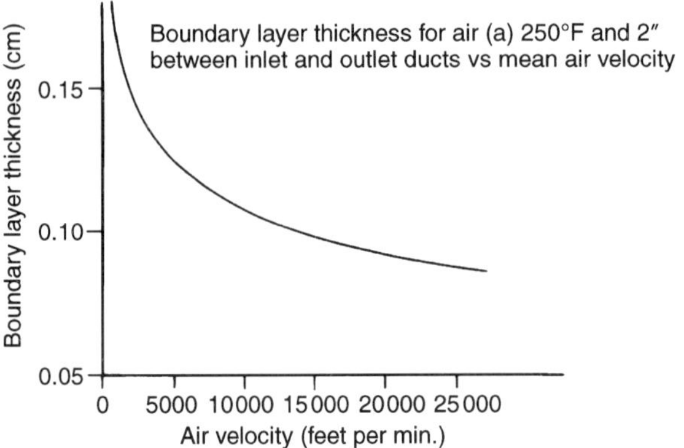

Figure 3.8 Relationship between air velocity and boundary layer thickness.

There is a practical limit of air velocity where the investment in equipment to achieve higher air velocities does not justify the results one gets. For example, calculated in Figure 3.8, this appears to be in the range 15 000 to 20 000 feet per min. A good deal of modern equipment is built to operate in the 10 000–15 000 feet per min range where a reasonable compromise between cost and boundary layer thickness seems to fall.

3.6.2 *Effect of air velocity*

The effect of air velocity on rate of heat transfer is shown in a qualitative way in Figure 3.9. The heat transfer rate rises rapidly and continues to increase with increasing air velocity past the point where the boundary layer thickness no longer seems to be very much affected by increasing air speed. Rate of heat transfer does not follow the same law relative to air velocity so that the boundary layer thickness is no longer strongly affected (say in the 20 000–25 000 feet per min range) but the heat transfer rate still is.

But drying is not solely a matter of heat transfer rate to the wet film. It also involves the problem of solvent transfer from the heated web back through the

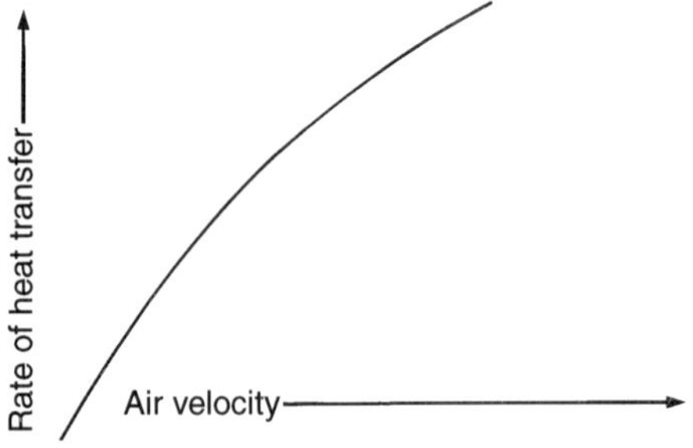

Figure 3.9 Effect of air velocity on rate of heat transfer.

boundary layer into the main air stream. This back-diffusion of solvent is more strongly affected by the boundary layer thickness than by the forward transfer of heat. Despite the improved heat transfer of solvent at very high air velocities, transfer of solvent through the boundary layer is probably not substantially improved beyond the 15 000 feet per min range air velocity with current dryer design. The improvement in heat transfer at the very high air speeds does not mean that solvent removal will be accelerated proportionately. In heated air drying, increasing air velocity produces these effects:

(a) Heat is transferred more rapidly to the wet ink film
(b) Solvent vapor is formed from the liquid solvent more rapidly
(c) The reduction in boundary layer thickness increases the rate of solvent diffusion from the web surface to the main air stream which carries the solvent into the dryer exhaust

3.6.3 *How dry is dry?*

All of this is very well, and there is no doubt that the introduction of high velocity heated air dryers has had a marked effect on production speeds and the overall quality of flexographic printing output, but one should still exercise some caution in speaking about 'dry' inks. Here we are concerned with: how dry is dry?

Recent work has shown that the ink film does not have all the solvent removed and that a considerable length of time is required to bring the ink to a 'bone-dry' condition. Once again we run into the essential non-linearity of nature. Figure 3.10 shows, in a qualitative way, residual solvent in a 'dried' ink plotted against time in the dryer. Beyond a certain point, extremely long drying (lower press speeds) is needed to make any significant change in the amount of solvent left in the ink. A compromise must be made between practical running speeds and an acceptable level of retained solvent. The ink formulator, within limitations, can help this somewhat by proper selection of resin binders and solvents, since some resins tend to retain some solvents more than others.

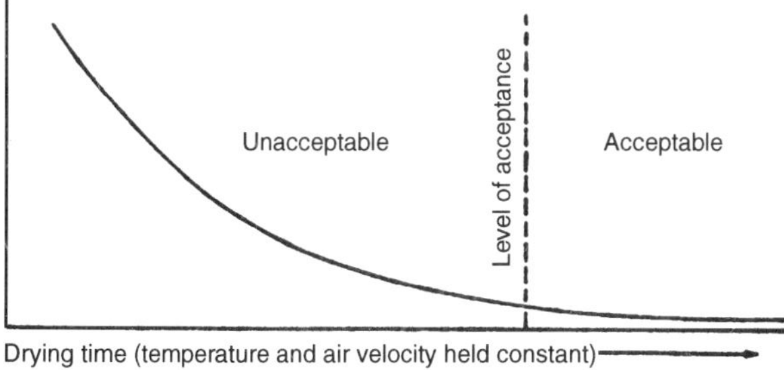

Figure 3.10 Relationship between residual solvent in a dried ink and time in the dryer.

Considerable progress has been made in developing resin–solvent combinations with improved solvent release that still retain all the other characteristics needed in a good ink.

The use of 'balanced' solvents is another way in which to improve drying without sacrificing other important ink properties. The customer should be educated, insofar as possible, as to his/her needs for adequate drying facilities if he/she is to achieve good quality at high speeds. Offset, odor, blocking and other problems resulting from inefficient drying are serious for the printer and are not always the fault of the ink.

Secondly, information concerning the kind of drying equipment a customer has should be made available. This information should include length of drier, air volume circulated, air velocity and air temperature.

3.6.4 *Residual solvent standards*

Thirdly, the ink user should be urged to establish standards of residual solvent with his customer. The standard of no residual solvent is all but impossible to meet. With the kind of analytical techniques available today (e.g. gas chromatography, infrared spectroscopy), solvent residues below the average human sensory levels can be detected. Levels as low as these may, nevertheless, in some cases, cause problems. Establishing a standard becomes important to the ultimate user of the printed web. Once such a standard is established and knowledge of the equipment to be used to produce the job is available, the technician has a far clearer understanding of what he/she must do in order to meet the requirements.

It is unfortunate that life tends to become more complicated with the passage of time, but the fact is that it does, and everyone has to adjust to the greater complexity of things. This has become true of even so simple a matter as flexographic inks.

3.7 Evaluation of anilox rolls

In this section some information will be given concerning how to evaluate anilox rolls, and what count and configuration to use, and when.

The engraving on the surface of an anilox roll has two main functions to perform. Firstly, it should measure or 'meter' to the printing plate surface, as accurately as possible, the precise ink film thickness needed for the ink, the copy and the substrate combination involved. Secondly, the 'land' or 'wall' surrounding each cell in the anilox roll surface should confine the flexo ink in the cells and prevent it from being squeezed outward toward the edges of the design when the 'kiss' impression between the surface of the plate and the surface of the anilox roll is too 'passionate'. These are the two main and unique functions of an anilox roll in a flexo ink distribution system. We will concentrate first on the ink metering function.

3.7.1 *Some standards for anilox rolls*

How can a flexographer tell how well a particular anilox roll will meet his needs with respect to metering the optimum ink film thickness for his particular job requirements?

Fortunately, he does not have to start from scratch. The optimum anilox roll surface count and configuration has been determined for most common flexo inks, printing copy and substrate surface combinations in the industry, and most of these have been standardized for some time. For instance, in the case of most 'average' copy consisting of line, type and solids printed on uncoated and relatively rough and absorbent paper surfaces with water-reducible flexo inks and conventional 'two-roll' flexo ink distribution system, the inverted pyramid configuration with 165 cells per inch in each direction was standardized about 40 years ago.

This standard '165-P' anilox roll surface has a total cell volume or capacity equal to about nine billion cubic microns per square inch of surface area, and the usual standard allows for a total maximum variation of plus or minus 10% in this volume as furnished either as original equipment or as re-engraved rolls. The standard nominal maximum cell depth of this engraving is 0.0018″ or about 45 μm, but the control emphasis on all anilox roll counts and configurations is on cell volume rather than depth because it is volume that is really the critical factor.

In the case of most plastic film and foil stocks with smooth, non-absorbent surfaces, where solvent inks are normally used, the anilox roll accepted as an industry standard for many years has a 180P engraved surface. As flexo inks were steadily improved with respect to 'strength' or color intensity at normal running viscosities, an increasing number of press manufacturers and flexo film printers standardized on 200P anilox rolls where fine halftones or color process printing was not involved.

For tone and process printing the anilox rolls used with conventional two-roll ink distribution systems are from 400 to 600 count. For the same work done with reverse-angle doctor blades, from 300 to 360 count quadragravure configurations are most widely used.

3.7.2 *The importance of cell volume*

In all cases, however, cell volume is the important consideration. Careful control of cell volume, plus close control of land width, smoothness and control are the key elements ensuring consistently uniform, predictable and dependable anilox roll performance. The importance of anilox cell volume versus a single measurement such as cell depth alone, is illustrated in Figures 3.11 to 3.13. Figure 3.11

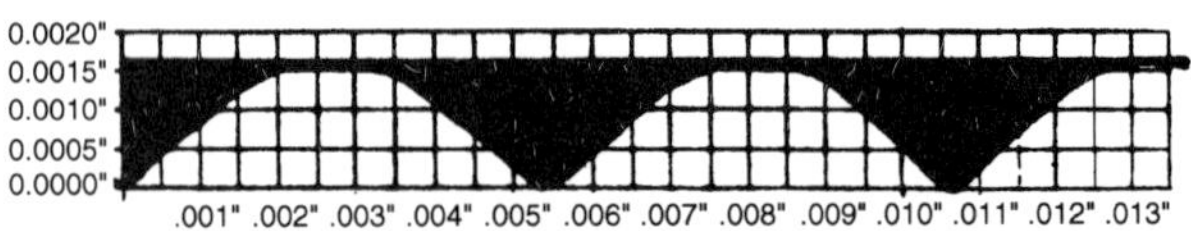

Figure 3.11 A scaled reproduction of profilometer tracing of a series of cells on the surface of a quality 180P anilox roll. Note the full cell depth and volume, the smooth, uniform walls and the rugged, fully formed lands. The solid black areas represent the ink capacity or volume and distribution provided.

shows a scaled reproduction of an actual tracing made of a series of cells in a standard 180P anilox roll by a very accurate and sophisticated profilometer. This tracing is for a 180P configuration anilox roll; a new top quality roll should not vary more than plus or minus 10% from this standard, with respect to volume.

Figure 3.12 shows a reproduction of a tracing of a non-standard, badly worn '180P' roll shown in Figure 3.11, but the profile speaks for itself. There are in fact, 180 cells per inch on the roll in Figure 3.12, and the cells are indeed shaped like inverted pyramids. But the cells shown in Figure 3.12 are too shallow and too narrow, metering less than half the volume of ink supplied by the cells shown in Figure 3.11. The lands in Figure 3.12 are wide, rough, burred and not fully formed. This roll cannot meter enough ink to print satisfactorily, and its service life is virtually ended. In addition, the rough, burred land areas will abrade the surface of both the printing plates and the fountain or wiping roller in contact with these rough land areas.

We have repeatedly emphasized cell volume rather than cell depth alone, as the most meaningful index of uniformity and value for anilox rolls. Figure 3.13 helps to show why this is true. When ink transfers from the anilox cells to the printing plate surface, this transfer is not complete. The ink film tends to 'split', most of it transferring to the rubber printing plate surface, which has greater affinity or attraction for the ink than does the chrome-plated anilox roll surface. Some of the ink always remains in the anilox roll cells, however, and it is especially true that the ink tends to cling to the 'corners', supplying essentially the same volume of ink to the plate, in spite of measurably less cell depth, and with easier cleaning, in addition.

In short, be sure to get proper anilox roll cell counts, configurations and ink volume to meet your various printing needs.

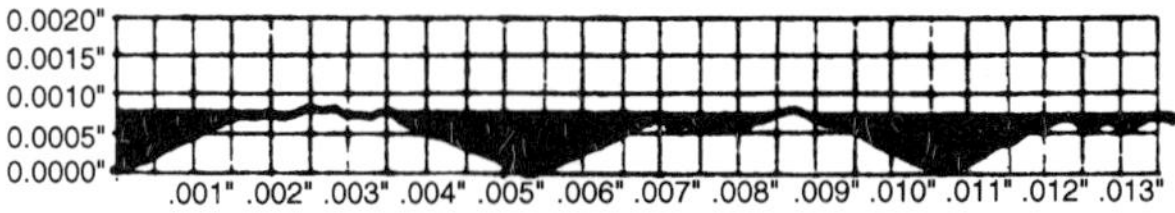

Figure 3.12 This profile, made with the same equipment as in Figure 3.11, shows the surface of a poorly engraved 180P anilox roll. Note the shallow cells and the wide, rough, burred and incompletely formed lands, with resulting inadequate and irregular ink distribution. These rough lands will also cause excessive wear on printing plates and rubber fountain roll.

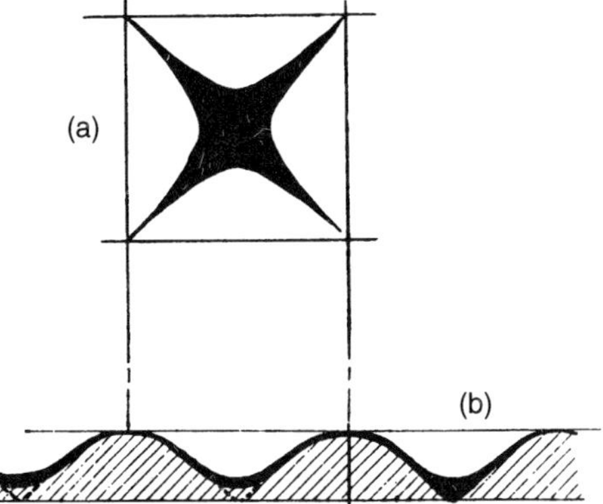

Figure 3.13 (A) Ink does not transfer completely to plate, but tends to cling to corners of anilox roll cells, shown looking down into a single cell, and in (B), in cross-section view of cells with rounded base (left) and pointed base (right). There is little difference in either the total or transferred ink volume between otherwise identical cells with shallower rounded bottom (left) and deeper, pointed bottom (right).

3.7.3 *What to use and when*

Now let us discuss the questions about what anilox roll surface counts, configurations and cell volumes should be used and when, for the best results in flexo printing. We will attempt to provide answers which are accurate and correct in 95 to 99% of all the cases covered.

It should be noted that no specific count given here is intended to be considered 'sacred', since there are so many variables involving differences among presses, inks, stocks and job requirements. Anilox rolls are readily available in designations which vary in capacity by increments of 5 to 10%, so that if production experience consistently indicates that for existing conditions, optimum results would more often be obtained by a slight deviation from the 'industry standards' such a deviation can easily be tried. When proven to be advantageous, the deviation can be established as a local or plant standard so long as those particular conditions and requirements prevail.

The flexographic printing industry almost universally uses anilox rolls with the (inverted) pyramid configuration for ink distribution systems utilizing a rubber-covered fountain or wiping roll. For reverse-angle doctor blade ink distributions, quadragravure configuration anilox rolls are the accepted industry standard. These same 'Quad' rolls are the most widely used rolls in rotogravure coating and laminating operations.

'Tri-helicoid' (triangle-helicoid) configuration rolls are rarely used in flexo printing operations although they are sometimes helpful in overcoming certain problem situations. Tri-helicoid rolls are mainly used in the application of heavy viscosity adhesives and surface-coating materials.

Tone-and-process anilox rolls, as the designation implies, are used for tone and color-process printing. Counts finer than 360 usually identify rolls primarily designed for use in the 'conventional' two-roll flexo ink distribution systems, while the tone and process designation rolls, with counts from 250 to 360, are usually intended for use with reverse-angle doctor blade systems.

3.7.4 *Anilox roll chart*

(a) *'P', inverted pyramid configuration.* Most common use: Anilox rolls in 'conventional' 2-roll (anilox + rubber-covered roll) flexo ink distribution systems.

120 to 140 count: For very heavy application on rough surfaced absorbent paper substrates, as for coating corrugated materials, and for heavy application of daylight fluorescent colors, etc. Seldom used in flexo presses except for special situations.

150 count: For heavier-than-usual applications on rough surfaced absorbent paper stocks. Furnishes about 10% more than standard 165P roll.

165 count: Established industry standard for printing on uncoated papers, including most corrugated linerboard stocks and most natural and bleached kraft bag stocks. Sometimes used for extra heavy application of first-down, opaque white ink and for silver and gold metallic inks on transparent films or for printing dense, opaque whites on aluminum foil surfaces.

180 count: Widely used standard for printing average line, type and solids, on films, foils and relatively smooth, non-absorbent (including coated) papers. Okay for fine-particle-size metallic inks.

200 count: Often used in those printing stations of flexo presses where medium-to-fine line, type and/or design elements are usually printed on film, foil and coated paper or glassine stocks. Also frequently used for printing fine line, type and design copy on paper envelope stocks. Not generally recommended for metallic inks compounded with any but the finest 'lining' powders, as coarser 'gold' (bronze) powders may 'pack' and 'fill' the cells in anilox rolls with counts higher than 165.

(b) *'Q', quadra-gravure configuration, sqare cell bottom.* Most common use: Anilox rolls in reverse-angle doctor blade flexo ink distribution units (and for most rotogravure coating operations).

120 to 140 count: Standard range in reverse-angle doctor-blade type flexo presses for corrugated printing.

150 count: Most popular count for rotogravure coatings on transparent films, foils and smooth surfaced paper stocks.

165 count: For heavy, opaque first-down white, etc., on transparent films, using reverse-angle doctor-blade units on flexo presses. Also recommended for metallic silvers and golds.

200 count: For most medium-to-light line, type and design printing on films, foils and smooth, non-absorbent paper stocks, using reverse-angle blade flexo units.

(c) *'T-H', tri-helicoid configuration.* Most common use: For application of heavy viscosity coatings and adhesives, usually in doctor-blade units. Popular counts range from 35 to 200, and for special needs go as fine as 600 to match job requirements.

180 to 230 count: Occasionally used for special flexo press anilox roll requirements, as to minimize such problems as striation and foaming with some types of 'problem' inks to meet unique job requirements.

(d) *'TP', tone and process anilox rolls.* Specifically designed for quality flexo tone and process printing.

350 to 400 count: For all quality tone and color-process printing on smooth surfaced non-absorbent surfaces, with conventional (2-roll) flexo ink distribution units. Suitable for plate screens from 65 and including 150 lines per inch.

550 to 600 count: For finest quality tone and color process printing on films with conventional two-roll systems or with very light blade pressures; has been used in reverse-angle doctor blade systems to print plate screens as fine as 200 line. These rolls are especially designed for use with 133-line and 187-line screen printing plates.

250 to 360 count: Designed for printing quality tone and color process work on films, from plates to 150-line screen, when used with reverse-angle blade inking units. Also for use in relatively coarse screen tone and process printing on paper stocks with conventional (2-roll) inking systems. Also for flexo-type letterpress and litho conversions for newspaper printing.

3.8 The reverse-angle doctor blade and anilox rolls

In this section aspects of reverse-angle doctor blade installations on anilox rolls used in flexographic ink distribution systems will be discussed. The features covered will relate to reverse-angle doctor blade materials, design, operation and press widths.

First, there appears to be a considerable range of opinions regarding the origin of the reverse-angle doctor blade concept. We are not certain of the exact date of the conception of the first commercially successful reverse-angle doctor blade unit, but around 1950 the inventor to whom the first U.S. patent was issued (who has otherwise left no imprint on the flexographic printing industry and whose name has been consigned to limbo), showed me an early unit, which had then been in operation for at least 2 years. This was one of several such units on a three or four-color rotogravure web press, about 12 to 18″ wide, and is different from the concept ultimately implemented in flexo.

The first flexo version of a reverse-angle doctor blade ink distribution system we recall was a laboratory unit designed and constructed for test purposes in 1959 and reported in some detail at the 1960 FTA Technical Forum. At the time, we believe, the principle of the recommended optimum blade angle of 30° to the tangent at the point of contact with the surface of the anilox roll, on the 'dry' side of the blade, was first established and published.

The flexo blade design and assembly described in the 1960 FTA Forum used a positive screw adjustment to set the position and pressure of the blade with respect to the anilox roll surface. Several subsequent reverse-angle blade flexo units followed the same basic design. However, because the importance of very light blade-to-anilox roll surface pressure was obvious, the design and use of spring tension automatically controlled pneumatic and hydraulic tension, and pressure controls soon followed.

The flexo industry soon learned that where reverse-angle doctor blade units were concerned very light blade pressure in contact with the anilox roll surface was extremely important to anilox roll life. And blade pressures from one-tenth to one-quarter of an ounce per linear inch were necessary to consistently ensure anilox roll life similar to that expected from an anilox roll used in a two-roll flexo ink distribution system.

Most flexo presses for printing flexible packaging materials in widths up to 60″ were designed and built with reverse-angle doctor blade assemblies rigid enough to provide uniformly light blade pressures over the entire width of the anilox roll surface, with no more than one spring, pneumatic or hydraulic control at or near each end of the assembly. However, with the advent of flexo corrugated printing presses in widths often from 120 to 240″ and in press weights and vibration conditions, multiple blade assembly controls became common. Figure 3.14 shows two spring-tension type doctor-blade assembly pressure controls placed near each end of a typical reverse-angle doctor blade flexo web press with anilox roll widths typically from 36 to 60″. These spring tension controls usually work very satisfactorily on presses of this size and at operating speeds up to 1,000 feet per min.

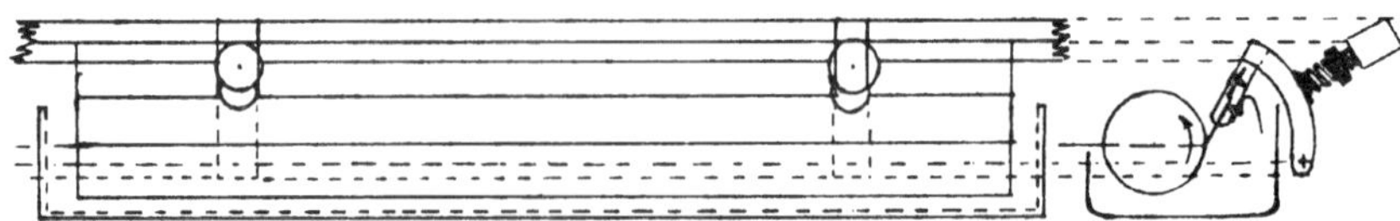

Figure 3.14 Reverse-angle doctor blade/anilox flexo ink distribution system, spring-tension blade assembly pressure.

3.8.1 *Pneumatics used on corrugated presses*

On flexo corrugated presses of widths from 86 to 240″ using reverse-angle doctor blade ink distribution systems, pneumatic pressure units are most commonly used. The original design used for these reverse-angle flexo corrugated presses employed a doctor blade assembly which consisted of two transverse elements mounted in a hinged relationship. The blade was fastened in one element and the second element contained a series of small air cylinders, each of which applied an equal force against the blade holder, thus causing it to exert a uniform pressure against the anilox roll surface. On the original press design, the lower portion of the anilox roll was below the surface of the constant ink level maintained in the ink fountain. As the anilox roll rotated, the reverse-angle doctor blade, located above the ink level, shaved the excess ink from the surface of the anilox roll and directed it back into the ink fountain by means of an apron provided for the purpose on the blade holder.

The pressman on this unit set the pressure of the blade to the anilox roll surface by adjusting a regulator knob which contained a pneumatic pressure gauge. The blade pressure was thus increased to the point where the doctoring or shaving effect was achieved uniformly over the entire length of the anilox roll surface. The blade pressure against the anilox roll surface was maintained at the minimum level to ensure an even doctoring effect across the anilox roll surface length. This pressure ideally should be less than half an ounce per linear inch. The blade assembly design is illustrated in Figure 3.15.

3.8.2 *Doctor blade forms fountain*

A subsequent flexo corrugated reverse-angle blade assembly design is similar to the one illustrated in Figure 3.15. Both of these designs employ a single anilox roll with a reverse-angle blade assembly and with no rubber-covered fountain roller. Both designs use multiple pneumatic blade pressure controls. The most obvious difference between the two designs is that in the design shown in Figure 3.15, the ink fountain is mounted under the anilox roll and the doctor blade assembly is mounted above the ink fountain and on the right side. In the second design, shown in Figure 3.16, the doctor blade assembly is mounted on the left side of the anilox roll and the ink fountain is formed by the doctor blade holder plus end plates.

Pneumatic pressure cylinders and their plungers vary a little in design but not in function. They are spaced about the same distance apart as the cylinders in

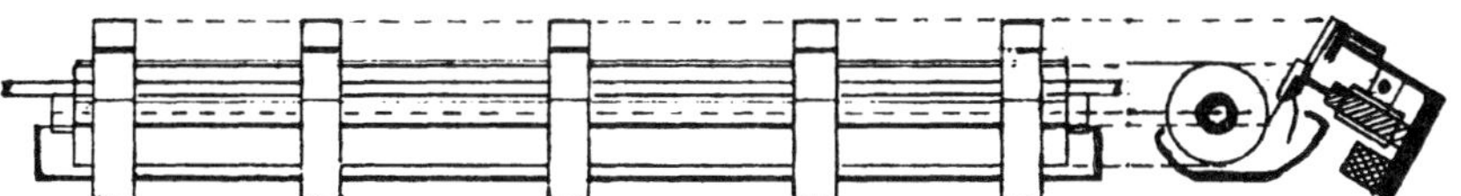

Figure 3.15 Reverse-angle doctor blade/anilox roll flexo ink distribution system, multiple location pneumatic blade assembly pressure control.

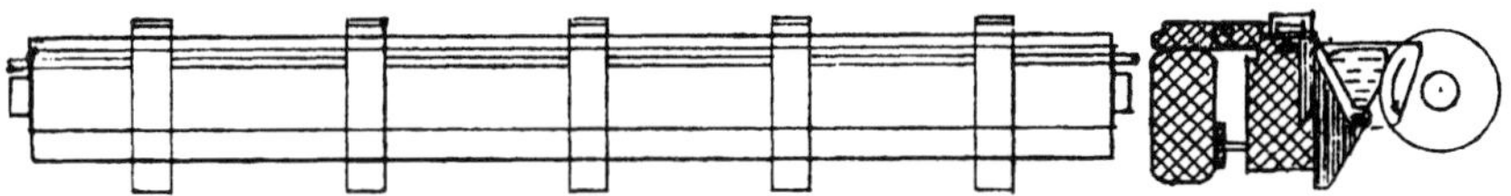

Figure 3.16 Reverse-angle doctor blade/anilox roll flexo ink distribution system, multiple location, pneumatic blade assembly pressure control. Ink fountain contained in blade assembly.

Figure 3.15, but since the ink fountain and doctor blade assembly plus the ink maintained in the fountain makes for greater weight of the total blade assembly in the design shown in Figure 3.16, the air pressure adjustment used in this design is set and maintained high enough to compensate for the added operating weight of the blade assembly.

3.8.3 *Pressure by expandable tube*

The designs shown in Figures 3.15 and 3.16 have been used satisfactorily in narrower width presses, ranging from 12″ to 72″ printing widths. In addition, another approach, based on compressed air, has been used for accurately adjusting and maintaining blade assembly pressures at the consistently low levels recommended for successful operation of flexo reverse-angle doctor blade ink distribution systems.

This additional option substitutes a flexible and expandable rubber or other elastomeric hose or tube, supported for the entire length of the doctor blade holder within a rigid casing permanently secured to the surface of the transverse element back of the blade holder, for about three quarters of this circumference, with the remaining one quarter of the casing, facing the back of the blade holder, open so that it allows the tubing, as it expands under the pneumatic pressure of air within it, to exert uniform pressure between the doctor blade and the anilox rolls surface (Figure 3.17).

The apparent and claimed advantages of this 'expandable tube' pneumatic pressure applicator are: (1) lower initial cost; and (2) even, uniform pressure applied against the entire length of the back of the blade holder. We have seen a limited number of these latter units used with apparent satisfaction.

3.8.4 *Advantages of the doctor blade system*

In spite of the growing popularity of flexo reverse-angle doctor blade ink distribution systems, many flexographers continue to ask what the specific

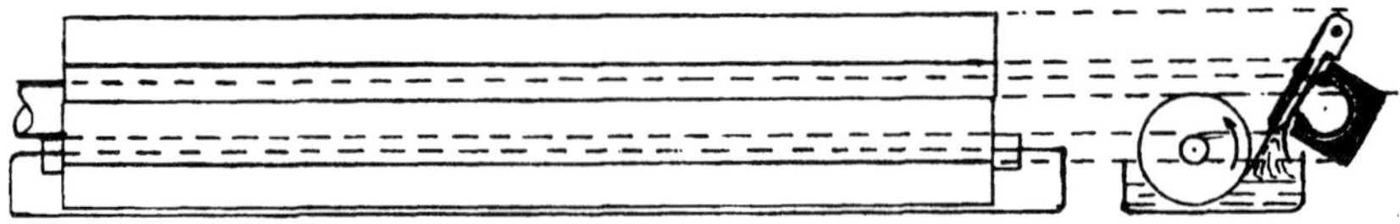

Figure 3.17 Reverse-angle doctor blade/anilox roll flexo ink distribution system, continuous pneumatic tube blade assembly pressure control.

advantages of these units are as compared to the older 'two-roll' anilox ink distribution systems. It may be appropriate to at least briefly review the flexo industry's experience with reverse-angle doctor blade ink distribution systems to the present time.

First, flexo reverse-angle doctor blade systems consistently provide a uniform film of ink to both the printing plate and substrate surfaces over a much wider range of ink viscosities.

Secondly, reverse-angle doctor blade systems consistently produce superior quality and more uniform printed results over a much wider range of operating speeds. For example, tests run by a group of FTA panel members, and reported at the FTA Forum in 1971, showed that using the same printing plate and ink a two-roller anilox press unit showed a 100% increase in ink density at an increase in operating speed from 100 feet per min to 400 feet per min. The reverse-angle doctor blade unit showed no increase in ink density at an increase in operating speed from 100 feet per min to 700 feet per min. Comparative runs made subsequently have shown no increase in printed ink density when run on the reverse-angle blade unit, at operating speeds from 100 feet per min to 1000 feet per min.

Thirdly, various tests over a period of years and numerous production jobs have shown drastic differences in the printed quality of four-color process flexo jobs produced by reverse-angle doctor blade flexo presses versus two-roll units. In fact, almost all the FTA top award-winning color process jobs during recent years have been printed on reverse-angle doctor blade flexo press units.

Finally, most of the recent developments involving new applications of the flexographic printing process to replace letterpress and lithographic graphic arts processes involve reverse-angle doctor blade/anilox roll ink distribution systems. These new flexo markets include newspaper and other publication printing applications, the folding box industry and various commercial printing applications. The new reverse-angle designs increasingly use one of the pneumatic blade pressure controls shown in Figures 3.15 to 3.17 so that such controls will become increasingly familiar to flexographers in the future.

3.9 How the reverse-angle doctor blade works, and a comparison with other metering systems

In reviewing the early history of reverse angle doctor blade flexo ink distribution systems, the first progress report on this development which we recall was covered by the Flexographic Technical Association in the early 1960s. Because so many of the points covered then are very similar to questions often being asked today by flexographers who are not yet using reverse angle doctor blade ink distribution systems on their flexo presses, the early F.T.A. reports on the subject seem to us an appropriate starting point for a discussion of reverse angle doctor blade technology in the graphic arts industries.

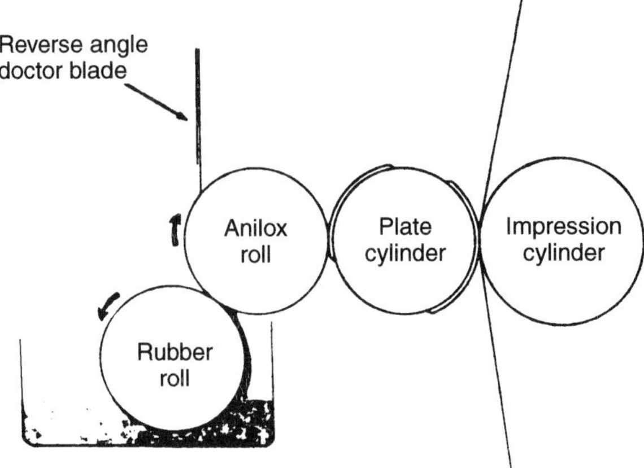

Figure 3.18 Reverse-angle doctor blade installation on an anilox roll.

Many flexographers today are familiar with reverse angle doctor blade flexo ink distribution systems, to the extent that they have one or more of these units installed and operating on flexo presses, but a number of flexographers are still looking forward to getting acquainted with the first reverse angle flexo installation.

Figure 3.18 shows schematically a reverse angle doctor blade installation as such a conversion might be made on the anilox form roll of a conventional two-roll flexo press. In this Figure, the blade is installed on the side of a flexo press where the web travel is upward, in order to ensure that the doctor blade installation will be readily accessible to the press operation. It will be noted that the 'reverse-angle' blade, in the position shown, 'shaves' or 'scrapes' the ink from the surface of the anilox roll surface, as would be done if the blade were mounted in a conventional position as in a rotogravure press.

Obviously, in Figure 3.18, the rubber-covered fountain roller would not be necessary if the object were simply to design a new flexo press with a reverse-angle doctor blade flexo ink distribution system, in which case the unit would look more like the schematic drawing shown in Figure 3.19.

3.9.1 *Development of flexo ink distribution systems*

Let us review the evolution of flexo ink distribution systems to see how and why we have arrived where we now are. The earliest flexo ink distribution systems consisted essentially of two smooth rollers, one or both of which were usually rubber-covered, rotating together with adjustable pressure, in the manner of an old-fashioned wash wringer. More pressure between these rollers supplied less ink to the surface of the printing plate and hence to the surface of the substrate to be printed, and increasing ink viscosity or press operating speed supplied more ink to the printing plate and substrate surface.

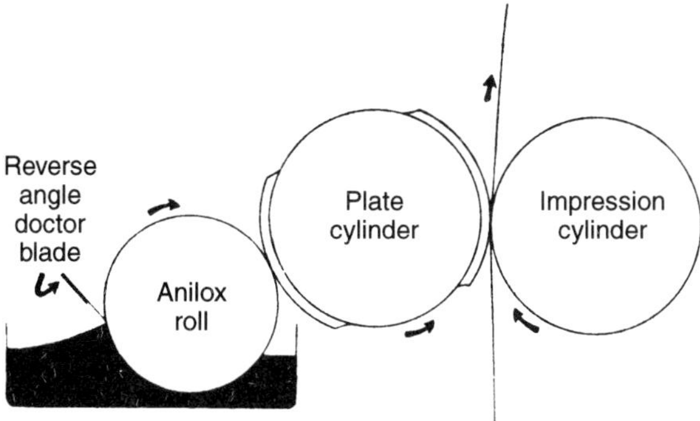

Figure 3.19 A flexo press with a reverse-angle doctor blade flexo ink distribution system.

Balancing these factors in order to maintain reasonable uniformity in the printed result during a press run was possible and reasonably practical at relatively slow press operating speeds, especially on absorptive paper substrates. However, at higher press operating speeds and on non-absorbent surfaces, such as cellophane or plastic films and metallic foil stocks, this balancing act became progressively less practical and more frustrating.

At about the end of the 1930s, the anilox ink distribution system was developed, soon after the development and increasingly widespread and growing use of 100% pigment inks for flexography, particularly for the printing of film and foil packaging materials. These two developments complemented each other, with the result that by the end of World War II, anilox rolls were almost universally being used as standard equipment in most flexo presses being produced in both the United States and abroad, but as two-roll inking systems.

Many flexographers today do not recall, however, that the anilox flexo ink distribution method originally introduced and promoted recommended a single anilox roll, rotating in the ink fountain, with a doctor blade which wiped the ink from the surface of the engraved, chrome-plated anilox roll. This system was essentially the same as that shown in Figure 3.19, except that the blade was mounted in the conventional rotogravure fashion, as shown in Figure 3.20.

(a) *Print quality.* The print quality produced on these presses was very good, but most flexo pressmen lacked experience and skill in mounting the doctor blades and it was feared the wear on the anilox roll surface might be excessive. In addition, the doctor blade units mounted on the side of the press where the web travel was downward were virtually inaccessible. For these reasons and because press operating speeds at the time anilox rolls were developed usually did not exceed 400 feet per minute, the two-roll anilox ink distribution system was most readily accepted and became the industry standard until the evolution of the reverse angle doctor blade.

The first two-roll anilox inking systems were designed with the fountain roll and the anilox roll rotating at the same surface speed, but as press operating

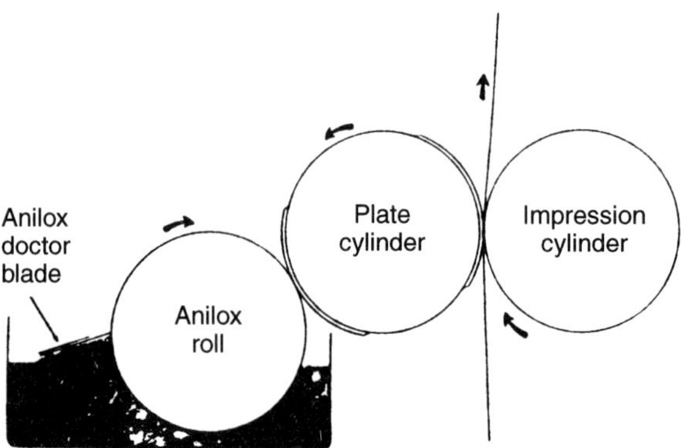

Figure 3.20 Original anilox flexo ink distribution method, with the blade mounted in the conventional rotogravure fashion.

speeds increased, differential surface speed fountain rollers became more common, gradually progressing from a 1 : 2 ratio to as much as a 1 : 4 geared ratio between the fountain roller and the anilox roll.

As press speeds increased, the need for greater surface speed differentials between the fountain roller and the anilox roll resulted in the development of a separate drive system for the fountain roller, which often provided a constant speed fountain roll, resulting in a variable ratio between the fountain roll and the anilox roll as the press speed increased. At operating speeds of 1000 feet per min or more, the ratio between the fountain and anilox roll might be 1 : 10 or more.

It was found, however, with two-roll systems, that, at surface speed ratios of 1 : 10 or more and at press operating speeds of 1000 feet per min or more, the ink film thickness applied to the surface of the printing plate, and in turn to the substrate surface, often became excessive. This, therefore, put more pressure on the need for flexo reverse angle doctor blade ink distribution systems, particularly since experimental work proved that reverse angle blade systems consistently furnished a uniform ink film thickness to the printing plate and substrate surface with virtually no variation from 'makeready speed', of 50 to 100 feet per min, to running speeds, from 1000 to 3000 feet per min and over a wide range of ink viscosities.

(b) *Contact angle.* A number of the earlier pioneers in reverse angle doctor blade technology determined that the best angle at which these blades should contact the surface of the anilox roll was 30° to the tangent at the point of contact, measured on the dry side of the blade. Another factor of major importance was found to be the lightest possible contact pressure between the blade and the surface of the anilox roll. The maximum pressure recommended is $\frac{1}{2}$ ounce per linear inch and preferably $\frac{1}{4}$ ounce per linear inch.

The schematic reverse-angle doctor blade assembly shown in Figure 3.21 was recommended in the one of F.T.A.'s early reports and is still considered

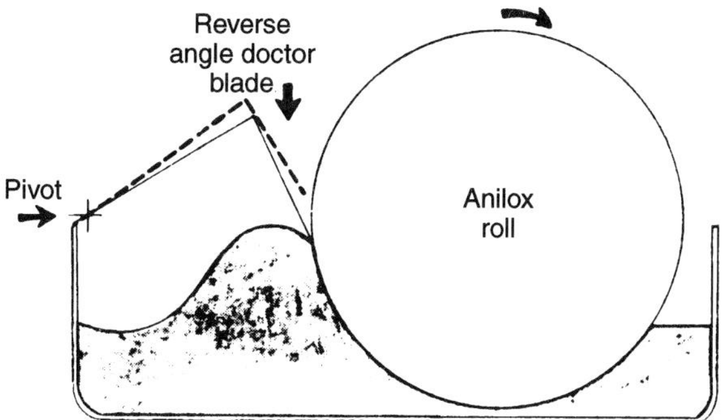

Figure 3.21 Reverse-angle doctor blade assembly recommended in an early F.T.A. report.

desirable, although the majority of units designed and still built are based on a screw-mounted rigid blade assembly, rather than a flexible pivoted mounting arrangement. A minority of reverse angle doctor blade ink distribution units in operation at the present time do have a pivoted blade assembly, with blade pressure controlled by pneumatic, hydraulic or spring tension or pressure. Such blade assemblies permit positive and continuous readings of blade contact pressure and most flexographers hope such arrangements will soon become universal.

Most reverse-angle doctor blades currently used on flexo presses are of Swedish blue steel stock, about 0.004″ in thickness, for presses up to 36″ printing width. The 'shaving' blade normally extends from $\frac{1}{2}$ to $\frac{3}{4}$″ from the blade mounting assembly. For wider press widths, the doctor blade stocks increase in thickness to from 0.006″ to as much as 0.010″ for presses up to 120″ wide.

For corrugated flexo printing presses, reverse angle doctor blades of high density polyethylene material up to $\frac{3}{32}$″ thick are commonly used, and polyester plastic blades from 0.015 to 0.030″ thick are occasionally but not widely used.

All reverse angle blades, whether of Swedish blue steel or plastic stocks, are normally ground and honed at a 45° angle on the 'shaving' surface. Stainless steel doctor blade stock is not commonly used or recommended for reverse-angle doctor blade ink distribution systems.

(c) *Lack of a 'back-up' blade.* A reverse angle doctor blade assembly does not employ a 'back-up' blade, such as is normally used in a conventional rotogravure assembly, because in the reverse angle or 'shaving' configuration ink tends to accumulate between the back-up blade and the shaving blade, where the accumulated ink often dries, resulting in an uneven shaving surface on the blade, as in Figure 3.22. Therfore, reverse-angle blades are mounted without a back-up blade.

Various suggestions have been made during the past 20 years as a means for keeping the thin 'shaving' reverse-angle blade and at the same time stiffening its 'spine', including the 'laminated' blade structure shown in Figure 3.22, and the

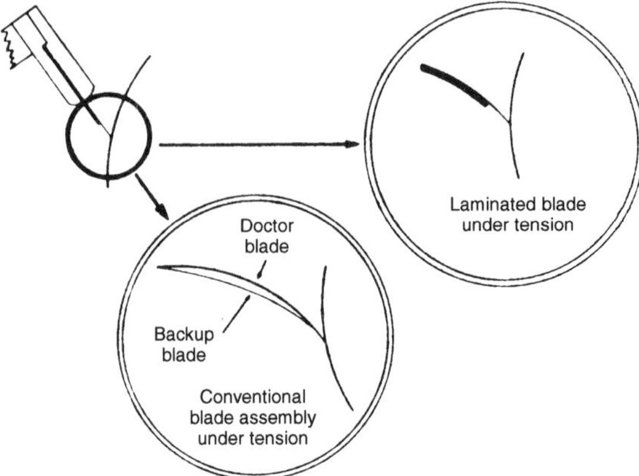

Figure 3.22 Laminated blade structure for keeping a thin 'shaving' reverse-angle blade.

tension assembly shown in Figure 3.23. Neither of these approaches has been widely adopted as yet, and the reverse-angle blade assemblies based on the single blade mounted in a holder, essentially as in Figure 3.19, represent the accepted designs in most common use in today's flexo presses.

3.10 The advantages and disadvantages of the reverse-angle anilox blade

In the last section the early development and evolution of reverse-angle doctor-blade ink distribution systems for flexo presses were discussed. In this section

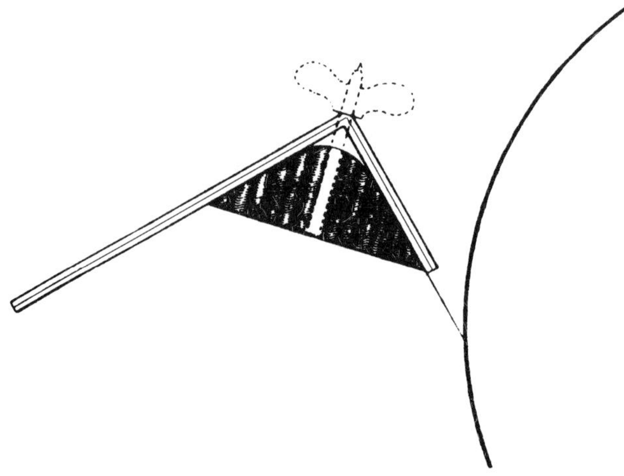

Figure 3.23 Tension assembly for keeping a thin 'shaving' reverse-angle doctor blade assembly.

some other designs which have been finding acceptance in reverse-angle blade applications in more recent years will be described. One such design, which was developed after the earliest reverse-angle doctor-blade systems, was a reverse-angle blade system specifically designed to operate on that side of a flexo press where the web travel was downward.

During the early development of reverse-angle doctor blade systems for adaptation to flexo presses on the side of the press where the web travel is downward, it was learned that the blade assembly was located back of the anilox roll, where the pressman could neither see it nor reach it conveniently, as shown in Figure 3.24. This condition essentially prevailed whether the inking unit on the side of the press where the web travel was downward and was a two-roll ink distribution system, as shown in Figure 3.24; or was a single anilox roll with the reverse-angle blade 'shaving' the ink from the upward-rotating anilox roll surface. In either event, the pressman found it inconvenient and often almost impossible to properly adjust the reverse-angle blade when it was mounted on the backward side of the anilox roll.

For these reasons, many flexo press manufacturers did not, for some years, recommend the use of reverse-angle doctor blade flexo ink distribution systems on the side of the press where the web travel was downward. However, by about 1976 one or two leading U.S. flexo press manufacturers introduced single-roll reverse-angle doctor blade units designed specifically for the 'downside' of the

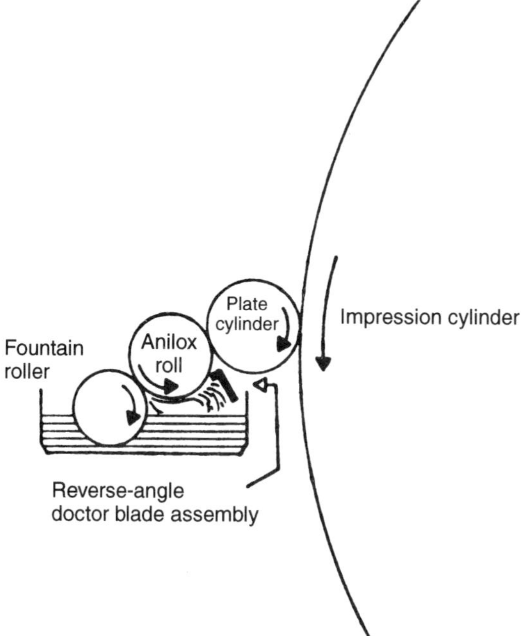

Figure 3.24 An early version of the reverse-angle doctor blade assembly.

press, similar to those shown in Figure 3.25. With this arrangement, the adjustment of the blade assembly and its attached ink fountain to the anilox roll surface was conveniently located and easily adjusted by the pressman. In addition, this adjustment lends itself readily to spring, pneumatic or hydraulic tension adjustment arrangements, making it easier to quickly and accurately read blade contact pressure on the anilox roll surface.

Since the evolution of this development, the use of reverse-angle doctor blade ink distribution systems on the 'downside' of flexo presses has steadily increased, and the earlier problems with the use of these blade systems on the first two or three stations of most flexo presses have steadily decreased. One happy result of this change has been an increasing growth in quality flexo color process printing, which is steadily reflected in the volume of color process

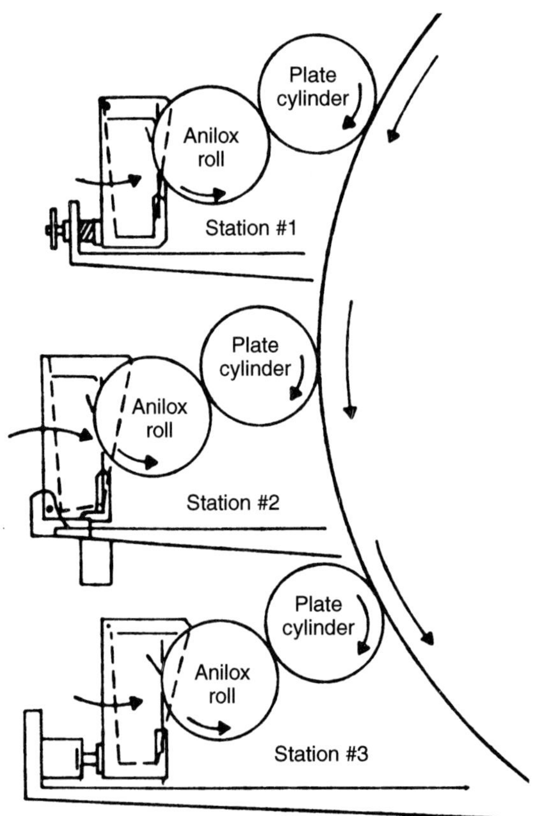

Figure 3.25 Single-roll, reverse-angle doctor blade unit, and ink fountain assembly on printing stations 1 to 3 of a six-color central impression cylinder flexo press.

flexible package printing produced by flexography, to be seen in a tour of the aisles of any modern supermarket.

3.10.1 *Applications in newspaper printing*

Another result, not initially anticipated by the flexo press manufacturers involved in the new 'downside', reverse-angle flexo design development, was the role this new design was to play in the ultimate application of the flexo-type reverse angle doctor blade anilox roll ink distribution system in the newspaper printing industry. In the simplified ink distribution system project activated by the American Newspaper Publishers Association, Research Institute, in 1976, the concept involved using conventional, unmodified letterpress news ink.

A.N.P.A. engineers originally anticipated that unmodified letterpress news ink could not be satisfactorily distributed by any existing flexo ink distribution system, but as a starting point it was decided to run a test with a standard letterpress newspaper printing ink on a flexo press with single anilox roll, reverse-angle blade ink distribution units. To the surprise of A.N.P.A. engineers the results of these trials were satisfactory in terms of print quality without any modifications at all in either the standard letterpress ink or the single-roll reverse-angle doctor blade flexo press units.

These initial trials were run on new design units on the downside web travel section of the flexo press used for this test; when the trials were repeated on the single anilox roll, reverse-angle doctor blade units on the opposite side of the press where the web travel was upward, the results showed what appeared to be fine ink particles scattered irregularly over the printed area. When the runs were repeated on both sides of the press, the results were the same, with no ink particles showing in the runs made on the downside of the press, but consistently apparent in the test runs made on the side of the press where the web travel was upward.

This phenomenon was puzzling to the engineers until careful studies disclosed that due to the different position of the reverse-angle blades contacting the anilox roll surface on the up and down web-travel positions, the relatively heavy viscosity of the letterpress newspaper printing ink resulted in traces of the letterpress ink getting under the blade's 'shaving' edge, with traces of ink accumulating between the back edge of the blade and the anilox roll surface. As the anilox roll rotated, tiny particles of this ink were attracted to the inked surface of the anilox roll, resulting in these tiny 'droplets' or particles of viscous ink being transferred to the surface substrate.

The obvious question, of course, was why did this phenomenon take place on the 'upside' units and not on the 'downside' units? The answer is indicated in Figures 3.25 to 3.28. Figures 3.25 and 3.26 show schematic sketches of the anilox roll, reverse-angle doctor blade and ink fountain assembly on printing stations 1, 2 and 3, of the six-color central impression cylinder flexo press on which the trials were run. Figure 3.27 is a schematic sketch of the corresponding

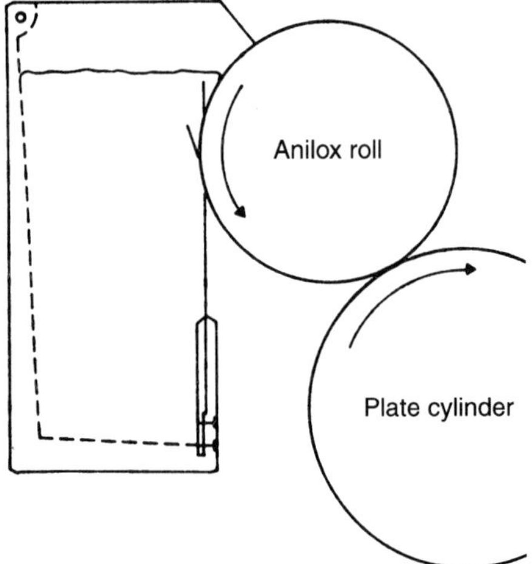

Figure 3.26 Enlarged version of part of Figure 3.25.

anilox roll, reverse-angle blade and ink fountain assembly originally used on printing stations 4, 5 and 6, on the same press.

Note that when particles of the heavy viscosity paste-bodied letterpress ink escape under the doctor blades in stations 4, 5 and 6, such particles tended to

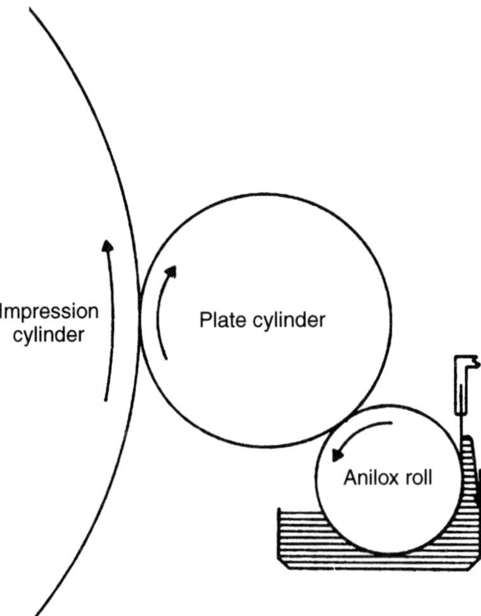

Figure 3.27 Similar to Figure 3.26, but for the ink fountain assembly in printing stations 4 to 6.

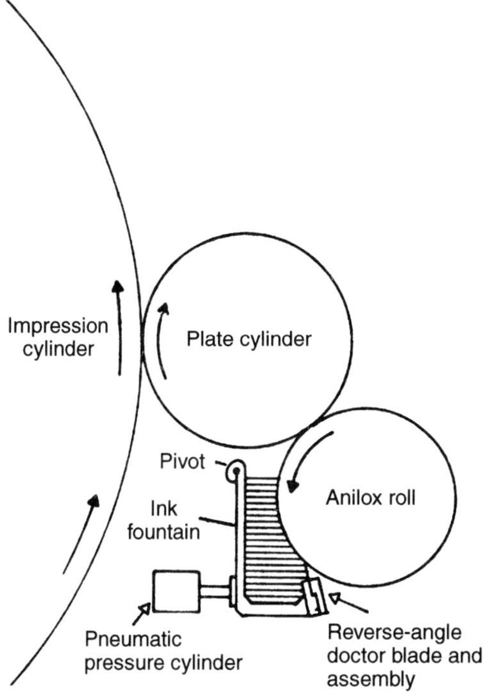

Figure 3.28 Similar to Figure 3.27, but with improved design to overcome 'specking'.

accumulate in the nip formed by the back of the blade and the 'shaved' or 'doctored' surface of the anilox roll. Gravity prevents the accumulation from escaping in any way except to the extent that these accumulated particles in contact with the anilox roll surface rotate with it until they are transferred to the plate surface, and thence to the newsprint surface, where they appear as printed 'specks'.

In the case of stations 1, 2 and 3 of this same press, the heavy viscosity, paste-bodied letterpress inks behave similarly up to the point where traces of the ink pass under the shaving edge of the doctor blade. However, at that point, because of the position and location of the back of the doctor blade, any ink particles which escape under the shaving edge of the blade tend to follow the back surface of the blade and because of gravity, the ink particles 'flow' downward along the blade surface and away from the anilox roll surface. The result is a clean print, free from specks.

By redesigning the number 4, 5 and 6 printing units similar to those on the 1, 2 and 3 stations, as shown in Figure 3.28, the 'specking' was overcome and has ceased to be a problem on the various retrofitted letterpress newspaper presses now in operation. Thus, the flexographic industry not only now has a range of reverse-angle doctor blade ink distribution systems ranging from narrow-web tape, label, envelope and business forms presses to corrugated presses as wide as 240″, but also a growing range and number of retrofitted letterpress and lithographic presses successfully using flexo-type anilox roll, reverse-doctor blade distribution systems. Because of the very light doctor blade pressures required for these systems, even the extremely wide corrugated presses do not pose a problem in doctor blade assembly design and construction.

3.11 Overcoming moire effects in flexography

This section is concerned with the problem of moire patterns in flexo printing. Perhaps we should begin by defining what 'moire' in printing actually is. A dictionary definition of moire is 'any silk, rayon, etc., fabric with a watery or wavelike effect', and in 'print', the dictionary describes moire as a 'smeared pattern of dots appearing in the print of process color'.

The Flexographic Technical Association in its book *Flexography: Principles and Practices*, defines moire as 'a pattern resulting from incorrect relative screen of the anilox roll and a halftone plate and/or two or more halftone plates'. And the Gravure Technical Association, in its book *Gravure Technical Guide*, defines moire as 'an undesirable checkerboard or plaid effect caused by the improper relationship of the screen angles used on the various positives'.

Obviously, about the only thing these definitions agree on is that the moire effect in printing is undesirable, and therefore what we are really interested in is getting rid of the moire effect. The various definitions make it clear that there are different causes of moire effects. The one cause peculiar to flexography is that based on the conflicting angles between the engraved anilox roll surfaces of the flexo ink distribution system and the color separation angles of each printing plate screen.

For some time during the early days of flexography's experimental period of color process printing, the finest anilox roll screens available were 200 to 220 cells per inch in each direction. These anilox rolls, as new, furnished too much ink for flexo color process printing, so resourceful flexographers learned that when a 200 count anilox roll was about one-half worn, it furnished about the proper ink density for 65 to 85 line screen color process printing plates.

Flexo printers also soon learned, however, that these partly worn 200 count anilox rolls usually caused moire problems in the print. It was generally assumed that this was because most anilox rolls were engraved with approximately a 45° angle, which obviously conflicted with the 45° angle at which the 'key' color, usually black in four-color process plates, was engraved. The conflicting angle seemed logically responsible for the resulting moire effect, and a number of tests seemed to confirm this. For some time, therefore, the flexo industry established the rule that in deciding on plate screen angles for flexo color process printing, the 45° angle should be avoided, since it was already used by the anilox roll screen. Some flexographers still follow this rule.

3.11.1 *Tone and process anilox rolls*

In the mid to late 1960s finer count anilox rolls were introduced, often referred to as 'tone and process' anilox rolls. These rolls ranged in cell count from 250 to as fine as 550 cells per inch in each direction, and the configuration of theis new generation of anilox rolls changed from the earlier 'inverted pyramid' configuration to what is essentially a 'quadra-gravure' configuration, with much narrower 'lands'. In addition to the initially narrower lands surrounding each cell, the cell walls of the tone-and-process rolls have a much steeper slope than that of the earlier inverted pyramid cells.

Because the lands surrounding the cells of the tone and process anilox rolls are initially as narrow as one-half thousandth of an inch, they have largely eliminated the automatic objection to using the 45° plate separation angle for one flexo color-process plate. Thus, there has been a marked and growing trend, during the past 10 years, toward using the same color separation angles for flexo printing plates as those traditionally used for litho and letterpress. So long as the lands do not wear to a point where they become wide enough to print as continuous lines or screens which generate objectional moire patterns, the anilox rolls do not constitute a problem due to the fact that they are engraved at a 45° angle.

3.11.2 *Creating moire patterns*

Typical moire patterns may be created in various situations, including:

1. By superimposing two different line patterns, one of which shows 32 lines per inch and the second 40 lines per inch, both at the same parallel, horizontal angle. If both of these line patterns were identical, i.e. either 32 or 40 per inch, and perfectly parallel, no moire pattern would result
2. When two identical, e.g. 40-per-inch, straight-line patterns are superimposed, one at an angle of 10° to the other, which is horizontal. Here the moire pattern is coarse and somewhat similar in nature to that in (1), but primarily oriented vertically
3. With two identical 40-count, straight-line patterns in which one is superimposed over the horizontal pattern at an angle of 30° to the horizontal, a much finer moire pattern appears with only a 20° increase between the two relative screen angles.

When one 40-count straight-line pattern is superimposed over an identical horizontally mounted 40-count pattern, at a 45° angle, the classic, objectionable moire pattern completely disappears and only a uniform, true 45° angle screen is visible. This 45° angle is essentially the same angle at which our anilox rolls are usually engraved, and which we have for some time been told we must avoid duplicating in the color separation angle of any flexo color-process printing plate.

Flexographers have learned from actual experience over the past 10 to 15 years that not only is this not always true but that there are times when the process color printed from a 45° angle engraved anilox roll surface and with a 45° angle engraved plate appears to show less tendency to develop a moire pattern than any of the other process color plates produced with standard angles other than 45°, provided that the anilox roll surface is not badly worn. For instance, most new 'tone and process' anilox rolls have a cell wall or land width at the surface of about 10 to 12 μm, which does not exceed one-half thousandth of an inch. As the anilox roll wears, the land width at the surface increases, and when that measurement increases to double its original width a moire pattern is likely to appear in a halftone or color-process print.

3.11.3 *Plate screen versus anilox screen*

It has been found that the development of a moire pattern in a flexo halftone or color-process print depends greatly on the relationship between the plate screen count and the anilox roll screen count. For example, where two identical line screens are superimposed at an angle of 10°, the moire band is extremely coarse and objectionable. When the same line screens are superimposed at a 30° angle, the resulting moire band, 'beat' or frequency is still noticeable and objectionable but much finer. Where these same identical line screens are superimposed at a 45° angle, the moire band is virtually invisible.

When two dissimilar screens are superimposed, similating the effect of printing a plate screen on a flexo press with a conflicting anilox roll screen count, even though the angles are the same, the pattern appears as double, coarse interwoven threads. The two screens used are a 50-line, 30% value, 45° screen and a 65-line, 30% value, 45° screen, superimposed at a 90° angle. When the same two screens are superimposed at a 45° angle, the pattern appears as 'diamond' shapes. The important difference causing the marked appearance in the moire band widths is in the interfering screen counts.

(a) *Nomograph of interfering screen.* The effect of interfering screen counts of plate screens versus anilox roll screen in generating objectionable moire patterns in the print was recognized and reported in a talk by Nason at the 1969 FTA Forum. In his presentation on the subject of 'Understanding Moire Patterns in Flexo Halftone Printing', he constructed a graph on which the horizontal axis was plotted showing the engraved anilox roll screen counts from 200 to 400 per inch. On the vertical axis of the graph was plotted the plate screen counts, from 65 to 125 lines per inch. On this graph, Nason then plotted four diagonal lines, each of which represented combinations of plate screen counts versus anilox roll screen counts found to produce moire band widths narrow enough to be virtually invisible. For instance, any combination of plate screen count and anilox roll screen count which falls on or very close to a point on any one of these lines will not produce an objectionable moire pattern due to the interference of the plate screen and anilox roll screen.

Nason's graph covered a range of plate screens from 65 lines per inch to 125 lines per inch and a range of anilox roll screens from 200 cells per inch in each direction to a maximum of 400 cells per inch in each direction. At the time Nason's work was done, this range of plate and anilox screen counts seemed adequate, but because of advances in flexo printing plates, anilox rolls and operational factors, we have extended the range of plate screen values from 65 to 205 lines per inch, and the anilox roll screen values from 200 to 600 lines per inch in each direction. Figure 3.29 shows the range of screen counts in terms of current flexo printing plate usages and anilox roll screen availabilities for color process printing.

(b) *How to use the nomograph.* In using the data in Figure 3.29, for example, if the flexographer were to run a color process job with plates made with a 100-line screen, he would want to know the best anilox roll to use in order to avoid objectional moire bands or patterns. He would find the 100-count plate screen identification of the left hand, vertical side of the chart, follow that line horizontally toward the right side of the chart to the point where it intersects the upward-sloping diagonal center lines of the bands within which any moire formed due to interference of the plate screen and the engraved anilox roll screen will be least visible.

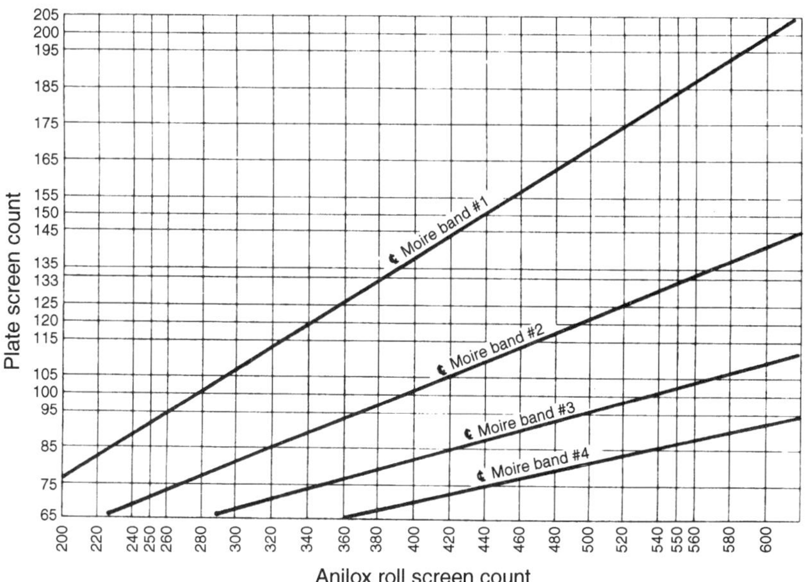

Figure 3.29 Variation of plate screen count with anilox roll screen count.

From that point, our flexographer will look for the vertical line indicating the anilox roll screen count which intersects the center line of the invisible moire band location. If the desired plate screen horizontal line intersects these other two lines, then that combination will not produce an objectional moire due to the interference of the plate screen and anilox roll screen. The farther the selected combination of plate screen and anilox roll screen intersection falls from the desired moire band center line, the larger and more objectional the moire pattern formed by the interference of these two screens will be.

Returning to our example, using projected 100-count plate screen, this intersects the center line of moire band number 1 at almost the same point as anilox roll screen count 280, and if such an anilox roll count were available it would be the obvious choice. Since our printer does not have a 280 anilox roll, he moves on to the intersection of the 100-count plate screen line and its intersection with the center line of moire band number 2. We find that the line for a 400-count anilox roll intersects the 100-screen plate line at a point very close to the center line of moire band number 2. Since a 400-line anilox roll is available and it is a finer screen and should give better quality print for color process work, it is a logical choice in this instance.

Using a popular 133-line plate screen for our second example, and proceeding in the same fashion, we find that a 380-count anilox roll would be excellent from the moire standpoint, but we do not have that count anilox roll available so we move on horizontally to the intersection of our 133-count plate screen line to its intersection with the center line of moire band number 2. We find that a 560-count anilox roll, which we do not have, would be almost perfect from the moire standpoint, but the 550 anilox roll, which we do have, is almost as close to the center line of moire band number 2 and should provide fine quality color process printing and therefore is an obvious choice.

(c) *Relating the screen to the roll and vice versa.* One of the most helpful things about Figure 3.29 is that it works in both directions. It not only helps the flexographer to find the best anilox roll screen to use with a given plate screen but also indicates a given plate screen to use with a given anilox roll screen. This is particularly important to a flexographer because most flexographers have a very limited number of different anilox roll screens on hand, but they can usually order a larger variety of screened plates, get them faster and at a much lower cost than anilox rolls.

For instance, if a flexo printer has 300-screen anilox rolls on hand, he need only follow the line for that anilox roll screen to the points where that anilox roll line intersects with the center line of either moire band number 1 or moire band number 3 to find that either a 105-line screen plate or a 65-line screen plate would ensure a virtually invisible moire band.

If this printer had a set of 360-screen anilox rolls, he would find that either a 65-line plate screen or a 125-line plate screen would ensure a virtually perfect invisible moire with a 360 anilox roll, a 75-line plate screen would be almost as good and a 95-line plate screen would probably be satisfactory. A 400-screen count anilox roll would be satisfactory for use with a 100-line screened plate, and a 550-count anilox roll would be excellent for use with a 185-line screened plate, and very good for either an 85-line screened plate or a 133-line screened plate.

3.11.4 *The basic principles of moire*

Historically, moire screen pattern has been a problem in the conventional letterpress and offset processes. By the same token, as flexography, essentially a letterpress process, goes further into the field of process reproduction, it also finds itself faced with the problems presented by moire.

Since the basic causes of moire can almost without exception be traced to the clashing of two or more regular patterns of different periods or frequencies, or of two or more regular patterns of the same frequency oriented at angles to one another, a discussion of the basic principles of moire, from the standpoint of the conventional halftone pattern will help in reaching some understanding of the problems as they occur in flexography.

(a) *Conditions for visible moire.* The conditions for visible moire pattern are present whenever two regular patterns are superimposed which differ either in frequency (that is, number of lines per inch), or in angle, or both. Let us first take the simplest possible case to illustrate the basic conditions.

This, of course, is the familiar case of two picket fences, one a short distance behind the other (Figure 3.30). Because of the difference in spacing between the pickets in the two fences from the viewpoint of the observer, the pickets in the farther of the two fences are alternately hidden by and then visible through those in the nearer fence. If, as shown in the figure, the pickets are black and the background white, we will see a very pronounced pattern of alternately light and dark bars, and these bars will be spaced much further apart than the pickets. The actual spacing depends on the relative spacing of the pickets in the two fences as seen from the observer's position.

Figure 3.30 The simplest situation for the creation of a moire pattern, as exemplified by two picket fences, one a short distance behind the other.

Now, how could we change the color of these pickets, or the material from which they are made, so that the moire pattern would no longer be visible? One obvious method would be to paint them so that they matched the background. Then, however, not only would the pattern be invisible but so would the fences. This solution, although simple, has little else to commend it, since ink which was indistinguishable from the paper on which it was printed would not have sufficient demand to keep you in business very long!

(b) *Two transparent colors.* There is another approach, however. If the pickets in both fences are of two different transparent colors, and are viewed against a white background, then it is only necessary that the mixture of the two colors when they are juxtaposed (that is when the pickets of the further fence are seen filling the spaces between the nearer fence) matches visually the mixture of the white background and the color resulting from the superimposition of the pickets (that is when the pickets in both fences line up). It can easily be shown that this condition is fulfilled when the pickets correspond to pairs of the ideal process colors, that is, colors which meet the following conditions:

1. Each one absorbs all the light in the region of the spectrum which it is intended to control and transmits all the light in all other regions of the spectrum
2. No color absorbs in the same region as any other color
3. There is no part of the visible spectrum where at least one of the colors does not absorb

This is all a complicated way of stating the characteristics of the ideal process colors, characteristics which can be shown far more simply in terms of spectrophotometric curves. Figure 3.31 shows curves for a set of process colors meeting these conditions.

(c) *Why moire will not be visible.* If we imagine the pickets shown in Figure 3.30 to consist of transparent films of a pair of these colors – for instance, the red or magenta, and the blue or cyan – and consider each region of the spectrum by itself, we believe it will be clear why the moire will not be visible. In the blue area, neither color absorbs or otherwise affects the light from the background, so it will make no difference visually whether the pickets are side by side (juxtaposed) or behind one another (superimposed); in either case 100% of the blue light will reach the eye. In the green region, when the pickets are juxtaposed, the magenta pickets absorb all of the light, but only half the area. The cyan pickets transmit all the light, but again, in only half the area. Since zero divided by two plus 100 divided by two equals 50, the net result is that one half the light in the green region reaches the eye when the pickets are juxtaposed.

On the other hand, when the pickets line up behind one another, no light is transmitted in the green region where they are superimposed, since the cyan

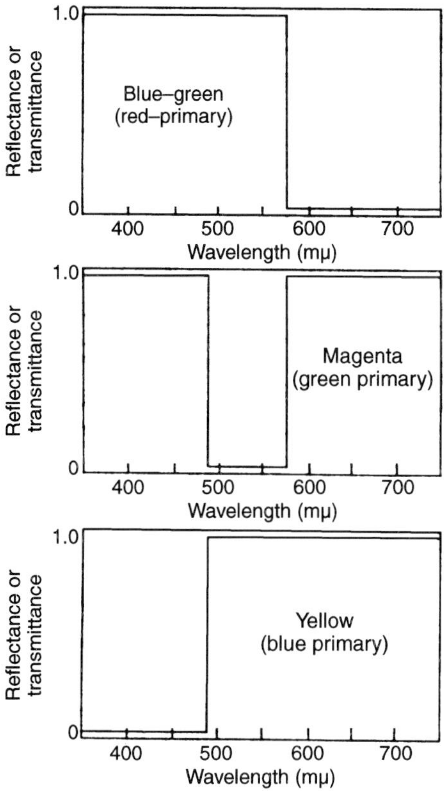

Figure 3.31 The spectrophotometric characteristics of ideal process colors.

pickets absorb all the light in this region. However, since there are equal areas of white light between the superimposed pickets, we will again arrive at 50% of the green light being returned to the eye. In a similar way, 50% of the light in the red region will enter the eye, regardless of whether the pickets are juxtaposed or superimposed. To sum up, the light that we will see will consist of 100% of the blue region and 50% of each of the green and red regions, regardless of whether the pickets lie side by side or lie on top of one another. Therefore no pattern will be visible, despite the fact that the geometrical conditions for moire are present.

Similarly, the visibility of a moire pattern in printing is directly dependent on the contrast between the two colors resulting from the superimposition of two inks and their juxtaposition.

(d) *Yellow causes least moire.* The lowest contrast occurs where the yellow is one of the colors. This is basically because yellow approaches most closely the ideal characteristics mentioned above and is, of course, the reason why yellow is invariably the color printed at the angle which would cause the most objectionable pattern. It also shows why screen angle changes are necessary, since otherwise the overall color of the picture would be extremely sensitive to very slight variations in register.

The cross-line screens used in the graphic arts have what has been called four-fold symmetry. This means that when the screen is rotated through a full circle, or 360°, there are four times that the pattern can be superimposed exactly. Therefore, there is only a quarter of a circle, or 90°, available for positioning the screen to minimize or avoid the pattern resulting from overprinting different colors.

It can be shown quite easily that the least objectionable pattern occurs when two screens of the same ruling are spaced 30° apart. However, this allows for only three colors, and by far the greatest amount of color printing employs four: yellow, red (magenta), blue (cyan) and black.

(e) *Screen angles are critical.* Although it would seem that the best solution for four colors would be to set the screen angles $22\frac{1}{2}°$ apart, this angle produces quite an objectionable moire in most cases. It is generally much more satisfactory to maintain the 30° angle between three of the colors and place the yellow halfway between two of them. Of course, this means that the yellow is at 15° to two of the other colors, and not only is the 15° moire large enough to be objectionable if the contrast ratio mentioned previously is high, but there is a secondary moire produced, also at a 15° separation. This comes about because the orientation of the moire produced by two cross-line screen patterns of the same ruling at 15° to one another is halfway between, or at $7\frac{1}{2}°$ to both.

The moire patterns resulting from a yellow placed halfway between a red and blue, spaced 30° apart, are themselves only 15° apart and cause a compound moire which can, in many cases, be even more objectionable than the primary, or two-color moire.

(f) *Screen ruling variable.* Fortunately, we have one other variable at our disposal, namely, the screen ruling. In overprinting two colors with different screen angles, the orientation as well as the period of the moire pattern varies with the screen ruling. Hence, if we can find a screen ruling for the yellow plate that will put the two primary moire patterns 30° away from one another, the secondary moire should be at a minimum.

If the red, blue and black plates are made with 120-line screens spaced 30° from one another, and the yellow is made with a 133-line screen oriented halfway between two of them, the primary moire patterns are almost exactly at 30° on either side of the yellow. The resulting compound moire is then at very nearly 60°, which is equivalent to 30°, because of the four-fold symmetry mentioned earlier.

Calculations show that the higher ruling should be actually 1.115 times that of the lower one to fulfill the above conditions exactly when the higher ruling is oriented between two lower ones angled 30° apart. If the lower of the two rulings is at 120 lines per inch this requires that the higher one be at 133.8 lines per inch (close enough to 133 for all practical purposes). To avoid the possibility of very low frequency beats under these circumstances, the lower ruling should be 0.707 times that of the higher ones.

(g) *A complication arises.* Although it might seem that this same ratio of two screen rulings could be used if the yellow printer were at a lower screen ruling than the other two instead of higher, a complication arises. In this case, the primary moire pattern will be at 15° on either side of the yellow, instead of at 30°. The two primary moires then will be 30° apart, which is effectively the same as in the first case and should result in the same reduction in secondary moire pattern. However, with the primary moires at 15° from the yellow, the yellow–red moire will line up exactly with the blue screen rulings and the yellow–blue moire with the red screen rulings.

Since the perception of moire is basically determined by the contrast between areas where the dots are superimposed and those where they are juxtaposed, there are many other factors which can affect it. To mention only a few: there are trapping, absorption characteristics of the stock, under-color removal, regular patterns in the printing surface of the stock itself, and differences in color rotation and in proofing versus production printing.

It is certain that most of you can think of additional variables which may have given you problems in your own operations. I do hope, however, that this discussion has given you something to think about and at least some clues that will help you when moire next raises its ugly head.

3.12 Drying power of inks

Flexographic inks are defined as 'fluid, fast-drying inks'. It is sometimes assumed by non-flexographers that this means that flexo inks dry 'immediately', and without the aid of drying equipment of any kind. As experienced flexographers understand, this is not true.

Flexo inks consist of colorants dispersed in a vehicle plus additives. The colorants may be pigments or dyestuffs. The flexo vehicle usually consists of one or more resins dissolved in a volatile solvent, typically alcohol or water. Additives may be various modifiers such as waxes, anti-foam agents, slip or anti-slip agents, etc. Such flexo inks do not serve their ultimate purpose until the volatile solvents are evaporated. This is the process we call 'drying'.

Drying of a fluid (alcohol or water soluble) flexo ink involves the technologies of flow, of heat transfer and of ink formulation. The technologies of flow and of heat transfer are equipment problems; ink formulation is of course the ink technician's problem.

In flexography, ink drying is usually accomplished by using hot air as the drying medium. The heated air flow must be controlled in terms of temperature, velocity and direction.

The inkmaker's concern with ink drying involves compromises between drying speed and leveling characteristics of ink. If the ink dries too fast, it may dry on the surface of the printing plate, especially when printing halftone or color process work. If the ink dries too slowly, the printer will be faced with problems of ink trapping, blocking and setoff. To further complicate the ink formulators' problems, flexo press dryers may vary from the most modern to minimal.

3.12.1 *Hot air drying systems*

Most modern flexo press dryers use high-velocity hot air drying systems engineered based on an understanding of the 'boundary layer' concept of the behavior of a fluid (gas or liquid) adjacent to the surface over which it flows. The significance of the boundary layer behavior in a high-velocity hot air dryer is that the boundary layer tends to act as a barrier to the transfer of heat from the fluid to the surface of the printed substrate and to the transfer of solvent vapor from that surface to the hot air.

The effects of the high velocity hot air in a modern flexo dryer are:

1. Improvement in the transfer of heat from the hot air stream to the printed web which is being dried
2. Removal of residual solvent from the surface of the boundary layer, permitting further diffusion of solvent from the printed substrate surface, through the boundary layer
3. Reduction in the thickness of the 'boundary layer'

As with most natural phenomena, the relationship between hot air velocity and boundary layer thickness is not linear, i.e. if the air velocity is doubled, the boundary layer thickness is not halved. Figure 3.8 shows the effect of dryer air velocity on the boundary layer thickness of air at a temperature of 250°F, and with 2″ between inlet and outlet ducts versus mean air velocity.

As the velocity of the heated air in the press dryer increases, its effect on a further decrease in boundary layer thickness is substantially reduced. It is apparent, from Figure 3.8, that there is an economically practical limit to the velocity of dryer air, above which the cost of maintaining higher velocities does not justify the results. Referring to the example in Figure 3.8, the practical limit to air velocity is apparently about 15 000 to 20 000 fpm, which represents the best compromise between cost and effectiveness. Many modern flexo press dryers are designed and built to maintain nozzle air velocities of from 10 000 to 15 000 feet per min, for the reasons outlined above.

From time to time flexographers have raised the question of why flexo press dryers do not have nozzle velocities in the range of 25 000 to 30 000 feet per min, since this would increase the drying rate of the applied ink of coating. This is a logical question, but the answer is a bit complicated. The effect of dryer air velocity on the rate of heat transfer is illustrated in a qualitative sense in Figure

3.9. Note that the heat transfer rate increases rapidly and continues to rise with increasing air velocity, past the point where the boundary layer thickness is no longer significantly affected by increasing air velocity. The rate of heat transfer obviously does not follow the same law relative to air velocity, which applies to boundary layer thickness. Therefore, beyond the point where boundary layer thickness is no longer strongly affected, as in the range of 20 000 fpm and up, heat transfer rate still is affected.

Drying is not only a matter of heat transfer rate to the wet ink film. It also involves solvent transfer from the heated substrate, back through the boundary layer and into the main air stream. This backward diffusion of solvent vapor is more strongly affected by the boundary layer thickness than by the forward transfer of heat. Hence, in spite of the improved heat transfer of solvent vapor at very high air velocities, transfer of solvent through the boundary layer does not seem to be substantially improved beyond the 15 000 fpm range of the air velocity in the case of most current press dryer designs.

The improvement in heat transfer at the very high air velocities does not mean that solvent removal will be acclerated proportionately. In high velocity hot air drying, increasing the air velocity results in these effects:

1. Heat is transferred more readily to the wet ink film
2. The volatile solvent vaporizes more rapidly
3. The reduction in boundary layer thickness increases the rate of solvent diffusion from the surface of the substrate to the main air stream from which it is then exhausted

3.12.2 *When is a flexo ink dry?*

So much for the theory and operation of typical high velocity hot air flexo ink dryers, which unquestionably have helped to increase the operating speeds, quantity and quality of flexo printing during the past 25 years. However, we should remember that we really do not have a simple and always accurate definition of when and under precisely what conditions is a printed flexo ink film really 'dry'?

Various tests during past years have proven that most flexo ink films at the rewind end of the press do not have all the solvent removed, and actually a substantial period of time is often necessary before the printed ink film attains a truly 'bone-dry' condition. This is related to the fact that so many 'laws of nature' are inclined to be nonlinear.

Figure 3.10 illustrates, in a qualitative fashion, the solvent remaining in a normally 'dried' printed ink film versus time in the dryer. Beyond a certain point, very long drying times, which usually translate into much lower press operating speeds, are required in order to make any worthwhile change in the relative volume or percent of solvent left in the ink film. As a practical matter, therefore, a compromise is necessary between practical press operating speed and an acceptable level of retained solvent.

Ink formulators can help flexographers by knowledgeable selection of resin and/or ink vehicle binders and solvents, since some resins will retain some solvents more than others. Continual progress has been and is being made in developing resin–solvent combinations with optimum solvent release which still retain all the other characteristics needed in a good flexo ink. When an ink formulator arrives at such an optimum vehicle resin/solvent combination, he/she can make an additional worthwhile contribution to the total flexo printing operation by passing along to the pressman the solvent mixture which should be used to reduce the ink to press running viscosity and to maintain the optimum printing qualities throughout the press run.

All too often, ink suppliers have provided flexo press operating personnel with little information except the vague impression that an alcohol-based ink should properly be thinned at press-side with alcohol only, and a water-based ink with water only. This practice frequently results in second-rate print quality which may vary throughout a given run and from one run to another.

It is important for the modern flexographer to recognize not only the importance of using balanced solvent mixtures for each different type of ink formulation but also the importance of properly designed, constructed and operated press dryers on which all ducts are kept clean and properly balanced or 'tuned'. Also, today's flexographer must recognize the extent to which he/she depends on all suppliers of inks, plates, press, dryer, papers, films, foils, laminants and certainly on FTA and its various activities, workshops, seminars, etc. for the latest and best industry technology and developments.

3.12.3 *Turbulence at the boundary layer as an aid to drying efficiency*

In the previous section, some of the facts and factors relative to medium and high-velocity hot air dryer design and operation for flexographic inks, coatings and presses were discussed. From the standpoint of cost, press operating speed and dryer efficiency, there are practical limitations which the flexographer must recognize, and in order to achieve the maximum cost effectiveness from his/her flexo press dryer design the flexographer will benefit from the intelligent cooperation of the ink formulator.

Figure 3.32 shows schematically the relationship between typical dryer high-velocity hot air impinging on the boundary layer, ink film and substrate surface of the moving web.

In the area of most direct impingement or maximum dryer air velocity, the boundary layer of air tends to be thinnest, especially at minimum press operating speeds and where the dryer is so designed as to cause the direction of the high velocity air jets to impinge the boundary layer of air at an angle opposite to the direction which the boundary layer and web are moving through the dryer. Such a dryer design tends to develop a 'scrubbing' action between the hot air jets and the boundary layer of solvent-laden air, increasing the dryer efficiency. As the high velocity hot air stream from the dryer jets mixes with the boundary layer of

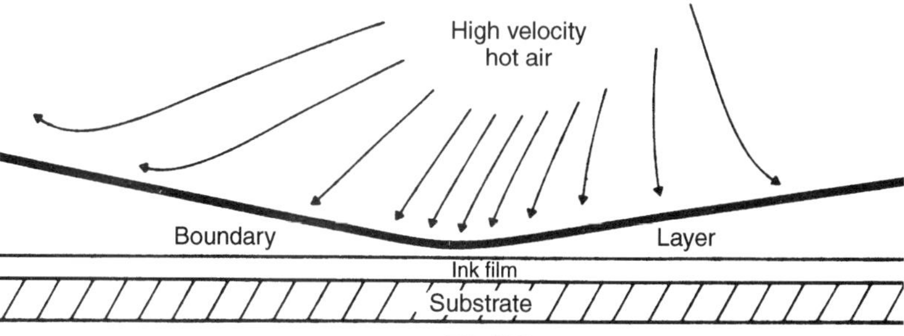

Figure 3.32 Relationship between high-velocity hot air impinging on the boundary layer, ink film and substrate surface of the moving web.

air and solvent vapors, the solvent vapors are diluted and removed as the solvent vapor saturated air is exhausted from the dryer hood. This drying process actually consists of the following four steps:

1. Diffusion of solvent from the interior of the ink film to its surface
2. Vaporization of solvent in the boundary layer of air at the interface of the ink film surface
3. Diffusion of solvent vapors through the boundary layer of air
4. Scrubbing and sweeping the solvent vapors away through the exhaust system of the dryer

The flexo press dryer is essentially designed to heat the printed ink film, vaporize its solvent, scrub, and eliminate the ink solvent vapors through the dryer's exhaust system. The effectiveness of these dryer functions depends on the composition of the solvent and resin portion of the ink formulation. Likewise, the vaporization of the solvent at the surface of the ink film at a given temperature depends on the composition of the ink. The remark often made by flexo pressmen, to the effect that one ink formulation is 'harder to dry' than another, results from the fact that one flexo ink solvent–resin combination may release its solvents more rapidly than another.

It is a proven fact that vapor pressures of solvents in equilibrium with resin and polymer solutions vary widely with the types of solvent and polymer used. For instance, solutions of different polymers at the same volume concentration in the same solvent give different vapor pressures. Solutions of the same polymer in different solvents at the same concentration and temperature give different vapor pressures. Vapor pressure is a measure of the tendency of a particular solvent to escape from a particular solution. It seems logical to expect that those resin–solvent combinations which give the highest vapor pressures at given concentrations and temperatures should be the easiest to dry. Those giving very low vapor pressures under similar conditions would be more difficult to dry.

In referring to a 'fast' or 'slow' drying solvent, what is normally meant is that the 'fast' drying solvent has a high vapor pressure at a given temperature and the 'slow' drying solvent has a low vapor pressure. Such a reference is usually made to the pure solvent, so with respect to flexo inks the vapor pressure of a single solvent provides only part of the story because an ink formulator and a flexo pressman are both really concerned with the vapor pressure of a given solvent over the particular resin solution to be used.

As an example of what this means as compared to the vapor pressure of a single solvent see Figure 3.33. In this example, we have used two solvents, carbon tetrachloride and benzene, neither of which is a flexo ink solvent, over rubber solutions in each of these solvents versus rubber concentration at a constant temperature of 25°C. Figure 3.33 shows the vapor pressures of pure carbon tetrachloride and pure benzene, as they vary with temperature. As noted from Figure 3.34, at every temperature the vapor pressure of carbon tetrachloride is higher than that of benzene. Thus, at every temperature, the tendency of carbon tetrachloride to evaporate is greater than that of benzene. From this, a logical conclusion would be that carbon tetrachloride is a 'faster' drying solvent than benzene, and so far as the pure solvents are concerned this is a valid conclusion.

(a) *Modifications to vapor pressures.* Figure 3.33 shows, however, what happens to these vapor pressures when a polymer, rubber in this case, is dissolved in these solvents. Up to a volume fraction of about 0.57 (57% by volume), the carbon tetrachloride vapor pressure is significantly higher than that of benzene, but as the rubber solution becomes increasingly concentrated, the

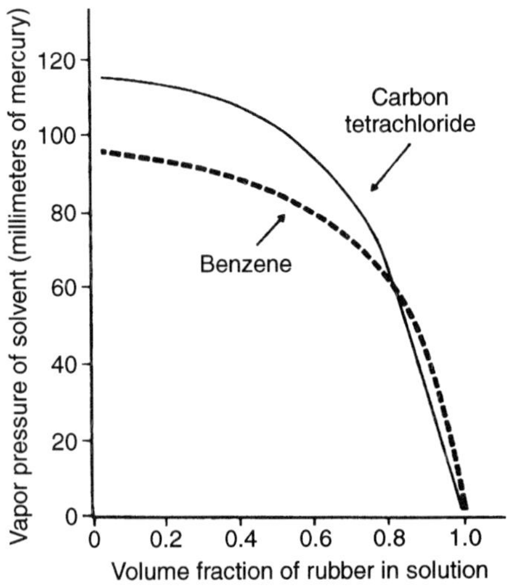

Figure 3.33 Vapor pressure of carbon tetrachloride and benzene over rubber solutions versus rubber concentration.

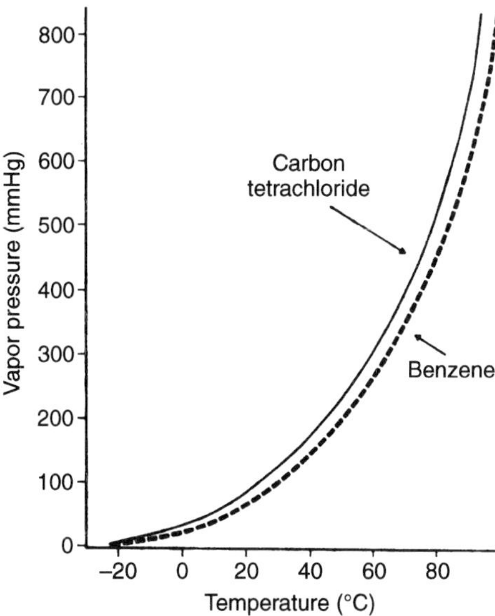

Figure 3.34 Vapor pressure of carbon tetrachloride and benzene versus temperature.

vapor pressure of the carbon tetrachloride becomes lower than the vapor pressure of the benzene, a rather unexpected result.

This means that with the same dryer, and using the same air velocity and temperature, a solution of rubber in carbon tetrachloride would begin to dry more rapidly than a similar rubber solution in benzene, but the final drying would proceed more slowly from the carbon tetrachloride solution than from benzene solution. A different resin or polymer dissolved in both of these same solvents could very well have the opposite effect.

From the flexo ink chemist's viewpoint, when he has a choice of solvent–resin combinations, he will obviously choose the combinations where the interaction of solvent and resin will promote rapid solvent release. In this way it is possible for the ink formulator to influence the drying rate of a flexo ink or coating without changing the drying equipment or its adjustments.

(b) *Drying rate.* The drying rate of an ink or coating is, of course, only one factor to be considered in its formulation. Such characteristics as gloss, pigment wetting, film strength, adhesion, etc. are also important considerations in ink formulation; and depending on individual requirements some of these factors may be more important than solvent drying rate.

Nevertheless, there are frequent situations where such factors as retained odor in printed packaging materials and complete freedom from traces of solvent in printed films are of primary importance in formulating as well as in printing flexo inks, coatings and laminants. In order to adequately discharge this aspect of his total responsibilities as a formulator, today's flexo chemist needs to know about the vapor pressures of solvents over resin and polymer solutions, the effects of various additives used in the whole range of flexo formulations on vapor pressures, and much more.

The accumulation of such data is not easy, since vapor pressures of polymer solutions are not easy to determine, but when the vapor pressures are known for a given solvent–resin combination it becomes possible to calculate an interaction parameter. The higher this parameter, the more rapidly will the resin or polymer release the particular solvent. The lower the parameter, the more likely it is for the polymer to retain the solvent. For instance, in the example involving rubber in benzene and rubber in carbon tetrachloride, the rubber–benzene interaction parameter has a value of 0.44, while the rubber–carbon tetrachloride parameter is very nearly 0.0.

It would be helpful to flexo ink and coating chemists and formulators if such interaction parameters were included in resin and polymer data, along with other useful information such as data of solubilities, compatibilities, viscosities of solutions, etc. to help the formulator select the best resin–solvent combinations for solvent release as well as for pigment wetability, gloss, solubilities, film strength and adhesion. Hopefully, we will soon see that day.

3.13 Variations in thickness of substrates

In a recent discussion of a group of Fellow Flexographers the subject of the increasing popularity and growth of various laminated structures being printed by flexography and used for both flexible and non-flexible packaging came up. Because the topic was of such obvious interest to so many flexographers present at the industry meeting, the discussion continued into the wee hours of the morning. We decided you might like to be included in at least some aspects of this discussion.

First, since we were surprised at how fast the discussion grew to include virtually a whole roomful of flexographers, many of whom were not flexible packaging materials printers, we began to ask about the different phases of printing and converting represented by each of these vocal participants. Because the conversation started among a group of flexo printers and converters of flexible packaging materials, the assumption was that probably most of the other flexographers interested in coated and laminated materials would also be flexible packaging-oriented.

Before long, however, it became obvious that a number of both listeners and contributors to the discussion involved flexographers primarily concerned with other areas. For instance it was apparent that a significant number of label printers considered themselves to be very frequently printing not only coated but a considerable variety of laminated materials. In addition to printing, many such label and other narrow-web flexo printers were involved in die-cutting, slitting, sheeting and other aspects of laminated structures. As the discussions continued, it was obvious that many of the flexo narrow-web printers were contributing valuable knowledge of flexo printing and converting of combined and laminated structures to their wide-web flexo printing friends in the flexible packaging field.

3.13.1 *Common problems*

One of the most common problems shared by all the flexographers involved in this discussion seemed to be concerned with variations in the total thickness of the combined laminated structures, and the effects these substrate thickness variations had on the resulting print quality. The total thickness variations, of course, varied considerably, depending on the number, composition and caliper of the individual plies in the laminated structures. Some flexo corrugated printers joined the group as it grew, and the flexo printers of preprinted linerboard surprised most of the narrow- and wide-web printers of flexible packaging materials when they mentioned that it is not unusual for some corrugating linerboard stock to vary in total caliper or thickness by as much as 0.010″ (ten-thousandths of an inch), which is unheard of in most laminated flexible packaging substrates.

At that point, however, several flexo printers of pre-formed corrugated containers contributed their own experience, explaining that it is not unusual for them to encounter pre-formed corrugated containers which vary in total caliper as much as 0.050″!

The total range of caliper or thickness variations in laminated or combined substrate structures encountered by this entire group of flexo printers, therefore, according to our survey, was from a minimum of about 0.001″ for the two- to four-ply flexible film and foil structures to a maximum of 0.050″ for pre-formed corrugated containers.

The most obvious problem which all the flexo printers in this particular group complained about was the effect which the caliper variations had on the printed quality results, particularly where full, four-color process printing was required.

3.13.2 *Compressibility of plate mounting material*

Regarding the steps which these flexographers had found most effective in overcoming the quality printing problems caused by the thickness variations of various combined and laminated structures, the most universally successful was the use of foamed two-way adhesive plate mounting materials. Of course there is no one foam-backed two-way adhesive plate mounting material which is suitably compressible from as little as 0.001″ to as much as 0.050″ with equally satisfactory results. Nor is there a single foam-backed plate-mounting material satisfactory for use with tape-printing flexo plates 0.062″ in thickness, wider width presses with 0.125″ overall thickness. Fortunately, however, a variety of foam-backed flexo plate-mounting materials are available today, from a number of different suppliers and with compressibilities ranging from about 0.005″ to more than 0.050″ so that the requirements of all flexo presses and all normal single- or multiple-ply, extruded, coated or laminated structures can be accommodated.

One of the earliest developers of foam backing materials for flexo printing plates was E.L. Harley, Inc., one of flexography's early and well-known pioneers. Earl Harley, early in his career, invented and marketed a machine for automatically grinding the backs of flexo printing plates, resulting in more uniform plate thickness, greater accuracy and improved print quality. These Harley plate grinders became well known and widely used throughout the world.

The second Harley contribution to the flexo industry was his development of the optical flexo plate mounting and proofing machine for mounting and pre-proofing flexo printing plates on plate cylinders prior to press installation and makeready. This development added to his reputation, and when a few flexo printing pioneers began to experiment with early types of two-way adhesive foam mounting materials, Earl Harley saw this as another opportunity for the development of a new material to improve the printed quality and expand the scope of the flexographic printing process.

He recognized that rubber and other similar elastomeric flexo printing plates were not compressible and therefore could not readily and accurately conform to uneven printing surfaces. At the same time, it was becoming increasingly obvious that a growing range of substrates being printed by flexography tended to vary in uniformity of thickness, due to factors such as composition and structure of coated combined, laminated and extruded packaging materials. There was, simultaneously, a growing demand for higher quality printed graphics, including full-color process reproductions. In order to meet the needs of the growing flexo industry, Harley developed an accurate, truly compressible, closed-cell foam, PVC, a flexo plate mounting material with a pressure-sensitive adhesive on each surface. Field trials using these and similar materials repeatedly proved that flexo printing of four-color process graphics, in plate screens up to 133 lines per inch, with total compression of 0.005″ consistently resulted in uniformly acceptable printed quality with minimum dot growth.

Closed-cell foam-cushion backing, according to our own latest information, in total thickness of 0.015″ and 0.020″, is intended primarily for all popular flexible packaging substrates (Figures 3.35 and 3.36).

A range of compressible backing materials in greater thicknesses have been developed in recent years by the Rogers Corporation, under the trade name of 'R/bak' to meet the needs of printers of heavier and thicker substrates, ranging from corrugated containers and folding boxes to preprinted linerboard and multiple-ply laminants. The R/bak material is an open-celled microcellular compressible polyurethane backing for molded rubber or photopolymer flexo plates, and it is available in thicknesses of 0.040 to 0.180″ in increments of 0.020″. Special modifications of the R/bak development are available in other

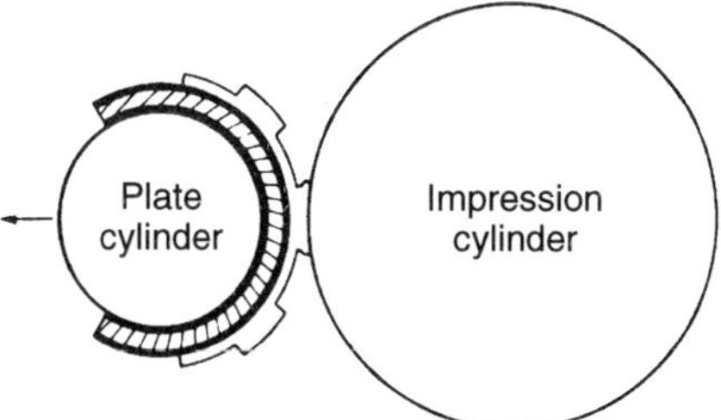

Figure 3.35 Owing to the fact that rubber is not compressible, spreading or mushrooming of the face of the printing plate to compensate for inaccuracies occurs. This action leads to lower quality printing and slower press operating speeds, due to plate cylinder bounce.

thicknesses and cast on polyester backings of 0.0005″ thickness. According to Warren Taylor, Rogers Corporation, corrugated printers who have used R/bak have reported the following benefits:

1. Prints lines and bars without distortion
2. Prints consistently scannable symbology
3. Less waste due to poor print quality
4. Gives much longer plate life
5. Prints both solids and screens on same plate

Whether today's flexographer is printing flexible packaging materials, corrugated containers, folding boxes or anything in between, he/she will benefit from the knowledge and use of compressible plate-mounting materials.

3.14 Flexographic printing on corrugated board

Flexo printing in the corrugated container industry is still in its infancy, and we are all pioneers at the present time. In fact, as flexo printing plate manufacturers have found, we have learned that flexo corrugated printing plants with flexo presses quickly accepted the philosophy that this equipment requires an extremely level flexo printing plate.

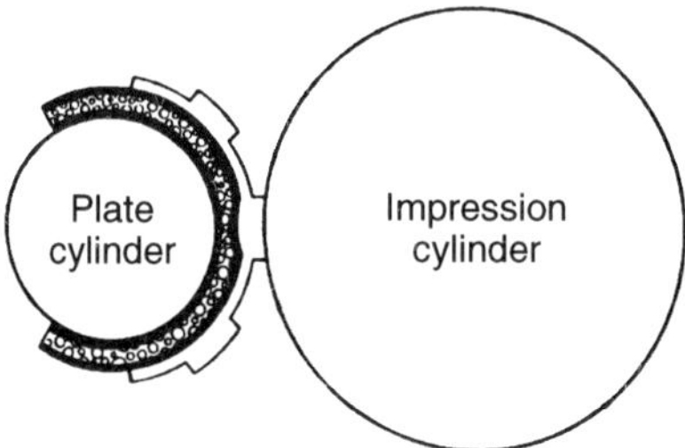

Figure 3.36 Variations of tolerances in the plate cylinder, plates and substrates are automatically compensated for by the cushioning effect.

Many flexo corrugated printing companies have established very close quality control specifications, and will reject printing plates unless they have extremely close tolerances, some in fact as tight as two or three thousandths of an inch. To meet these demands, those in the flexo printing plate business have installed new molding equipment which will mold printing plates to these rigid specifications, but many believe you do not need a perfectly calibrated, level flexo printing plate.

The sole purpose of the level plate concept is to assume that the surface or printing face of the plate is level, so that it will rightly accept ink accurately from the surface of the anilox form roller.

There is no argument about this fact, in flexo printing, because of the extreme accuracy of the anilox roll surface. The amount of contact roll pressure between the anilox roll surface and the printing plate is critical, in order to ensure an even ink film on the surface of the printing plate. If too much pressure is applied, the ink is squeezed onto the plate, causing filled-in printing and halos.

Not only will too much anilox roll pressure affect printing, but it will also wear out plates prematurely. We have personally seen a new set of plates worn out during its initial run on a flexo press by too much anilox roll pressure to the 'die' or plate surface, but if there is not enough pressure, a slight variation in the plate surface is enough to keep the low areas from picking up any ink.

3.14.1 *When ink gets to the plate*

We might also ask what happens after the ink gets to the surface of the plate? For instance, can the operator print, with sharpness and clarity, large solids, together with detail and fine type, on the same plate that is perfectly level and therefore, which exerts equal pressure in all areas? Ideally, the best solution to printing a complex layout of this sort, with solids and detail, would be to split the color and run it on two different cylinders. Then you could print all heavy coverage, requiring more pressure in one station, and all the detail in another. In most cases, this would be impractical because very few flexo presses are capable of printing more than two colors. Therefore, this would be impossible on two-color jobs. It would create two-color printing on single color jobs, requiring two plates, longer setups, and longer washups.

Not many flexo corrugated printing plants have shown a willingness to accept such an approach and for almost a hundred years, the skilled craftsmen of letterpress printing—and flexography is a part of letterpress—have known that it takes more pressure to print solids than it does to print fine detail. Pressmen have used the technique called 'makeready', which in essence adjusts the amount of plate pressure to the board or paper at the point of impression, so that fine printing areas are relieved and will not 'distort'. Some printers have felt that this technique is wrong in flexo, because it is easy to print with a light 'kiss-impression'.

3.14.2 *How level is the substrate?*

But we must ask, how level is the corrugated board stock to be printed when you receive it from the corrugator? For instance, is there washboarding? Is there warping? What are the tolerances of the printing cylinders that the plates are mounted upon?

If inaccuracies exist in any of these areas, and of course we know that they do, particularly in the corrugated board stock, additional plate pressure must be added, to compensate for uneven surfaces which we must print upon. And of course, how much pressure must be used? This is difficult to determine. But we know, with non-printing areas set back as much as 0.105″, we will still get background printing occasionally.

Now let us discuss the variable impression flexo printing plate and how it can potentially improve flexo plate printing quality. This plate has built-in make-ready, which is a component part of the plate, rather than an attachment and it is done during the vulcanizing operation.

All areas of the plate, except those requiring a softer impression, will be level, at 0.250″. The areas of fine printing and small type, such as an underwriters certificate, will show as a depression on the back of the plate, relieved as much as 0.015″. The operator must remember that the surface of the plate should remain level at all times. The relief is on the back side.

Since the relieved areas are small, there remains enough support to bridge the gap, and the printing surface, even in these relieved areas, will be at about 0.250″. A small air pocket is then formed behind the desired relief area, so that in making the impression, this printing will ‘give’, thereby creating a softer impression, and consequently a smoother print.

Those of you who mount plates by using double-faced two-way adhesive backing can accomplish the same results, by cutting holes in the backing in the desired areas, if they are cut in the right spots. Care must be used, however, in building up areas behind the plate, to relieve other areas and to ensure the same results. This will instead cause high and low surfaces, and ensure accurate inking by the anilox roller.

3.14.3 *VIP plate performs efficiently*

The variable impression plate (VIP) has performed effectively on conventional oil-type printer–slotters for a number of years. The only difference, in fact, between flexographic printing and the printer–slotter method, is the type of ink and the way in which the ink is transferred to the plate. From there on, the transfer method is identical.

In many cases, the VIP plates which were expected to be run on oil-type printer–slotter presses but which were later diverted to flexo equipment, were found to have varied only with respect to the type of ink and in the way in which the ink was transferred to the flexo plates. It was later found that these VIP plates actually performed very well.

In many instances in the early VIP plate development, operators found that the VIP plates actually performed surprisingly well, so it was learned that a perfectly level plate, in terms of calibration, is often not needed by flexo and it was found that the earlier proven theories of the oi-ink printer–slotter letter-presses could apply to flexo equally. The industry is now regularly providing these same type plates for flexo. Many test runs have been made, and the results have been found to be very encouraging.

3.14.4 *How much relief?*

So far, the main problem has been: How large an area can be relieved without affecting the accuracy of the printing surface? We know that for instance, one line, 10″ long, with 1″ type, presented no immediate problem because there is currently enough rubber around it to support it. But, if this area were to be increased to as much as five to 10 lines, this could cause a potential sagging problem, with insufficient ink volume.

Even though eventually perfected, the VIP plate may still be found to be less than perfect, because not all print-copy requires VIP printing. That is, all small type obviously cannot be made from a level flexo plate and conversely, printing all solids cannot be done from a level plate.

As long as we continue to print on uneven surfaces, such as linerboard with washboarding, we must continue to expect problems. This is a good reason that the industry must use the softer 20 and 30 Shore A durometer hardness rubber compounds. We feel confident that the flexo corrugated industry can improve print quality and also extend plate life, if the industry can continue to learn to use harder durometer rubber compounds.

3.14.5 *Harder plates may be better*

A research project conducted by the National Printing Ink Research Institute at Lehigh University some years ago found that harder durometer rubber plates transfer ink much cleaner than softer rubbers, and they leave less ink on the plate after impression. However, on uneven surfaces, such as corrugated board, harder plates were found not to lay ink as well as the softer plates.

With respect to ‘print copy’, we have seen a number of cases in which the corrugated stock has accepted print layouts that, in our opinion, many of the finest lithographers would not tackle. Many corrugated printers have two strikes against them before they ever get to the press, because they do not give enough attention to factors such as unrealistic register; printing through scores; halos that are caused by reverses in the middle of large solids, ink-trapping problems; type and detail that are too fine; unrealistic stretch factors, and color sequence problems.

4 Pigments

C. DURGAN and E. REICH

4.1 Pigments for printing inks

Pigments are colored powders that are either organic or inorganic in nature and which are used to color many substrates. Properly chosen, pigments impart brilliant color, gloss, transparency or opacity and resistance properties which are necessary to the finished ink. The pigment chosen for an ink also provides a unique drawing power for the finished package, promoting the product and making it stand out among other competitive packaging. For this reason making the proper pigment selection is invaluable for the ink formulator.

Pigments are the most expensive raw material utilized in a printing ink formula. They are also the most widely handled component and probably the least understood material in the ink formula.

Colored pigments impart the color of the printing ink by themselves or in conjunction with other materials such as metallic, pearlescent or fluorescent components. Pigments also exhibit many different properties such as lightfastness, solvent resistance, alkali resistance, dispersion, opacity and compatibility in various ink systems. Due to the many differences in properties and the wide availability of pigments, both organic and inorganic, the ink formulators must take it upon themselves to learn as much as possible about their properties. Establishing a sound relationship with their pigment suppliers will aid in properly obtaining the optimum pigment selection for their application.

With the proper understanding and selection of pigments an ink can be formulated to not only achieve the correct hue, but also give lasting performance in the finished ink film.

4.2 Classification of pigments

4.2.1 *Color index system*

Pigments are identified using an international system for classification called the Colour Index System developed by the Society of Dyers and Colourists (SDC) [1] and the Association of Textile Chemists and Colourists (AATCC) [2]. This system designates a generic name and constitution number which is given to all pigments according to their chemical composition. The generic name assigned to the pigment is based on its chemical composition, hue and sequence in which the color was registered.

A separate coding system names the category, color, identification and chemical class of the material [3].

The first letter indicates the category:

A — acid dye
B — basic dye
P — pigment
F — food pigment
L — lake
N — natural pigment
V — vat pigment

The second letter indicates the color:

B — blue	O — orange
Bk — black	V — violet
Br — brown	R — red
G — green	W — white
M — metal	Y — yellow

The Colour Index number identifies the color and the sequence of registration. The five-digit number designates the constitution number which discloses the structure of the pigment. The first one or two of these digits discloses the chemical class of the pigment [4]. An example can be seen in the coding of C.I. Pigment Red 1 (12070). C.I. Pigment Red 1 indicates that the color is a pigment rather than a dye; the hue is red and it was the first registered red pigment with a corresponding C.I. name. The chemical class is a monoazo-3-hydroxy-2-naphthanlide type.

Pigments which are chemically similar but differ in the metal, acid or crystal form used in salt formation use the same generic name but are differentiated by a colon and an additional number. An example of the different salt forms of Permanent Red 2B are as follows:

C.I. Pigment Red 48 : 1	Permanent Red 2B (barium)
C.I. Pigment Red 48 : 2	Permanent Red 2B (calcium)
C.I. Pigment Red 48 : 3	Permanent Red 2B (strontium)
C.I. Pigment Red 48 : 4	Permanent Red 2B (manganese)

The Colour Index System, with constitution numbers, is outlined in Table 4.1. When composing a constitution number where different salt forms are available, a sixth number is issued to distinguish the individual salt forms. An example of this can be seen in Blue 1 as listed in Table 4.2.

The constitution number signifies the structure of the pigment. Only one constitution number is assigned to a pigment with a particular molecular structure; however, several generic names can be associated with a particular constitution number. An example would be C.I. Pigment Blue 15 which is classified as an insoluble copper phthalocyanine blue pigment. The constitution

Table 4.1 Colour Index chemical class categories organic pigments

Chemical class	Constitution number
Insoluble azo	
Acetoacetyl	11640-11790
Heterocyclic hydroxy	12600-12825
Disazo	20000-29999
2-Naphthol	12050-12211
3-Hydroxy-2-naphthanilide	12300-12520
Precipitated azo	
2-Naphthol (sulfonic) acid	15500-16815
Other	(Two classes)
Precipitated non-azo	
Xanthene	45000-45999
Triphenylmethane	42000-44999
Other non-azo	(Three classes)
Insoluble non-azo	
Phthalocyanine	74000-74999
Anthraquinone	58000-72999
Quinacridone	73900-73999
Methene	48000-48999
Other insoluble non-azo	(Five classes)
Inorganic pigments	
Colored salts	77000-77999, for all inorganic pigments
Colored oxides	
Inorganic blacks	
Metallic pigments	
White pigments	

number associated with this pigment is 74160. There are several crystal forms of C.I. Pigment Blue 15, all carrying the 74160 constitution number but differing in their crystal modifications. Some of the red shade crystal forms also carry an additional constitution number 74250 which indicates mono-chlorination of the copper phthalocyanine, used to stabilize the alpha crystal form. A similar situation is seen with a few other pigments as listed in Table 4.3.

The Colour Index System identifies the pigment regardless of manufacturer. Those manufacturing a pigment under a particular Colour Index Code must maintain the proper chemical structure to be considered under that Colour Index Code.

Table 4.2 Examples of constitution numbers for different salt forms

Generic name	Constitution number	Common name
C.I. Pigment Blue 1	42595 : 2	Victoria blue (PTMA)
C.I. Pigment Blue 1 : 2	42595 : 3	Victoria blue (SMA)
C.I. Pigment Blue 1 : X	42595 : X	Victoria blue (PMA)

Table 4.3 Examples of pigments with the same constitution number but different crystal forms

Generic name	Constitution number	Common name
C.I. Pigment Blue 15	74160	Copper phthalocyanine blue α RC (Cu)
C.I. Pigment Blue 15 : 1	74160, 74250	Copper phthalocyanine blue α R, NC (Cu)
C.I. Pigment Blue 15 : 2	74160, 74250	Copper phthalocyanine blue α R, NCNF (Cu)
C.I. Pigment Blue 15 : 3	74160	Copper phthalocyanine blue β G, NC (Cu)
C.I. Pigment Blue 15 : 4	74160	Copper phthalocyanine blue β G, NCNF (Cu)
C.I. Pigment Violet 15	77007	Ultramarine violet
C.I. Pigment Blue 29		Ultramarine blue
C.I. Pigment Black 6, 7	77266	Lamp, carbon black

4.3 Pigment composition

This section is limited to those pigments whose wide commercial use is in the packaging ink field. The classification of pigment types can be seen in Figure 4.1 [5]. A summary of azo couplers and amines and the composition of the aforementioned pigments can be found in Table 4.4.

4.3.1 *Azo pigments*

Azo pigments comprise a large group of pigments containing the —N=N— azo chromophore within the molecule. The azo group is formed by the reaction of a

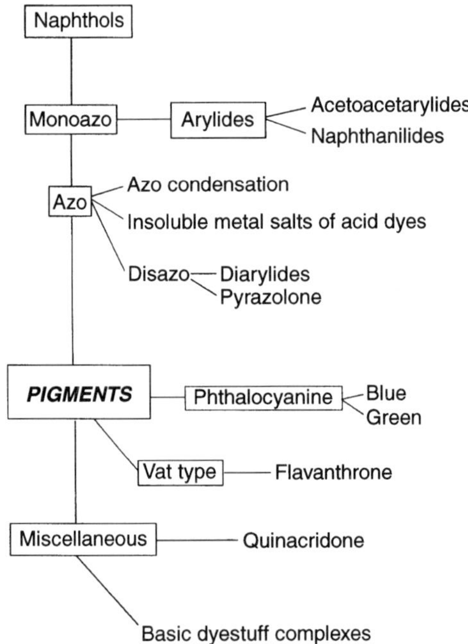

Figure 4.1 Classification of organic pigments used in the printing ink industry.

Table 4.4 Summary of azo couplers and amines

Industrial name	Chemical name
AAA	Acetoacetanilide
AADMCA	Acetoacet-2,5-dimethoxy-4-chloranilide
AAMX	Acetoacet-*meta*-4-xylidide
AAOA	Acetoacet-*ortho*-anisidide
AAOCA	Acetoacet-*ortho*-chloranilide
AAOT	Acetoacet-*ortho*-toluidide
AAPP	Acetoacet-*para*-phenetidide
AAPT	Acetoacet-*para*-toluidide
2B acid (2B amine)	*Meta*-toluenesulfonic acid, 6-amino-4-chloro
4B acid (4B amine)	4-Aminotoluene-3-sulfonic acid
Beta-naphthol	2-Naphthol
Beta-oxy-naphthoic acid (BONA)	2-Hydroxy-3-naphthoic acid
C acid (C amine)	*Para*-toluenesulfonic acid, 2-amino-5-chloro
Ethyl C amine	3-Amino-6-chloro ethyl-benzene-sulfonic acid
MNDA	2-Methoxy-4-nitro-aniline
Naphthol AS	2-Napthalenecarboxamide, 3-hydroxy-*N*-phenyl
Naphthol AS-BS	2-Napthalenecarboxamide, 3-hydroxy-*N*-(3-nitrophenyl)
Naphthol AS-D	2-Napthalenecarboxamide, 3-hydroxy-*N*-(2-methylphenyl)
Naphthol AS-OL	2-Napthalenecarboxamide, 3-hydroxy-*N*-(4-methylphenyl)
Naphthol AS-PH	2-Napthalenecarboxamide, 3-hydroxy-*N*-(3-ethoxyphenyl)

primary aromatic amine with nitrous acid, forming a diazonium salt, which is reacted with a coupler to form the pigment or the sodium salt.

Generally, azo pigments are divided into two classifications, metallized and non-metallized. While azo pigments are found in various shades of yellow, orange and red, there are a few exceptions which include dianisidine blue and some browns and violets. Metallized azo pigments contain one or more acid groups in the molecule which are reacted with a metal salt to render the pigment completely insoluble. The most commonly used salts are those of barium, calcium and strontium; however, other salts are also used.

Non-metallized azo pigments contain no acid groups in the molecule and, therefore, do not require metallization to completely precipitate the pigment. Non-metallized reds are divided into two groups depending on the number of amino groups and nature of the amine used. Mono-azo pigments contain one amino group in the molecular structure. The diazotized amine is reacted with a coupler to form the pigment. Disazo pigments are formed from aromatic primary diamines, which are diazotized and reacted with a coupler to form the pigment [6].

4.3.2 *Metallized azo pigments*

(a) *Metallized azo red pigments.* Metallized azo pigments comprise the largest category of red pigments used in packaging inks (Table 4.5). These pigments are based on four coupling components:

Table 4.5 Composition of the major metallized red and orange pigments

Pigment	Diazotizing agent	Coupler	Amine	Salt
C.I. Pigment Red 49 : 1	Sodium nitrite	Beta-naphthol	Tobias acid	Barium chloride
C.I. Pigment Red 49 : 2	Sodium nitrite	Beta-naphthol	Tobias acid	Calcium chloride
C.I. Pigment Red 53 : 1	Sodium nitrite	Beta-naphthol	C acid	Barium chloride
C.I. Pigment Red 52 : 1	Sodium nitrite	Beta-oxy-naphthoic acid	C acid	Calcium chloride
C.I. Pigment Red 57 : 1	Sodium nitrite	Beta-oxy-naphthoic acid	4B acid	Calcium chloride
C.I. Pigment Red 48 : 1	Sodium nitrite	Beta-oxy-naphthoic acid	2B acid	Barium chloride
C.I. Pigment Red 48 : 2	Sodium nitrite	Beta-oxy-naphthoic acid	2B acid	Calcium chloride
C.I. Pigment Red 48 : 3	Sodium nitrite	Beta-oxy-naphthoic acid	2B acid	Strontium chloride
C.I. Pigment Red 48 : 4	Sodium nitrite	Beta-oxy-naphthoic acid	2B acid	Manganese chloride
C.I. Pigment Orange 46	Sodium nitrite	Beta-naphthol	Ethyl C amine	Barium chloride

beta-naphthol
beta-oxy-naphthoic acid (BONA)
2-hydroxy-3-naphthanilides (Naphthal AS)
naphthalene sulfonic acid derivatives

Metallized reds are used widely in packaging inks due to their strong color and tinctorial strength. Metallized reds offer a large variety of shades due to the phenomenon of crystal modification. Available color ranges from yellow shades such as C.I. Pigment Orange 46 and the barium lakes of Lithol Reds and Permanent Red 2B to the bluer shades of calcium lakes of Lithol Red, Permanent Red 2B, BON Red and Lithol Rubine. Reds based on BONA are bluer in shade, whereas reds based on beta-naphthol are yellower in shade. Metallized reds possess moderate fastness properties to solvent, lightfastness and alkali resistance. These pigments are very economical and lend themselves well to the competitive pricing seen in the ink industry. Metallized reds containing calcium salts based on a benzenoid structure such as C.I. Pigment Red 48 : 2 or C.I. Pigment Red 57 : 1 exhibit hydration in an alkali system if not stabilized. Dehydration of water, or crystallization, produces a color shift from a yellower shade to a bluer shade upon heating. Reds containing calcium salts based on a naphthalene structure such as C.I. Pigment Red 49 : 2 remain stable in an alkali system. Reds based on barium or strontium salts are stable in alkali systems. Due to the packaging trend towards barium-free colors, many of the barium-containing pigments are being replaced with barium-free blends of calcium reds and barium-free orange or yellow.

(b) *Metallized orange pigments.* Metallized orange, C.I. Pigment Orange 46, is also used widely in packaging inks due to its strong yellow shade, and as an

Table 4.6 Composition of monoazo yellows

C.I. Pigment	Diazotizing agent	Coupler	Amine
C.I. Pigment Yellow 3	Sodium nitrite	AAOCA	PCONA
C.I. Pigment Yellow 73	Sodium nitrite	AAOA	PCONA
C.I. Pigment Yellow 74	Sodium nitrite	AAOA	MNOA
C.I. Pigment Yellow 65	Sodium nitrite	AAOA	ONPA

inorganic orange replacement. Metallized orange offers strong color and tinctorial strength, moderate fastness properties to solvent, lightfastness and alkali resistance. As is the case with the reds, metallized orange is economical in price.

(c) *Mono-azo yellow pigments.* This group of pigments has historically been used in the coatings industry. The composition of mono-azo yellows is given in Table 4.6. However, with the increased requirements demanded in the packaging ink field these pigments have been implemented in many formulations for increased lightfastness and as an alternative to diarylide yellow pigments. Some mono-azo yellows such as C.I. Pigment Yellow 1 exhibit weak color strength in packaging ink systems but the very clean green shade is needed for certain shades and color matches. Other mono-azo yellows such as C.I. Pigment Yellows 73, 74 and 65 possess a strong color for packaging where increased lightfastness is required and as a lead-free replacement for inorganic yellows.

(d) *Mono-azo red pigments.* Mono-azo red pigments are produced by coupling diazonium salts with naphthol AS (Table 4.7). The pigments produced are of two groups (Table 4.8), based on their structure and amide groups.

Mono-azo reds are used in the paint and ink industries due to the color strength, tinctorial strength and moderate price and fastness properties. Reds in the Group 2 category possess a greater degree of resistance to solvents and

Table 4.7 Composition of major monoazo red and orange pigments

Pigment	Diazotizing agent	Coupler	Amine
C.I. Pigment Red 2	Sodium nitrite	Naphthol AS-D	5-Nitro-*o*-toluidine
C.I. Pigment Red 17	Sodium nitrite	Naphthol AS-D	5-Nitro-*o*-toluidine
C.I. Pigment Red 22	Sodium nitrite	Naphthol AS-D	5-Nitro-*o*-toluidine
C.I. Pigment Red 23	Sodium nitrite	Naphthol AS-BS	5-Nitro-*o*-anisidine
C.I. Pigment Red 170	Sodium nitrite	3-Hydroxy-2-naphtha-*o*-phenetidine	*p*-Amino benzamide
C.I. Pigment Red 184	Sodium nitrite	Naphthol AS-PH	*p*-Amino benzamide
C.I. Pigment Red 210	Sodium nitrite	Naphthol AS-OL	*p*-Amino benzamide
C.I. Pigment Red 238	Sodium nitrite		
C.I. Pigment Orange 5	Sodium nitrite	Beta-naphthol	2,4-Dinitroaniline

Table 4.8 Division of mono-azo red pigments, based on their structure and amide groups

Generic name	Common name
Group 1	
C.I. Pigment Red 2	Naphthol red, medium
C.I. Pigment Red 17	Naphthol red, medium shade
C.I. Pigment Red 22	Naphthol red, yellow shade
C.I. Pigment Red 23	Naphthol red, blue shade
Group 2	
C.I. Pigment Red 170	Naphthol red, medium shade
C.I. Pigment Red 210	Naphthol red, medium shade
C.I. Pigment Red 238	Naphthol red, rubine shade

lightfastness than Group 1 mono-azo reds. Both categories exhibit moderate (Group 1) to good (Group 2) alkali resistance.

(e) *Dinitroaniline orange.* This pigment is used in the coatings and ink industries due to its strong color and shade. It is often used as a replacement for inorganic orange or used in blending to produce barium-free colors. C.I. Pigment Orange 5 exhibits moderate lightfastness, solvent resistance, as well as weather fastness.

4.3.3 *Non-metallized disazo pigments*

(a) *Disazo yellows.* These pigments are produced by coupling acetoacetarylides or pyrazolones with diazonium salts (Table 4.9). Other disazo pigments are produced by the diazotization of aromatic amines which are coupled to bisacetoacetarylides. Diarylide pigments comprise the largest group of organic yellows currently in use. Diarylides are used primarily in the printing ink industry due to their strong color, high tinctorial strength, low cost and variety of shades available. Moderate properties for lightfastness, heat and solvent resistance are typical of diarylide pigments; however, these properties are well within the specification requirements for much of the packaging industry.

Table 4.9 Composition of major disazo yellow and orange pigments

Pigment	Diazotizing agent	Coupler	Amine
C.I. Pigment Yellow 12	Sodium nitrite	AAA	DCB
C.I. Pigment Yellow 13	Sodium nitrite	AAMX	DCB
C.I. Pigment Yellow 14	Sodium nitrite	AAOT	DCB
C.I. Pigment Yellow 17	Sodium nitrite	AAOA	DCB
C.I. Pigment Yellow 55	Sodium nitrite	AAPT	DCB
C.I. Pigment Yellow 83	Sodium nitrite	AADMCA	DCB
C.I. Pigment Orange 13	Sodium nitrite	PMP	DCB
C.I. Pigment Orange 16	Sodium nitrite	AAPP	PCONA
C.I. Pigment Orange 34	Sodium nitrite	TMP	DCB

Table 4.10 Composition of phthalocyanine pigments

Pigment	Phthalocyanine blue crude
C.I. Pigment Blue 15	Copper phthalocyanine blue crude
C.I. Pigment Blue 15 : 1	Copper phthalocyanine blue crude
C.I. Pigment Blue 15 : 2	Copper phthalocyanine blue crude
C.I. Pigment Blue 15 : 3	Copper phthalocyanine blue crude
C.I. Pigment Blue 15 : 4	Copper phthalocyanine blue crude
C.I. Pigment Blue 16	Copper phthalocyanine blue crude
C.I. Pigment Green 7	Copper phthalocyanine green crude
C.I. Pigment Green 36	Copper phthalocyanine green crude

(b) *Disazo orange pigments.* The major disazo oranges used in the printing ink industry are C.I. Pigment Orange 13, 16, and 34. These pigments are based on *o*-dianisidine and pyrazolone. They offer a clean yellow shade and are used as a monopigment and in blending with other pigments to produce a variety of color shades. Disazo oranges are moderate in resistance properties and priced very economically which makes them very popular for printing ink systems.

4.3.4 *Phthalocyanine pigments*

The composition of phthalocyanine pigments is given in Table 4.9.

(a) *Phthalocyanine blue.* Copper phthalocyanine blue (Table 4.10) is the most widely used blue in the ink industry. A copper (II) complex of tetra-azatetrabenzoporphine, this molecule exhibits a planar, completely conjugated structure with excellent stability. Phthalocyanine pigments were first produced as iron phthalocyanine by the Scottish Dye Works in 1928 by the reaction of iron, phthalic anhydride and ammonia. Cuprous chloride was then substituted producing the copper phthalocyanine blue complex. The first phthalocyanine blue pigment was introduced by ICI in 1935 and the first US production was carried out in 1940.

Phthalocyanine blue is today produced in several crystalline forms. The most important are the alpha, beta and epsilon forms. Several types of each crystalline form are produced which exhibit different properties for crystallization resistance and nonflocculation to strong solvents. The alpha crystal produces a very red shade product which is, however, unstable to excessive temperatures and strong solvents (C.I. Pigment Blue 15). To compensate for this, a noncrystallizing type (C.I. Pigment Blue 15 : 1) and noncrystallizing nonflocculating type (C.I. Pigment Blue 15 : 2) have been produced. The alpha crystal blue is used mainly in paints and plastics, with some use in the ink industry. The beta crystal produces a greener pigment which exhibits greater stability to heat and solvents. The two types of pigment produced are the noncrystallizing (C.I. Pigment Blue 15 : 3) and the noncrystallizing, nonflocculating (C.I. Pigment Blue 15 : 4). The beta blues are used to a large extent in the ink industry and comprise the major blue used. Although C.I. Pigment Blue 15 : 1 and C.I. Pigment Blue 15 : 3 are primarily used in waterbase applications, the nonflocculation additives of C.I. Pigment Blue 15 : 2 and C.I. Pigment Blue 15 : 4 do not prevent their use in water, as well.

Phthalocyanine blues exhibit a strong color and tinctatorial strength and excellent resistance properties to acid, alkali, solvent, lightfastness and weatherability. Their clean shade and good economics make them very popular for packaging applications.

Recently, due to environmental concerns the use of copper phthalocyanine blue is being replaced to a small degree with metal free blue (C.I. Pigment Blue 16), primarily in the corrugated industry. The metal free product exhibits a far greener shade and weaker color and tinctorial strength and is considerably higher priced that C.I. Pigment Blue 15.

(b) *Phthlocyanine green.* Phthalocyanine green is the halogenated form of copper phthalocyanine blue. Progressive substitution of chlorine on the copper phthalocyanine molecule produce phthalocyanine green (C.I. Pigment Green 7). The higher the degree of chlorine substitution the yellower the product.

The yellow shade of phthalocyanine green (C.I. Pigment Green 36) is produced from a mixed halogenation technique. Phthalocyanine blue crude is reacted with bromine and chlorine to achieve the yellow shade green product. Increasing the bromination of the molecule yields the yellowest product.

The phthalocyanine greens are the major greens used in the ink industry due to their clean shade, strong color and tinctorial strength and excellent resistance properties to acid, alkali, strong solvents, lightfastness and weatherability.

4.3.5 *Dioxazine violet pigments*

The major dioxazine pigment used in the ink industry is C.I. Pigment Violet 23, carbazole violet (di-indolo[3,2-6:3′,2′-*m*]triphenodioxazine-8,18-dichloro-5,15-diethyl-5,15-dihydro-). Carbazole violet provides a very clean, blue shade which is used for shading blues and where a deep violet shade is necessary. C.I. Violet 23 is very resistant to light, solvent, alkali, acid and weatherability. C.I. Violet 23 yields a very small particle size which makes it difficult to disperse, requiring considerable energy to develop optimum color strength and transparency.

4.3.6 *Triphenylmethane pigments*

Pigments of this class are known for their clean, bright shades. The composition of major triphenylmethane pigments is given in Table 4.11. Their resistance properties to polar solvents are poor and resistance to acid, alkali, lightfastness are from poor to moderate. Due to the inferior resistance properties and economics of these pigments, many formulators try to match the shade by

Table 4.11 Composition of major triphenylmethane pigments

Pigment	Composition
C.I. Pigment Blue 1	4,4′-Bis(diethylamino)benzophenone condensed with *N*-ethyl-1-naphthylamine in toluene with phosphorus oxychloride and converted to the phosphotungstomolybdic acid salt
C.I. Pigment Green 1	Benzaldehyde condensed with *N,N*-diethylaniline in the presence of hydrochloric or sulfuric acid, oxidized, and converted to the phosphotungstomolybdic acid salt
C.I. Pigment Red 81 : 1	3-Ethylamine-*p*-cresol condensed with phthalic anhydride, esterified with ethanol and a mineral acid, and converted to the phosphotungstomybdic acid salt
C.I. Pigment Red 169	3-Ethylamine-*p*-cresol condensed with phthalic anhydride, esterified with ethanol and a mineral acid, and converted to the copper ferrocyanide complex
C.I. Pigment Violet 1	*m*-Diethylaminophenol condensed with phthalic anhydride and converted to the phosphotungstomolybdic acid salt
C.I. Pigment Violet 3	*N,N*′-Dimethylaniline oxidized with cupric chloride and converted to the phosphotungstomolybdic acid salt
C.I. Pigment Violet 27	*N,N*′-Dimethylaniline oxidized with cupric chloride and converted to the copper ferrocyanide complex

utilizing various pigments; however, the bright clean shade cannot be duplicated. Due to the poor to moderate alkali stability properties of some of these products, waterbase inks using these pigments must be very carefully formulated. High pH systems can exhibit poor stability in concentrated bases and finished inks.

4.3.7 *Benzimidazolone pigments*

An example of this type of pigment is C.I. Pigment Orange 36 (Table 4.12). Orange 36 has historically been used primarily in the coatings industry but it is finding increasing favor in the ink industry due to its shade and resistance properties. C.I. Pigment Orange 36 is very similar in shade to moly orange and exhibits excellent opacity, resistance to acid, alkali, solvents and excellent lightfastness. C.I. Orange 36 performs well in solvent and water systems alike; however, its high price limits its use in low-cost systems.

4.3.8 *Quinacridone pigments*

Quinacridone pigments (Table 4.13) have historically enjoyed low to moderate use in printing ink systems due to their economics and relatively weak print

Table 4.12 Composition of benzimidazolone orange

Pigment	Diazotizing agent	Coupler	Amine
C.I. Pigment Orange 36	Sodium nitrite	Acetoacet-5-amino-benzimidazolone	4-Chloro-2-nitro-aniline

Table 4.13 Composition of quinacridone pigment

Pigment name	Pigment intermediate
C.I. Pigment Red 122	2,5-Di-*para*-toluidino-terephthalic acid
C.I. Pigment Red 202	2,5-Di-*para*-chloranilino-terephthalic acid
C.I. Pigment Red 209	*Meta*-2,5-di-*para*-chloranilino-terephthalic acid
C.I. Pigment Violet 19	2,5-Dianilino-terephthalic acid

strength. Quinacridones can also exhibit heavy body in base concentrates and finished inks. These pigments are excellent in lightfastness, solvent fastness, heat resistance and weatherability. Due to their excellent resistance properties, quinacridones are used where a high degree of permanency is required. Quinacridones can be utilized in both solvent and waterbase systems; however, a rigorous dispersion method must be employed to achieve an adequate dispersion.

4.3.9 *Inorganic pigments*

C.I. Pigment Blue 27 or milori blue has been used for many years in solvent packaging inks due to its shade and very low cost. Ferric ammonium ferrocyanide is precipitated by reacting a ferrocyanide with a sulfate and then reacting a bi-chromate. Historically, milori blue has not been widely used in waterbase systems; however, pigment grades are today manufactured for use in waterbase inks. Milori blue exhibits moderate properties for fastness with the exclusion of alkali resistance which is poor.

Historically, many other inorganic pigments have been used in the packaging field such as yellows and oranges. The use of many of these products has been discontinued, however, due to the content of heavy metals such as lead, chromium and cadmium.

4.4 Manufacture of azo pigments

Azo pigments, both metallized and nonmetallized, share a common manufacturing process as seen in Figure 4.2 which is based on the azo coupling reaction to produce the azo group. In the case of a metallized azo pigment, metallization completes the reaction and renders the pigment insoluble. Inasmuch as the reaction differs among monoazo, disazo, metallized and nonmetallized pigments, the differences will be addressed at the appropriate stage of manufacture.

4.4.1 *Diazotization*

Diazotization involves the reaction of a primary aromatic amine with nitrous acid to form a diazonium salt. This reaction product is, by itself, unstable and

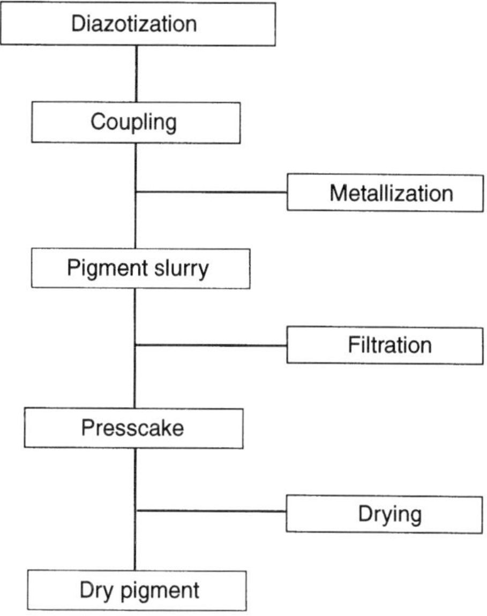

Figure 4.2 Outline of the azo manufacturing process.

requires low temperatures, approximately 0°C, throughout the reaction. Due to the thermal sensitivity of diazonium salts, free ice is often present to control and maintain the proper reaction temperatures.

Surfactants are sometimes used during this stage of manufacture to wet out and control the fineness of the precipitate and accelerate the azo coupling reaction.

The primary amine differs depending on the pigment type being produced. In azo yellow manufacture a substituted aniline is used for monoazos, whereas, diamines (e.g. 3,3′-dichlorobenzidine) are used for disazos. The amine used in production of azo reds will also vary, based on the pigment being produced.

4.4.2 *Coupling*

Immediately following diazotization, the diazonium salt is reacted with the coupler to produce the pigment, or the sodium salt of the pigment in the case of metallized azos. Couplers used in monoazo yellow production are based on one equivalent of acetoacetarylide coupling component while disazo products are based on two equivalents of acetoactarylide. The coupler used in red pigment production is based on beta-naphthol, BONA or a naphthanilide derivative.

During the coupling process, surfactants may be used to aid in the dispersion of pigment particles as they are being formed. This can be seen in the production of yellows which are precipitated in the pigmentary form, or the coupling of sodium salts to control salt formation as in the case of metallized reds.

4.4.3 *Metallization*

In the case of metallized reds the partially soluble sodium salt is of little use as a pigment and needs further processing to render it insoluble. Therefore, the addition of the salt of an alkaline earth metal such as barium, calcium, strontium or manganese is used to precipitate the partially soluble salt into an insoluble azo pigment.

4.4.4 *After-treatment*

After coupling/metallization is completed, most azo pigments are given some type of after-treatment in the form of rosination or surface treatment to aid in dispersion, prevent aggregation and to control the growth of the crystal particle. Surface treatments with fatty amines may also be used to create Schiff base stabilization resulting in more easily dispersing diarylide yellows.

Treatments with long-chain carboxylic acids are also used to produce hydrophilic anchor groups which are attached to the surface of the pigment. The hydrophilic anchor groups are attached to the pigment surface with lipophilic groups extending outward.

Rosination is a commonly used surface treatment which aids in inhibiting crystal growth, producing small pigment particles and promoting a very transparent pigment. Rosination also aids in dispersion by encapsulating the pigment particle preventing flocculation and enabling the pigment to wet out easily during the premix.

4.4.5 *Filtering, drying and grinding of azo pigments*

After the manufacturing process is completed, the pigment slurry is pumped into a filter press or other type of filtration equipment where the excess water is drained and the presscake is washed free of excess salts and impurities. The presscake is in the low solids form and can be dried by several methods including spray drying, oven drying, vacuum drying or belt drying.

A variety of mills can be utilized for grinding pigments. Specific mills can be chosen to produce finely ground pigment for easier dispersion. Other grinding methods produce coarse pigments or small lumps for low dusting pigments. Care must be taken in milling certain pigment types which are prone to dust explosion.

4.5 Pigments produced from crudes

4.5.1 *Phthalocyanine blue*

(a) *Manufacture.* The manufacture of copper phthalocyanine blue pigment involves two stages. First a crude pigment must be formed through a condensation reaction process known as the Wyler–Riley process. This involves the reaction of a heated phthalic acid derivative, urea, cuprous chloride and ammonium molybdate catalyst in a high boiling solvent such as kerosene or alkylbenzenes.

Another process for the manufacture of blue 'crude' is the phthalonitrile process which was developed in England and Germany. This process involves heating the phthalonitrile with a copper salt by means of a baking process or solvent process. The phthalonitrile route, although the more expensive of the two processes, produces a purer copper phthalocyanine without large amounts of by-products.

The phthalocyanine blue manufactured from these processes is termed a 'crude', with a very large particle size which displays weak strength and poor color development. The word 'crude' thus refers to tinctorial, rather than to chemical purity. To further develop the color and strength and reduce the particle size of the crude, additional conditioning and finishing are necessary. This can be carried out by a variety of methods. An illustration displaying the difference in particles between blue crude and the finished pigment can be seen in Figure 4.3.

(b) *Finishing methods*

Salt attrition. The crude is ground in salt with an organic solvent, such as glycol, in a double arm mixer to produce the beta-phthalocyanine blue pigment.

Solvent free milling. The crude is milled in salt to produce the alpha-phthalocyanine blue pigment.

Acid pasting. The crude is dissolved in an excess of concentrated sulfuric acid, then re-precipated under controlled conditions from the acid solution into water which produces the crystallizing grade of alpha-phthalocyanine blue pigment.

Acid swelling. The crude is treated with a low concentration of sulfuric acid, and re-precipitated into water, as in the acid pasting process, which produces the alpha-phthalocyanine blue product.

Acid kneading. The crude is milled with urea and sulfuric acid to produce the very red shade epsilon phase of phthalocyanine blue (C.I. Pigment Blue 15 : 6).

(c) *Stabilizing the pigment*

Crystallization resistant (NC). Further modification of the pigment can be carried out to produce a 'non-crystallizing' product by a small degree of chlorination of the copper phthalocyanine molecule. This stabilizes the alpha crystal and prevents shifts from the alpha to the beta phase due to heat.

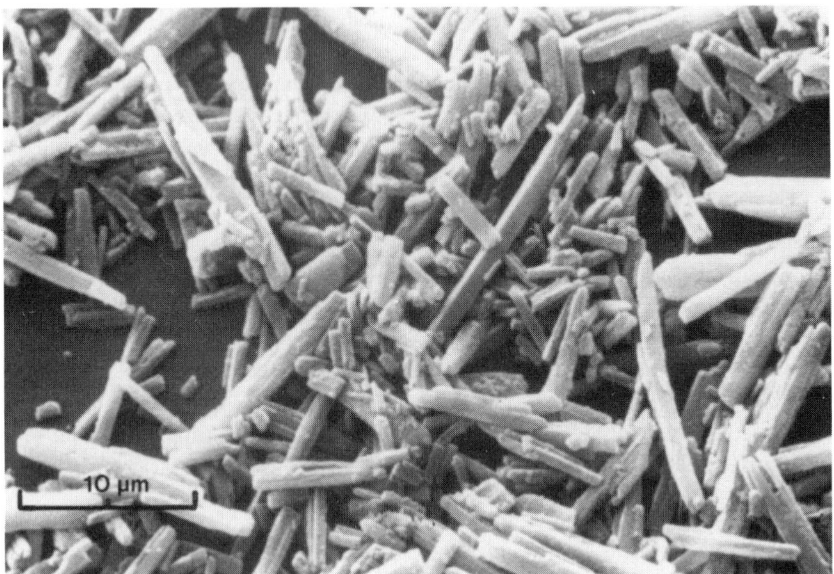

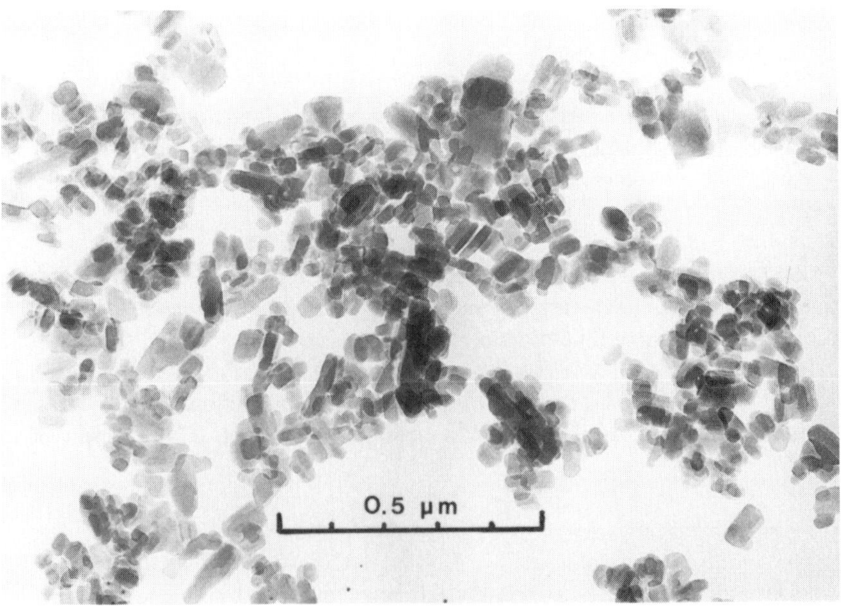

Figure 4.3 Scanning electron photomicrograph of copper phthalocynanine crude (top) and transmission electron photomicrograph of copper phthalocyanine blue pigment (bottom).

Flocculation resistant (NF). Modification of the pigment particle to resist flocculation is another common stabilization technique performed on phthalocyanine blue pigments. Examples of flocculating and non-flocculating inks can be seen in Figures 4.4 and 4.5. Stabilization can be carried out by a variety of methods including partial sulfonation, partial sulfonamide formation or the partial introduction of dialkylaminomethyl groups into the aromatic ring of the molecule.

4.5.2 *Metal-free phthalocyanine blue*

Metal-free phthalocyanine crude is manufactured by the phthalonitrile process to form the sodium, calcium or magnesium salt. Treatment with dilute acid then forms the metal-free crude. The crude produced from these processes requires finishing by acid pasting, producing the slpha (red shade) product, or by grinding with a polar solvent to produce the beta (green shade) product.

4.5.3 *Phthalocyanine green*

Phthalocyanine green is a halogenated form of phthalocyanine blue. The blue crude is halogenated with sulfur chlorides, thionyl chloride or in a heated aluminium chloride/sodium chloride eutectic mixture, or in chlorosulfonic acid. The blue shade green product, C.I. Pigment Green 7, is halogenated with chlorine while the yellower shade product, C.I. Pigment Green 36, additionally

Figure 4.4 Optical photomicrograph of liquid ink exhibiting flocculation of pigment particles.

Figure 4.5 Optical photomicrograph of liquid ink exhibiting no flocculation of pigment particles.

contains bromine. The crude pigment is 'finished' by means of salt attrition with glycol, or by solvent swelling to produce the final pigment.

4.5.4 *Carbazole violet*

Like phthalocyanine blue, carbazole violet is produced in two steps. First, production of the crude involves condensing amino ethyl carbazole with chloranil in a high-boiling solvent with the use of a condensation agent such as benzenesulfonylchloride or *p*-toluenesulfonylchloride. The crude is then filtered, washed and dried. Additional finishing is required to reduce particle size and develop strength and coloristic properties.

4.5.5 *Quinacridone reds and violets*

Manufacture of quinacridone reds and violets involve a three step process: production of the intermediate, reaction of the intermediate into a crude, and conditioning of the crude into a pigmentary form.

Two methods are typically used for the synthesis of quinacridone crude: the oxidation of dihydroquinacridone or cyclization of 2,5-diarylaminoterephthalic acid. Cyclization of 2,5-diarylamino terephthalic acids is carried out by a condensation reaction in poly-phosphoric acid. Crude quinacridone has a very small particle size, thus requiring further conditioning in a solvent to grow the particle to an appropriate size for the final pigment. The shade of quinacridones is controlled by the use of different crudes. Thus, the gamma crude of C.I. Pigment

Violet 19 produces quinacridone reds while the beta crude produces quinacridone violets. The crystallinity of the pigment can also be controlled by 'mixed crystals' of solid solutions. In this process two or more pigment crystals are conditioned together, the minor component being accommodated in the crystal lattice of the major component. This process not only extends the available shades of quinacridone but also improves the lightfastness of the pigment.

4.5.6 *Pigments produced from dyes*

A general method for the manufacture of PMA and PTA basic dye or triarylcarbonium pigments can be seen in the production of C.I. Pigment Blue 1.

A precipitation tank is used to prepare a 1.0% solution of victoria pure blue BO dye using water and steam. In a smaller tank, disodium phosphate and sodium molybdate are dissolved in water. A mineral acid is added to form a complex phospho-molybdic acid solution (PMA) [7]. Sodium tungstate can be substituted for sodium molybdate to form a phospho-tungstic acid solution (PTA). The complex acid solution is then added to the stirred dye solution in the strike tank. The pigment which is precipitated can be modified by additional treatments such as heat or surfactant treatments. The pigment is then filtered, washed free of impurities, dried and ground.

4.5.7 *Inorganic pigments: manufacture of C.I. Pigment Blue 27 (milori blue)*

The manufacture of milori blue is based on the reaction of a solution of sodium ferrocyanide and ferrous sulfate in the presence of ammonium sulfate forming Berlin white. The Berlin white is digested in sulfuric acid and oxidized with sodium bichromate or chlorate to form the milori blue precipitate. The finished pigment is then filtered, washed free of water soluble salts, pressed and dried. Surface treatments are sometimes added to improve the texture and processing [8].

4.6 Advantages and disadvantages of dry pigments in a water based system

There are many advantages and disadvantages associated with the use of dry pigments. This section addresses areas of proper pigment selection, resination, pigment/vehicle compatibility, production of a high quality dispersion using dry pigments, ease of dispersion and handling of dry pigments.

4.6.1 *Proper pigment selection*

When formulating an ink system several decisions must be made concerning the pigment form which will be utilized, and the long-term stability of the pigment for the formulation. Will the best choice for the pigment selection be in the form of a dry toner, presscake, or dispersion? Is the long-term stability of the pigment chosen for the formula adequate for the needs of the product? An advantage of choosing a dry pigment over other pigment forms is the fact that dry pigment is 100% dry product. Inks and concentrated bases can be formulated without estimating the percentage of pigment which is necessary when formulating with a presscake or dispersion. The dry pigment form can also be easier to handle in production than other pigment forms. The dry form is stable over an extended period of time, even several years, without change in the pigment or its properties. Another advantage of choosing a dry pigment in formulating is the wide range of dry pigments offered by most manufacturers.

When formulating a water based ink, the proper pigment choice is of primary importance. Most pigment manufacturers offer several pigment choices which may include various degrees of transparency, end treatments, shades, easily dispersible pigments or pigments developed specifically for a particular system. The hue of the pigments chosen is only the first consideration for a properly formulated ink. Other properties such as lightfastness, chemical resistance, ease of dispersion, compatibility within the ink system need to be addressed prior to pigment selection. The type of milling equipment available to the formulator has a major impact upon choosing an easily dispersible pigment over a pigment which would require considerably more energy in grinding to achieve the ultimate dispersion for color and strength development. When the most efficient grinding equipment is not available to attain a properly dispersed ink base or finished ink, the formulator may choose another form of pigment such as a presscake or dispersion as the optimum choice for the colorant.

Another area which requires addressing when selecting a pigment for the application is the economics. Cost is always an important factor when formulating an ink which is both competitive and will fulfill the properties which are required by the end user. Choosing a low-cost pigment which fulfills the coloristic parameters required in the finished ink may not fulfill other resistance properties which are also required in the formulation. Sometimes more expensive pigments formulated in the proper combination will not only satisfy the color and resistance properties which are required but will also be very economical as well. Most pigment manufacturers publish data describing their products and have technical departments available to aid ink formulators in the optimum pigment selections.

Finally, after the proper pigments have been selected and ink formulations are completed, the finished ink should be tested for the end use application. Final testing by the formulator for all required resistance properties and specifications could mean the difference between a successfully formulated finished ink and a possible product claim by the customer.

(a) *Dry pigment toners.* Dry pigment toners are the most widely utilized form of pigment. Presscakes are dried, ground and blended to produce a consistent product which is used in many different ink systems. The dry pigment form enables ink formulators to use the toners in highly concentrated ink bases, chips and many additional applications.

Dry pigment is the most difficult pigment form to disperse due to particle agglomeration. As water is removed from the presscake in the drying process, the primary pigment particles can come together forming aggregates, agglomerates and flocculates (Figures 4.6, 4.7 and 4.8) and may become very compacted thus requiring considerable energy to disperse. Typical dispersion equipment such as a standard media mill is usually suitable to grind dry toner with an adequate premix.

Opaque pigments. Opaque pigments can be easily used in water based ink formulations where a high degree of transparency is not of primary importance. Many water based ink formulations used for line printing on paper, board, polyethylene, polypropylene, corrugated and label substrates can easily utilize opaque pigments which will impart the necessary print strength and gloss for the end product.

Opaque pigments are often used in ink formulations for corrugated printing. The opacity provided by these pigments offers excellent hiding power for printing on corrugated stock or lower quality paper substrates used in some gift wrap printing.

Opaque pigments can be dispersed more easily, requiring less energy due to the larger particle size associated with an opaque pigment. The larger particle

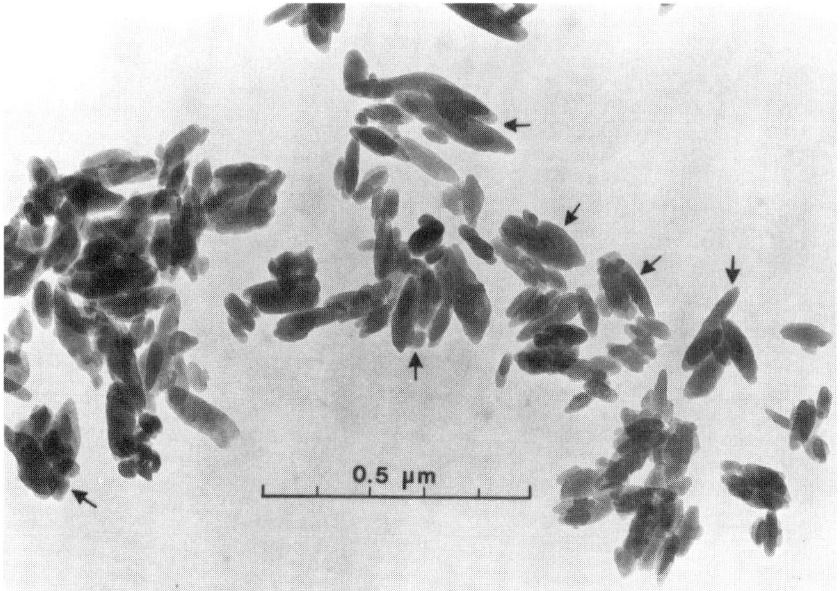

Figure 4.6 Transmission electron photomicrograph of pigment particles exhibiting agglomerates.

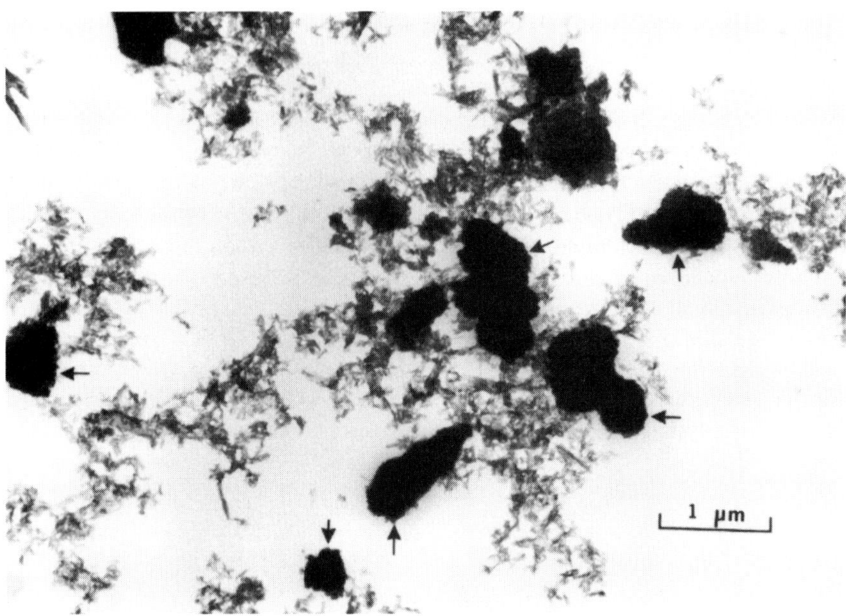

Figure 4.7 Transmission electron photomicrograph of pigment particles exhibiting aggregates.

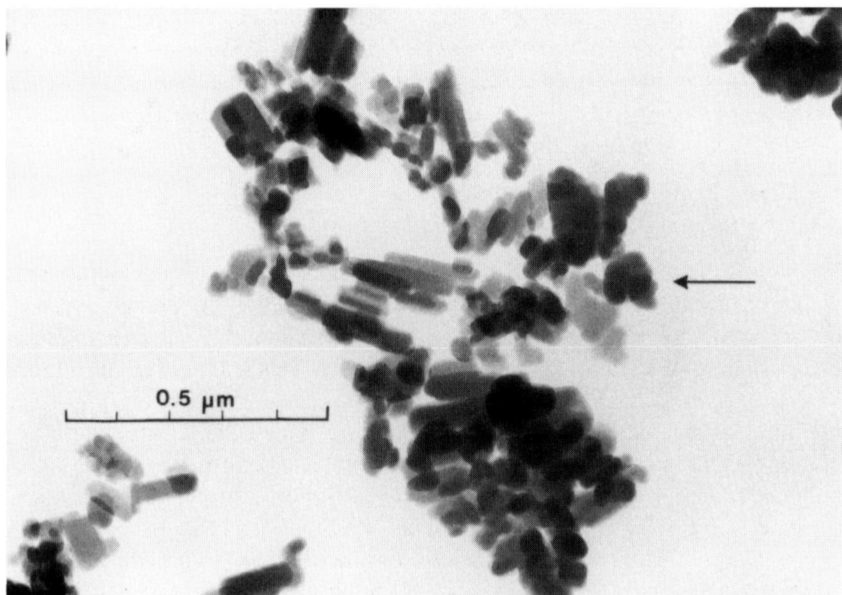

Figure 4.8 Transmission electron photomicrograph of pigment particles exhibiting flocculates.

size will also impart better lightfastness and lower rheology in the finished ink base.

Transparent pigments. Small particle pigments are normally utilized where transparency and gloss are of major concern. Application areas using transparent pigments include process printing, printing on high quality substrates such as foil and high quality label and paper stocks, applications requiring a high gloss, high color strength colorant or applications requiring a small particle product.

Transparent pigments offer good gloss, high print strength and tinctorial strength for many packaging applications. A transparent pigment will exhibit a lower degree of lightfastness than a larger particle pigment of the same class due to the smaller particle size and transmission of light. Transparent pigments can be more difficult to disperse and can produce a heavier bodied ink due to its small particle size.

4.6.2 *Resination*

The degree of resination of the pigment is of utmost importance when formulating a water based ink. Most pigments have resin incorporated during the manufacture for various reasons. Resin added during or after the coupling reaction helps to promote transparency due to its ability to inhibit crystal growth and aids in the dispersion of the pigment. The resin also encapsulates the pigment particle enabling the pigment to more easily 'wet-out' during the premix stage of dispersion. A resinated pigment can produce a more transparent ink over a nonresinated pigment. Also a pigment ground in a solvent ink system will normally produce a more transparent ink than the same pigment ground in a water based system. However, many pigments are being developed for use in water based systems which will impart excellent transparency, gloss and low rheology which cannot be captured as fully in a solvent system.

4.6.3 *Pigment/vehicle compatibility and stability*

When choosing a pigment for a water based system it is very important to look at the compatibility between the pigment and the vehicle system in which it will be employed. In a water based ink system, the use of a highly resinated pigment can produce an ink with poor rheological stability. This is due to an incompatibility between the resin used in the pigment manufacture and the acrylic resin used in grinding and formulating water based systems. The incompatibility can be seen as an ink with an initially heavy body, or as an ink which exhibits a building or gelling of the viscosity upon aging. Highly resinated pigments, such as some C.I. Pigment Yellow 83 pigments, should be chosen very carefully to avoid this problem. A pigment which is considered non-resinated usually lends itself best for water based systems. Not all water based systems exhibit incompatibility with resinated pigments. Moderately resinated pigments often work well in many water based systems. Adequate testing of resinated pigments in the various rosin systems which the pigments will be employed should ensure the formulator of proper rheological stability and compatibility within the ink system.

Instability is also seen with the use of metallized reds in a water based ink system. Commonly, in a water based ink system, metallized azo red pigments can produce both rheology and shade instabilities. Rheological instabilities are generally caused by an incompatibility between the rosin in the ink system and the rosin which is incorporated into the manufacture of the pigment. Most metallized azo reds carry a degree of rosination, therefore, proper selection and testing within the vehicle system is essential for a rheologically stable formulation.

Another form of instability which is seen when using metallized azo reds in a water based ink system is shade stability. Shade instability is seen when using metallized azo reds with calcium salts, where the amine portion of the molecule is based on a benzenoid structure (Figure 4.9) such as in C.I. Pigment Red 48 : 2, Permanent Red 2B, or C.I. Pigment Red 57 : 1, Lithol Rubine. In these pigments, a shade shift is observed due to the loss of water of crystallization from the molecule. When these pigments are formulated in the presence of water, the color shifts to a yellower shade as they absorb water, then shifts to a bluer shade as they lose water in drying. If the pigment is stabilized, the color will shift back to the original shade. Pigments with calcium salts where the amine portion of the molecule is based on a naphthalene structure (Figure 4.9) such as C.I. Pigment

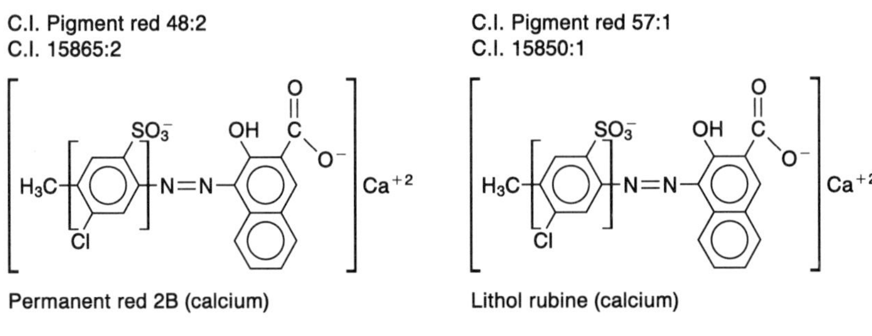

Figure 4.9 Structures of Pigment Red 48 : 2 and Pigment Red 57 : 1 with the benzene portion of the amine bracketed and Pigment Red 49 : 2 with the naphthalene portion of the amine bracketed.

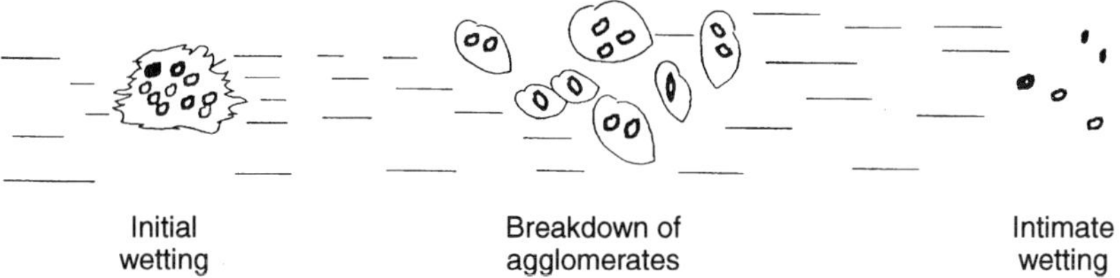

Figure 4.10 Stages of the wetting, or premix, phase.

Red 49 : 2, calcium lithol red, do not exhibit this phenomenon. Reds based on barium or strontium salts are stable in alkali systems.

4.6.4 *Production of high-quality dispersions using dry color*

The production of a pigmented dispersion is the separating of particle agglomerates in a liquid medium by means of a standard milling technique. This process requires three phases.

(a) *Phase 1, 'wetting-out' of the pigment.* The 'wetting-out' phase consists of premixing the pigment with the appropriate vehicle system and mixing under moderate shear to wet the pigment agglomerates. The wetting or premix phase in effect encapsulates the agglomerates with vehicle in combination with moderate shear, allowing the agglomerates to begin to break up into smaller agglomerates or clusters of pigment particles (Figure 4.10).

Pigments are hydrophilic or hydrophobic with respect to the ease of wetting. Hydrophilic pigments are water attracting and are not readily wetted. Hydrophobic pigments are water repelling and are more easily 'wet-out' by the vehicle system. The use of a variety of dispersants or a vehicle system which allows better wetting will aid in this stage, if needed.

Dry pigments typically have a slightly charged particle. Metallized pigments often possess a higher charge than nonmetallized pigments, with black exhibiting the lowest charge. Similarly, charged particles are important in the dispersion process to produce a repulsion between particles and to promote dispersion. The charge aids in the stabilization of the particle by creating an electrical repulsion which surrounds the pigment particle. An electrically charged double layer surrounds the particle creating a repulsion to other similarly charged particles to prevent reagglomeration. The repulsion extends into the liquid medium thus stabilizing the dispersion [9].

Another form of stabilization is steric or entropic stabilization. This creates steric hindrance resulting from the adsorption of a dispersing agent surrounding the particle, creating an effective steric barrier which repels the surrounding particles due to a loss of entropy. Charge and steric stabilization is shown in Figure 4.11.

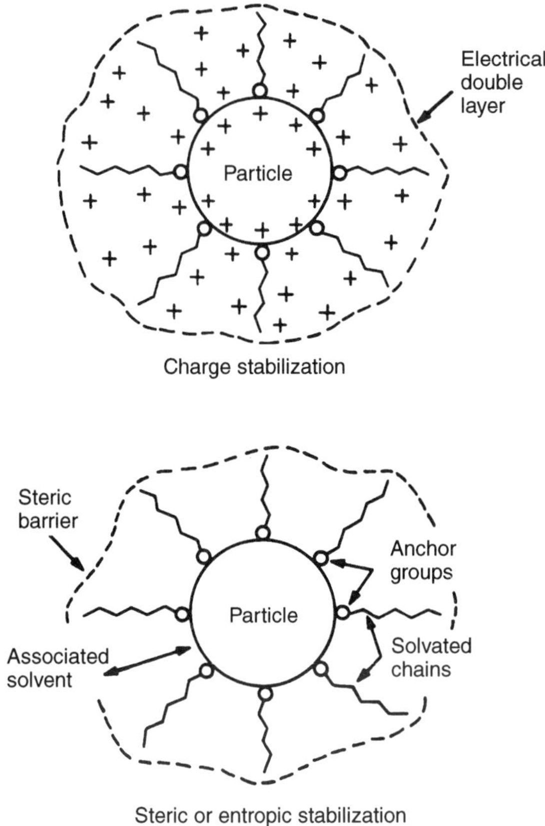

Figure 4.11 Charge and steric or entropic stabilization.

In choosing a dispersant, the primary objective is to stabilize the pigment particle by adsorption or anchoring the dispersant on the pigment particle and matching the polarity of the solvent in the vehicle system. For a water based system the proper dispersant should be adsorbed or anchored (Figure 4.11) onto the pigment particle and be soluble in and compatible with the solvents in the vehicle system. Dispersants and surfactants lower the interfacial tension between the liquid and the pigment thus improving pigment wetting. In water based systems a surfactant should be chosen which possesses a high HLB value (hydrophile–lipophile balance), which exhibits solubility in water. Obtaining technical information from the suppliers of vehicles and dispersants as well as pigment suppliers can aid in dispersant selection.

(b) *Phase 2, breaking down of agglomerates.* In the premix, after initial wetting of the agglomerates has taken place, by displacing the air surrounding the agglomerates with vehicle, the agglomerates begin to break down into smaller clusters. As the smaller clusters are agitated, further wetting takes place, encapsulating the clusters and in turn causing the clusters to continue to break down. This process continues until only small groups of agglomerates remain. The premix phase should continue until the pigment clusters are reduced to an appropriate size as indicated with a premix grind gauge. The premix phase of dispersion is highly important. Without proper wetting and reduction of the pigment agglomerates any 'unwetted' pigment clusters can become compacted under the high shear of milling and proper dispersion of the pigment will not take place.

(c) *Phase 3, grinding.* In the grinding phase of dispersion, the premix is contacted with grinding media such as sand, glass, ceramic or steel beads in a media mill or other type of milling equipment, to enable further reduction and wetting of the agglomerates to take place. Dispersion takes place by shear and the forcing of the premix through the media. This procedure is typical of grinding which takes place in a shot mill. The premix is passed through the mill one or more times until the proper dispersion is achieved as indicated by a 'fineness of grind' gauge. Several passes may be required depending on the hardness and particle size of the pigment.

Proper grinding is an essential final step in the dispersion process. Utilizing the appropriate grinding equipment and proper grinding time is essential to achieve a fully developed dispersion. In the initial grinding process, color and strength development take place. As grinding continues, gloss and transparency are developed and a stable dispersion is produced. Undergrinding of the pigment will result in a dispersion which is poorly developed, low in gloss, or lacking in tinctorial strength. Undergrinding or overgrinding can cause reagglomeration of the pigment, resulting in a dispersion which initially produces good strength and later appears weak in print and tinctorial strength. Generally a large particle pigment will be easier to disperse requiring lower energy for proper dispersion. On the other hand, a very small particle, such as carbazole violet, will require more energy for dispersion due to its hardness and small size of the particle.

4.6.5 *Ease of dispersion*

The ease of dispersion of various pigments can be as diverse as the pigment form chosen for grinding. A regular solids presscake will disperse more easily than a high solids presscake, which will in turn disperse more easily than a dry pigment. The type of milling equipment available and the degree of energy which the milling equipment is capable of producing, will determine the pigment form which is most suitable for use in manufacture.

4.7 Handling dry pigments

When handling dry pigments several areas which must be addressed are those of health and physical hazards, industrial usage and environmental concerns. The proper handling and usage of dry pigments protects not only those in contact with the products but also the environment in which we work and live.

4.7.1 *Physical and health hazards*

Most dry pigments (excluding certain metals, ferriferrocyanides and sulfides) are considered stable chemicals for storage and use under normal conditions [4]. The Federal Hazardous Substance Act defines materials having a median lethal dose (LD_{50}), the most common measure of acute toxicity, of greater than 5000 mg/kg as being 'non toxic'. Pigments are, in general, considered to have low levels of acute toxicity according to studies performed on over 4000 pigments by ETAD (Ecological and Toxicological Association of the Dyestuffs Industry), the National Printing Ink Research Institute and a Nifab Symposium. Most of the oral LD_{50} values were above 5000 mg/kg; none was below 2000 mg/kg [10].

Under normal conditions, dry pigments are considered to be nonflammable and unreactive substances. Pigments will, however, burn if involved in a fire. Additionally, a few pigments may exhibit a phenomenon known as self-heating. To date, two types of pigments have shown this behavior. The first type involves highly resinated pigments which have been micropulverized and packed out and stored at high temperatures [11]. Pigments such as resinated versions of C.I. Pigment Yellow 13 and C.I. Pigment Black 11 have shown evidence of self-heating properties, thus requiring special labeling to conform to new transportation regulations. Self-heating powdered organic pigments are classified in Division 4.2, Packing Groups II or III, UN No. 3088 and self-heating powdered inorganic pigments are classified in Division 4.2, Packing Groups II or III, UN No. 3190. A self-heating material of Class 4, Division 4.2 is defined as a material which is liable to self-heat and exhibit spontaneous ignition when in contact with air and without an external energy supply [12]. This same phenomenon is also seen in products such as flour and powdered chocolate [11].

Another type of physical hazard with pigments can be seen in the mixing and heating of monoazo pigments with lead chromate. In this case the organic pigment smoldered and burned. This could also possibly occur with the accidental introduction of tramp metal into organic pigments while drying or pulverizing [11].

4.7.2 *Industrial use/hazard communication*

Another area for concern is dusting. Any dusting material requires care in handling in any operation. Any operation which generates considerable dust could be subject to dust explosion. Details involving dry pigments are supplied on material safety data sheets. Due to the dusting and chemical nature of many dry pigments, proper ventilation and the use of protective coverings such as face masks and respirators should be worn when dealing with dry pigments in ink manufacture and testing. Proper training should be made available for all persons involved with pigments in laboratory and manufacturing areas.

In addition to training, the appropriate material safety data sheets for all pigments and raw materials associated with ink manufacture should be made available for information purposes and inspection.

4.7.3 *Environmental concerns*

Pigments are normally considered inert in nature and insoluble materials which pose no threat to the environment. However, care must be taken when disposing of organic and inorganic pigments, waste generated from laboratory or manufacturing sites or other chemicals used in ink production. The federally mandated disposal requirements of these and many other materials, both hazardous and nonhazardous, can be found in Title 40 of the Code of Federal Regulations [13]. Questions concerning the disposal of various pigments should be directed to the pigment supplier for information regarding proper disposal of such products.

4.8 Marketplace analysis for dry pigments

In predicting the marketplace for dry pigments several factors must be considered. A few of these considerations include changes in the demand for pigments, increased difficulty in obtaining permits to build or enlarge existing pigment manufacturing sites, changes in the classification and labeling of pigments and environmental movements in the industry.

Currently, the largest demand for organic pigments is in the printing ink industry, utilizing 40% of the pigments manufactured as reported at the Color Pigments Manufacturers Association annual meeting in 1992 (formerly DCMA, Dry Color Manufacturing Association) [14]. As of 1992, the world consumption of organic pigments was reported at approximately 180 000 tons [15]. A breakdown by product group as seen in Figure 4.12. This consumption is up from 158 000 tons in 1990 (Figure 4.13), the annual growth equaling the growth of the gross national product of the industrialized countries, making the pigment market relatively stable [15]. However, there are factors which may alter the future mix of pigments which will be demanded in the marketplace.

Recently, changes have been seen in the demand for pigments containing barium, lead and other heavy metals as well as the trend to investigate alternatives to other pigments. Many forms of packaging have increased specifications for lightfastness and product resistance, thus shifting the demand from lower cost pigments such as red lake C and lithol rubine to higher cost naphthols and quinacridones. The use of leaded pigments is decreasing throughout the industry, as they are being replaced with alternative products such as organic oranges and yellows. These factors will shift the production and consumption of pigment types for the future.

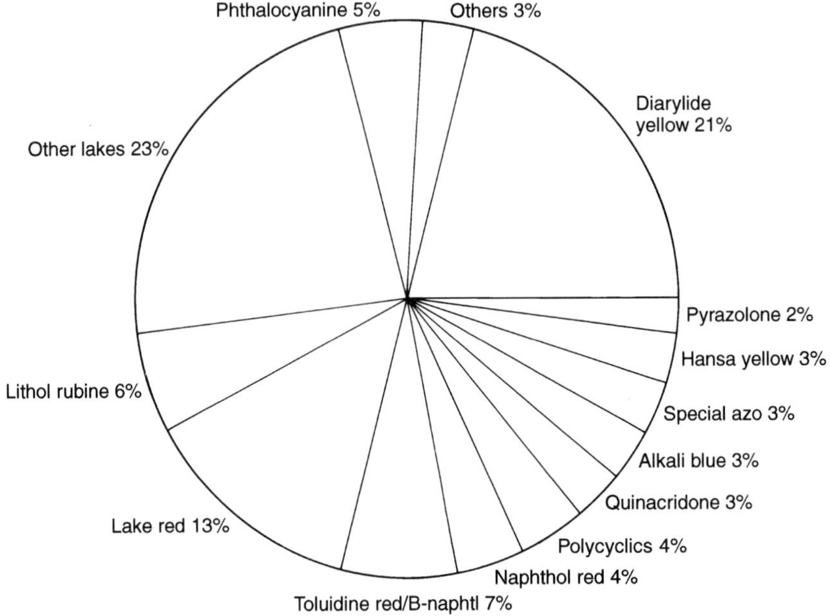

Figure 4.12 World consumption of organic pigments by product group (powder basis).

Another factor affecting the future production of pigments is the increasing difficulty in obtaining permits for the building or expansion of pigments producing facilities [14]. Increasingly stringent regulations concerning the use of certain raw materials, emission levels and effluent controls make permitting more difficult in the US and Europe. Obtaining permits can take considerable time in countries such as France, Portugal, Spain or Germany [14]. The tightening of regulations will, in all probability, continue making better planning a necessity.

The classification of pigments is also becoming more stringent. Producers of pigments must comply with ever-increasing laws and regulations in production and packaging. Regulations regarding emission limits were established in 1964, 1974, 1984 and 1986; standardization of forms in 1990 and regulation of effluent in 1992 [16]. Special packaging and labeling have been implemented in the US and Europe. Products are classified as to their potential danger in shipment and use. Material safety data sheets are becoming standardized in the US and abroad.

Environmental movements are also changing the industry practices. Recycling and the reuse of containers has become a widely used practice. The 'Green Dot' concept for packaging has been initiated in Germany, where from January 1, 1993 all packaging must be taken back by the shop [17]. Recycling

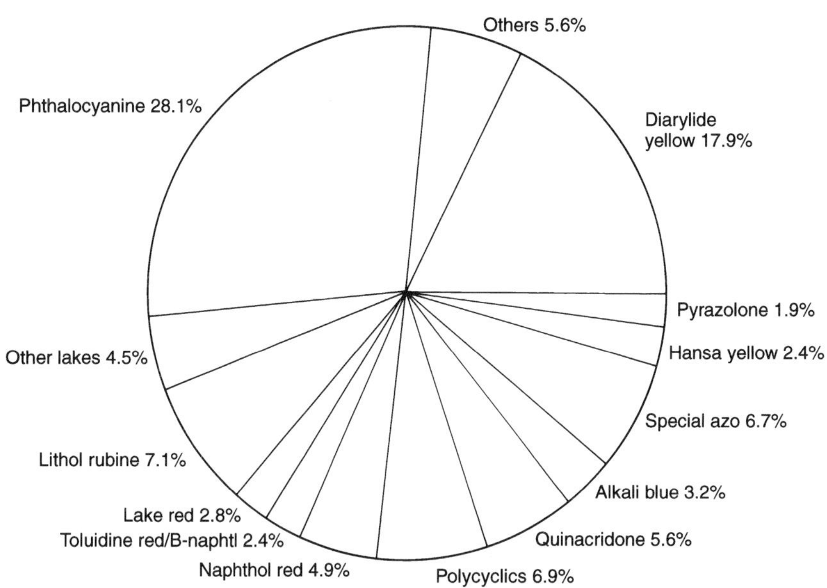

Figure 4.13 World consumption of organic pigments by product group (value basis).

has changed the industry from the packaging standpoint, with many producers reverting to larger, reusable or recyclable containers.

With the ever-increasing globalization of the industry, more conforming regulations and practices are likely to be seen. A change in the mix of products demanded is likely, but even with these changes, there is an optimistic future for the industry.

4.9 Water based pigment dispersions

Aqueous pigment dispersions are fairly new, but have seen outstanding improvements concerning rheology, compatibility and stability in water based inks. Aqueous dispersions for latex paints have been in use for some time, but are usually formulated using nonionic surfactants and or glycols to slow drying and in most instances contain on resin. The lack of resin binder makes aqueous pigment dispersions specifically formulated for latex paints, more universally compatible with latex binders in paints. In some instances these dispersions can be used in flexographic inks, but it is up to the formulator to determine its feasibility.

Aqueous pigment dispersions for water based flexographic and gravure inks in most instances contain acrylic resins as binders, they may contain some anionic, nonionic or amphoteric surfactants, or a combination of these, depending on the type of pigment, their concentration and the end use in a finished ink.

Pigments with the same pigment C.I. number from various suppliers are not necessarily interchangeable with one another. One pigment may have good flow and excellent rheology, while an alternative may increase in viscosity or gel even in a premix. Some pigments are extremely heat sensitive and gel if the temperature is not properly controlled during the mixing and milling process. Others may need higher temperatures to develop a low particle structure for more intense tint strength and gloss required in high quality flexo and gravure inks. Many commercially available organic pigments are resinated. The types of resins used may vary from one supplier to another. As an example, one lithol rubine pigment may be completely stable in an aqueous dispersion, while another is totally incompatible.

Besides acrylics, other resins, such as polyesters, epoxies, urethanes, etc. may be used to manufacture aqueous or water reducible dispersions. Depending on the resin system, it may be necessary to incorporate solvents other than water and/or amines that could be objectionable to some printers due to objectionable odors or higher VOCs. Again, with different resin systems, pigments that are suitable for acrylics may not be with these resins.

It is of the utmost importance that the dispersion chemist has a thorough knowledge of the end use of the product being formulated in order to avoid unnecessary complications with the customer. Depending on the pigment to be dispersed along with the quality and end use of the final ink product, aqueous pigment dispersions may be manufactured using high speed dispersers and mixers with variously configured dispersing blades, kinetic dispersion mills, ball mills (some manufacturers still use them) or media mills.

The current state-of-the-art equipment for such operations are media mills, using steel or ceramic media. The quality of the dispersion to be made is dependent on the size of medium, the load percentage of medium in a given size vessel and the through-put rate (pounds or gallons per hour) of the product. Using the smallest size medium possible (depending on the screen size in the mill), the highest percentage medium load in the grinding chamber with the slowest through-put will achieve the best quality dispersion. Even then the quality of the product may not be up to standard and a second or even a third pass through the mill may be required.

Other important factors which determine the quality in the final dispersion product are the percentage of pigment, resin and the required amount of dispersing aid or surfactant. The formulator (usually by trial and error) needs to determine the percentage levels, keeping in mind the specified requirements the ink maker is looking for. Depending on the ink maker and his customer's requirements, various solubilizing agents or amines can be used to make the resin in the dispersion water soluble or reducible. Chemicals such as ammonia, caustic soda, monoethanolamine or other amines can be used. The main purpose of these additives is to make the resin binder water soluble. The pH of aqueous dispersions is usually 8.5 to 9.5. A pH lower than 8.2 may result in thickening of the dispersion over a period of time. Also a low pH can produce incompatibility with the customer's water based ink vehicles, plugging of the cells in the anilox rollers. Poor transfer of the ink to the substrate being printed. A pH higher than 9.5 in the dispersion could cause problems in the formulators finished printing ink. Excessive foaming may also occur in ink systems with very high pH levels.

4.10 Pigment structures

The chemical structures, C.I. classification, and common names of 53 pigments commonly used in ink manufacture, are given in Figure 4.14 at the end of this chapter.

References

1. Society of Dyers and Colorists, Bradford, Yorkshire, BS1 2JB, UK.
2. American Association of Textile Chemists and Colorists, Research Triangle Park, NC, USA.
3. Patton, T.C. (ed) (1973) *Pigment handbook, volume I*, John Wiley and Sons, New York, p. 983.
4. *NPIRI raw material data handbook, volume 4, Pigments* (1983) Lehigh University, Bethlehem, PA.
5. Lewis, P.A. (1988) *Organic pigments*, Federation of Societies for Coatings Technology, Philadelphia, PA.
6. Forbes, R. Personal communication.
7. Oliver, V. (1973) Basic dye pigments, in *Pigments handbook, vol I*, (ed Patton, T.C.), John Wiley and Sons, New York, pp. 605–16.
8. Sistino, J.A. (1973) Ferriferrocyanide Pigments, in *Pigments handbook, vol I* (ed Patton, T.C.), John Wiley and Sons, New York, pp. 401–7.
9. Vernardakis, T.G. (1991) Pigment dispersion, in *Coatings technology handbook* (ed Satas, D.), Marcel Dekker, New York, pp. 529–50.
10. Color Pigments Manufacturers Association, Inc. (1992) *Safe handling of color pigments*, p. 4.
11. Color Pigments Manufacturers Association, Inc. (1992) *Safe handling of color pigments*, p. 7.
12. Vernadakis, T.G., Personal communication.
13. Code of Federal Regulations (1991) *Protection of Environment, Code CFR, 40 Parts 260–299*, Office of the Federal Register National Archives and Records Administration as a Special Edition of the Federal Register, Washington, p. v, vi, vii, ix.
14. Mueller-Kaul, W. and Ratcliff, C. (1992) The world market for organic pigments with special emphasis on Europe, *American Ink Maker*, **11**, 39.
15. Mueller-Kaul, W. and Ratcliff, C. (1992) The world market for organic pigments with special emphasis on Europe, *American Ink Maker*, **11**, 38.
16. Mueller-Kaul, W. and Ratcliff, C. (1992) The world market for organic pigments with special emphasis on Europe, *American Ink Maker*, **11**, 40.
17. Mueller-Kaul, W. and Ratcliff, C. (1992) The world market for organic pigments with special emphasis on Europe, *American Ink Maker*, **11**, 46.

Bibliography

Code of Federal Regulations (1991) *Protection of the Environment, Code CFR, 40, Parts 260-299*, Office of the Federal Register National Archives and Records Administration as a Special Edition of the Federal Register, Washington.

Color Pigments Manufacturers Association, Inc. (1993) *Safe handling of color pigments.*
Forbes, R. (Sun Chemical Corporation). Personal communication.
Hunger, K. (1993) *Industrial organic pigments*, VCH, New York.
Lewis, P.A. (1988). *Organic pigments*, Federation of Societies for Coatings Technology, Philadelphia, PA.
Morgan, W.M. *Pigments for paints and inks physical and chemical properties*, Trade and Technical Press Ltd., Morden, Surrey, UK.
Mueller-Kaul, W. and Ratcliff, C. (1992). The world market for organic pigments with special emphasis on Europe, *American Ink Maker*, **11**.
Patton, T.C. (ed) (1973) *Pigments handbook, volume I*, John Wiley and Sons, New York.
Vernadakis, T.G. (1984) Pigments particle size and effects on printing ink performance, *American Ink Maker*, **62**(2), 24.
Vernadakis, T.G. (1991) Pigment dispersion, in *Coatings technology handbook* (ed Satas, D.), Marcel Dekker, New York.
Veradakis, T.G.. (Sun Chemical Corporation). Personal communication.

C.I. Pigment blue 1
C.I. 42595:2

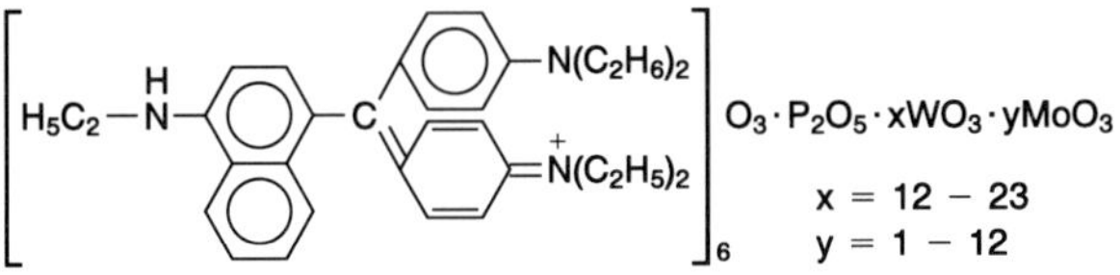

H_5C_2-NH- ... $N(C_2H_6)_2$ ... $\overset{+}{N}(C_2H_5)_2$]$_6$ $O_3 \cdot P_2O_5 \cdot xWO_3 \cdot yMoO_3$

x = 12 − 23
y = 1 − 12

Victoria blue (PTMA)

Figure 4.14 Pigment structures.

C.I. Pigment blue 15
C.I. 74160

Copper phthalocyanine blue (R.C.)

C.I. Pigment blue 15:1
C.I. 74160 C.I. 74250

X = 0 − 1 Cl
Copper phthalocyanine blue (R, NC)

C.I. Pigment blue 15:2
C.I. 74160 C.I. 74250

X = 0 − 1 Cl
Copper phthalocyanine blue (R.NCNF)*

C.I. Pigment blue 15:3
C.I. 74160

Copper phthalocyanine blue (G, NC)

C.I. Pigment blue 15:4
C.I. 74160

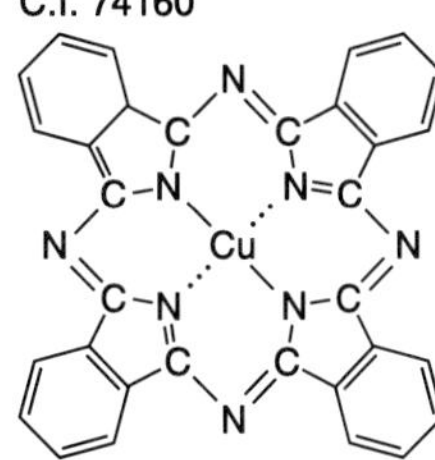

Copper phthalocyanine blue (G, NCNF)*

C.I. Pigment blue 16
C.I. 74100

Phthalocyanine blue (metal free)

C.I. Pigment blue 27
C.I. 77510

$Fe(NH_4)Fe(CN)_6 \cdot xH_2O$

Milori blue, iron blue

C.I. Pigment green 1
C.I. 45040:1

$[\ldots N(C_2H_5)_2 \ldots \overset{+}{N}(C_2H_5)_2]_6 \; O_3 \cdot P_2O_5 \cdot xWO_3 \cdot yMoO$

x = 12 − 23
y = 1 − 12

Brilliant green (PTMA)

* Specially treated to prevent flocculation.

Figure 4.14 *Continued*

C.I. Pigment green 1X
C.I. 42040:X

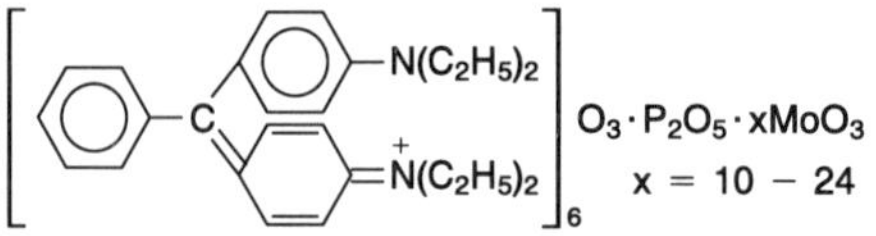

$O_3 \cdot P_2O_5 \cdot xMoO_3$
$x = 10 - 24$

Brilliant green (PMA)

C.I. Pigment green 7
C.I. 74260

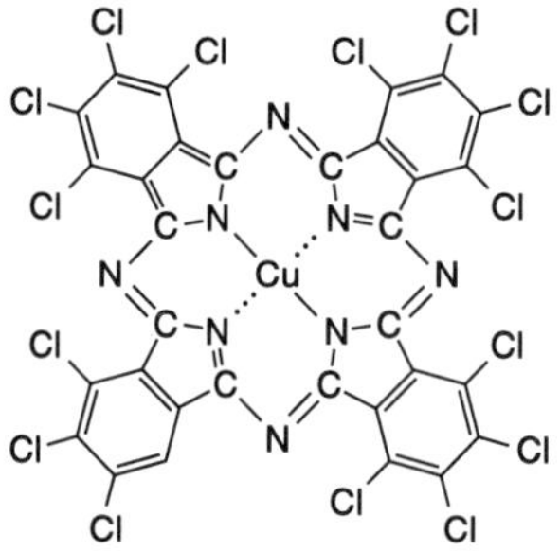

Copper phthalocyanine green

C.I. Pigment green 36
C.I. 74265

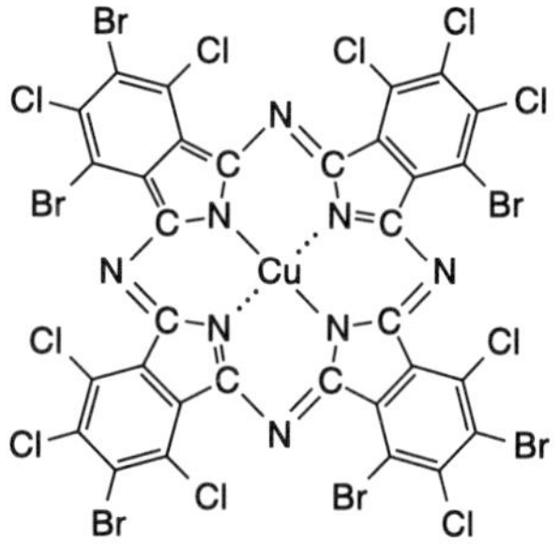

Copper phthalocyanine green (CT, BR)

C.I. Pigment orange 5
C.I. 12075

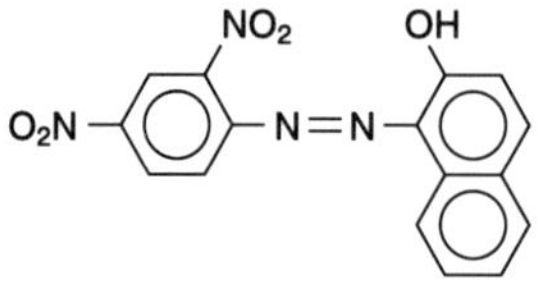

Dinitraniline orange

C.I. Pigment orange 13
C.I. 21110

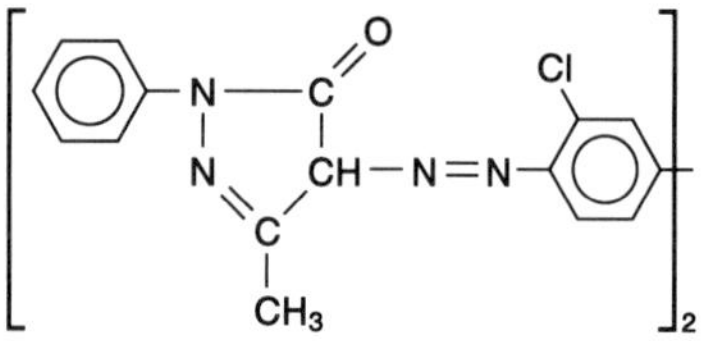

keto

Pyrazolone orange

C.I. Pigment orange 16
C.I. 21160

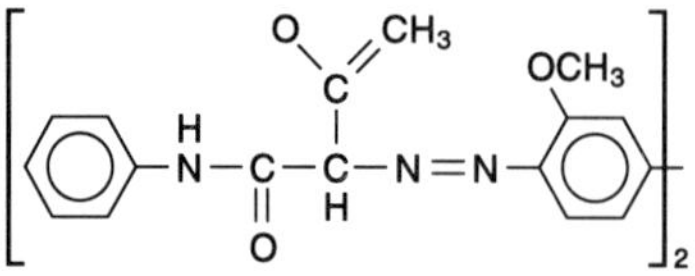

keto

Dianisidine orange

C.I. Pigment orange 34
C.I. 21115

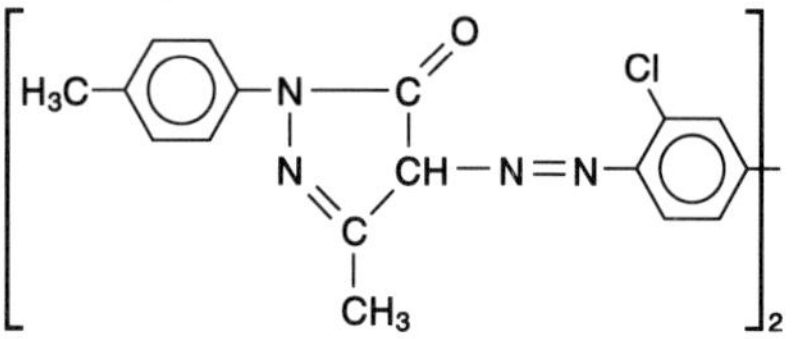

keto

Diarylide orange

Figure 4.14 *Continued*

C.I. Pigment orange 36
C.I. 11780

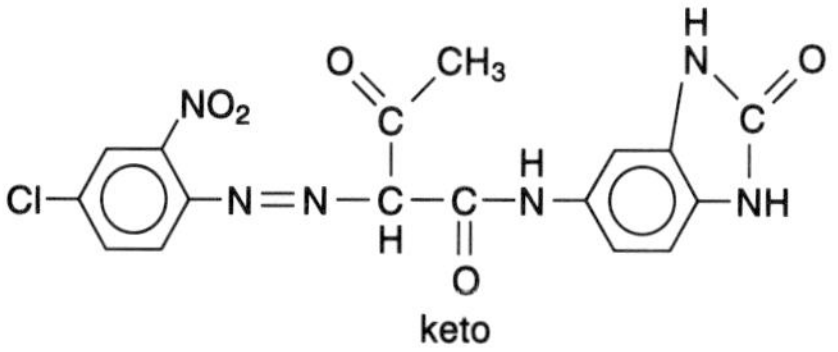

Benzimidazolone orange (barium)

C.I. Pigment orange 46
C.I. 15602

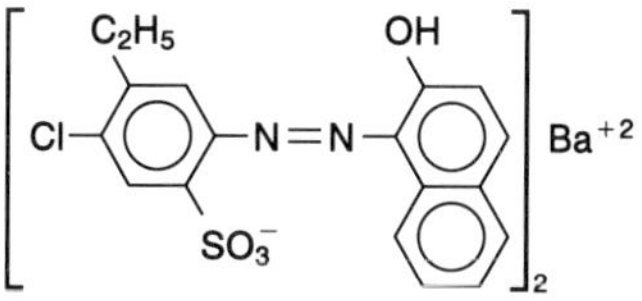

Ethyl red lake C

C.I. Pigment red 2
C.I. 12310

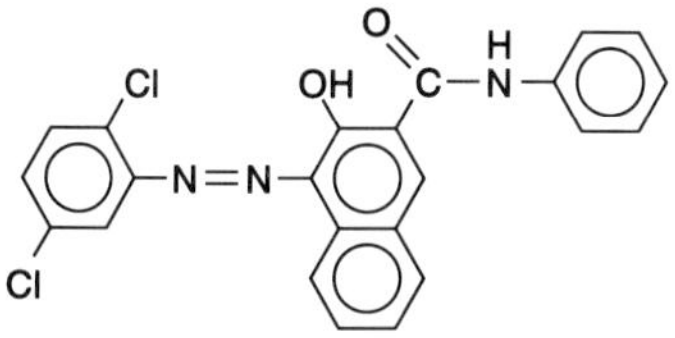

Naphthol red FRR (medium shade)

C.I. Pigment red 17
C.I. 12390

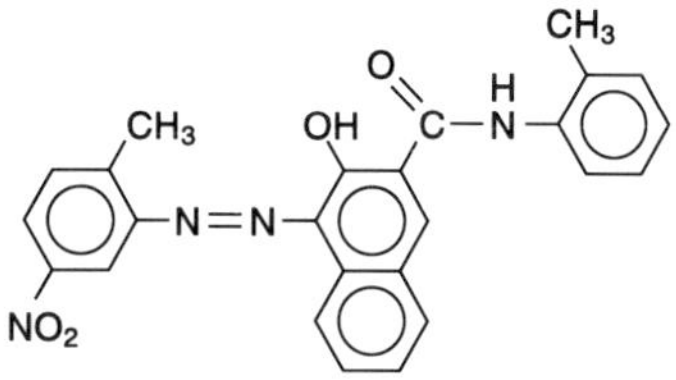

Naphthol red

C.I. Pigment red 22
C.I. 12315

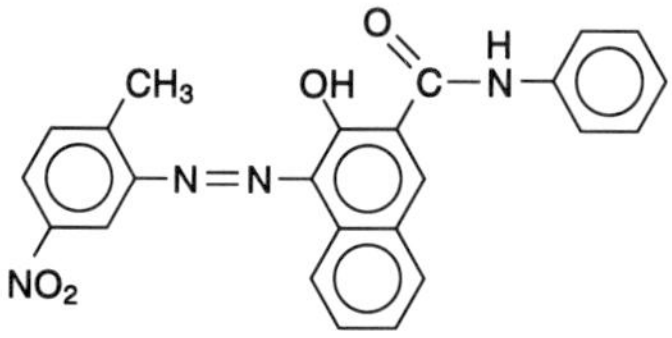

Naphthol red (light shade)

C.I. Pigment red 23
C.I. 12355

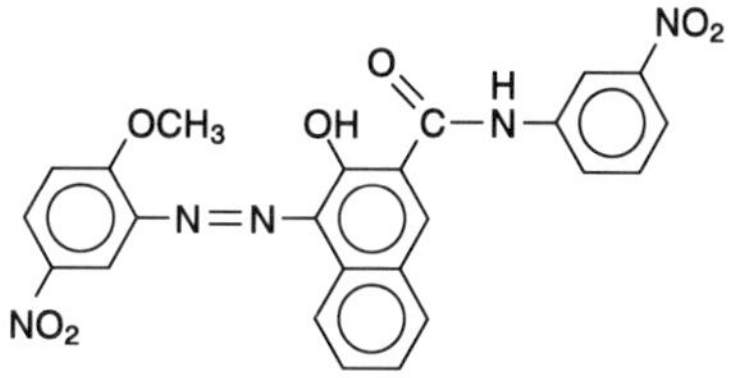

Naphthol red (blue shade)

C.I. Pigment red 48:1
C.I. 15865:1

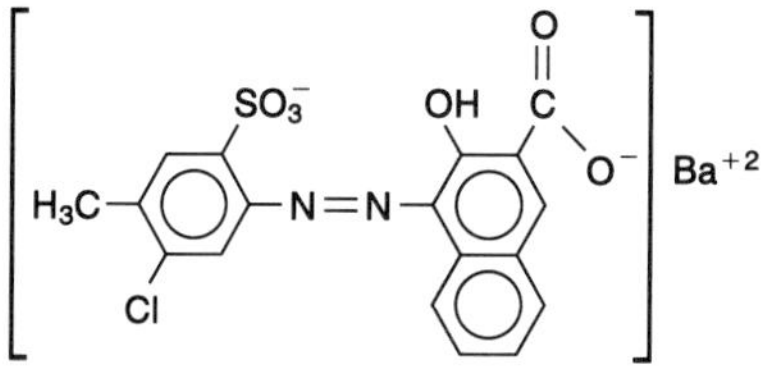

Permanent red 2B (barium)

C.I. Pigment red 48:2
C.I. 15865:2

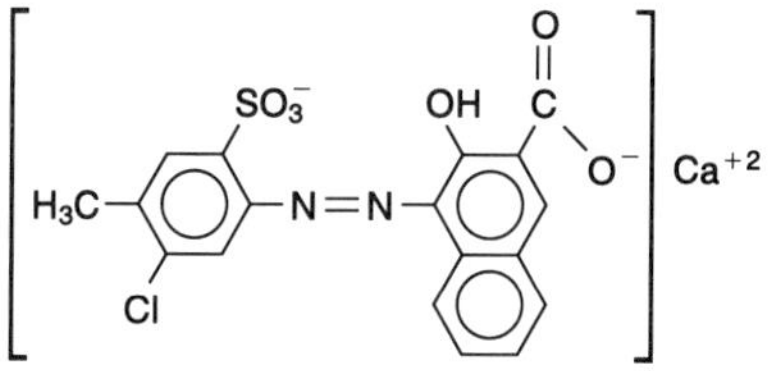

Permanent red 2B (calcium)

Figure 4.14 *Continued*

C.I. Pigment red 48:3
C.I. 15865:3

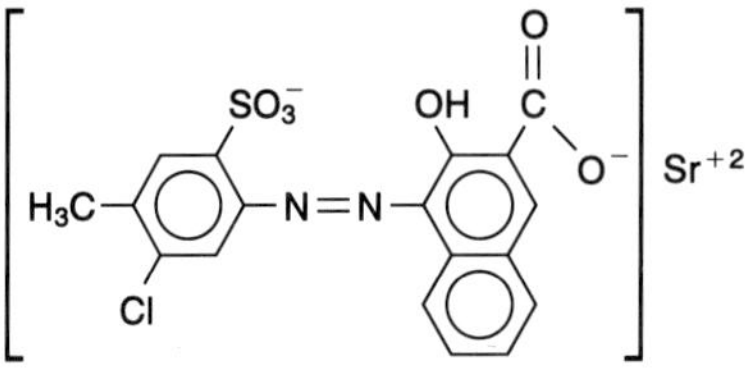

Permanent red 2B (strontium)

C.I. Pigment red 49:1
C.I. 15630:1

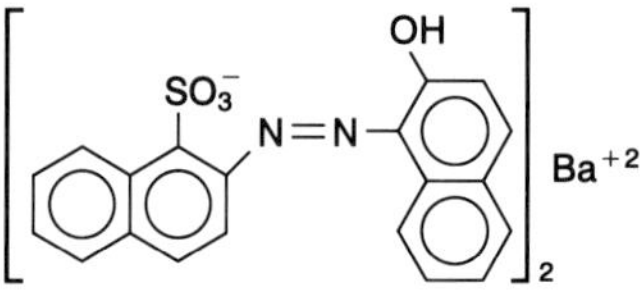

Lithol red (barium)

C.I. Pigment red 49:2
C.I. 15630:2

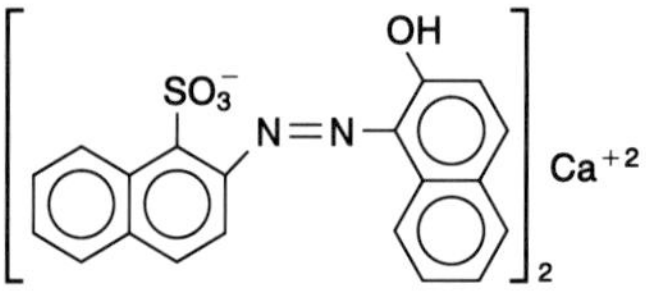

Lithol red (calcium)

C.I. Pigment red 52:1
C.I. 15660:1

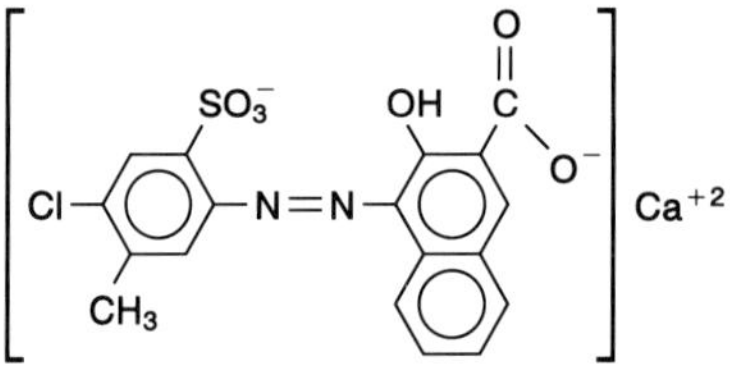

Bon red (calcium)

C.I. Pigment red 53:1
C.I. 15585:1

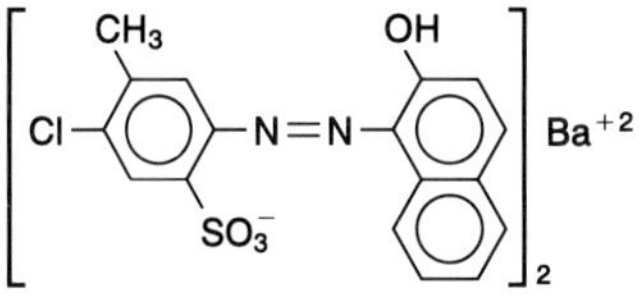

Red lake C (barium)

C.I. Pigment red 57:1
C.I. 15850:1

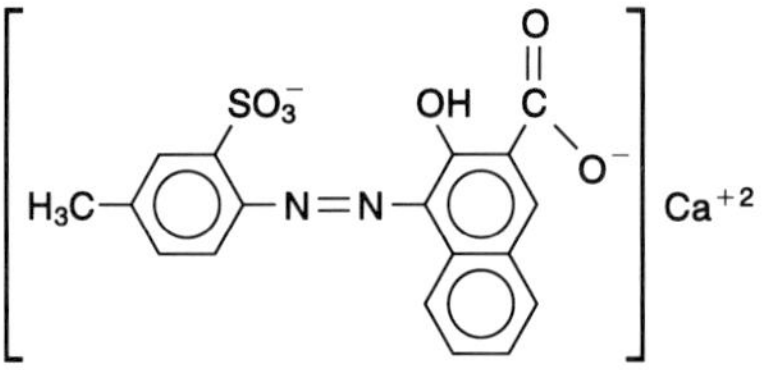

Lithol rubine (calcium)

C.I. Pigment red 81:1
C.I. 45160:1

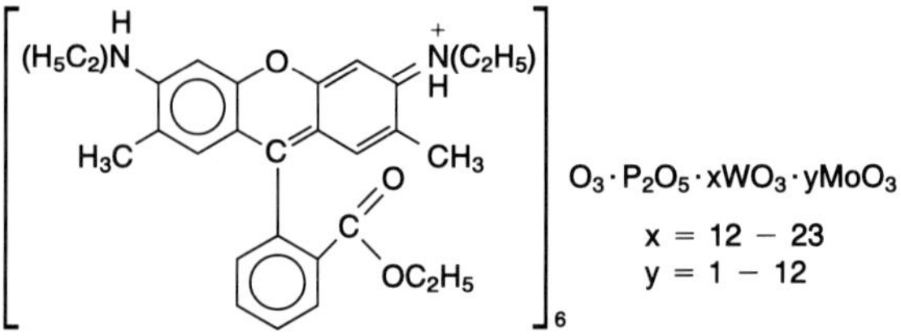

$O_3 \cdot P_2O_5 \cdot xWO_3 \cdot yMoO_3$

$x = 12 - 23$
$y = 1 - 12$

Rhodamine Y (PTMA)

Figure 4.14 *Continued*

C.I. Pigment red 122
C.I. 73915

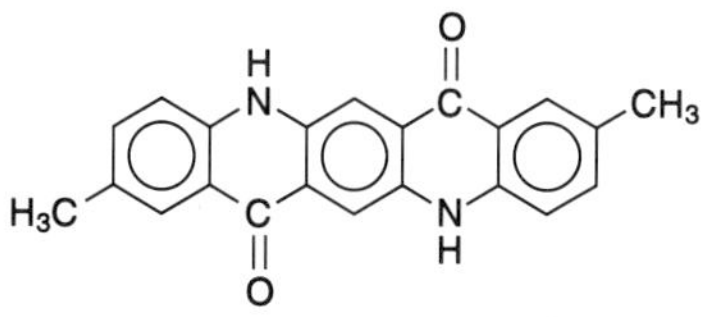

Quinacridone magenta

C.I. Pigment red 202
C.I. Unassigned

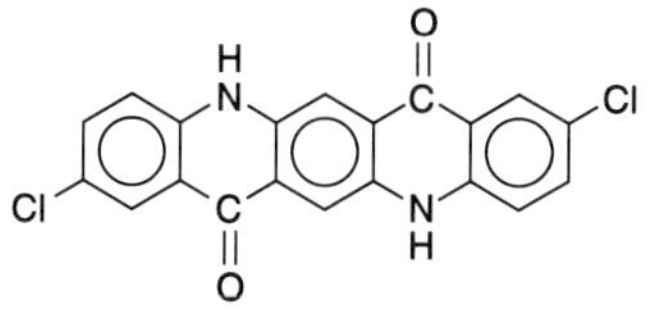

Quinacridone magenta B

C.I. Pigment red 209
C.I. 73905

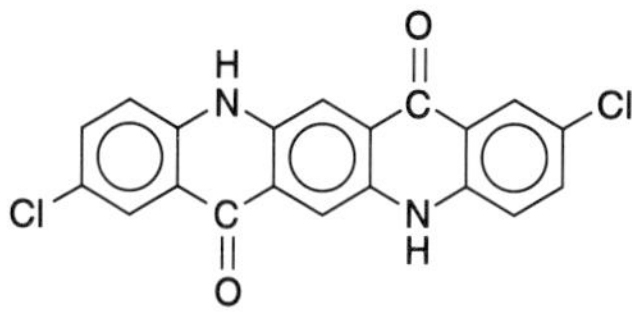

Quinacridone red

C.I. Pigment red 169
C.I. 45160:2

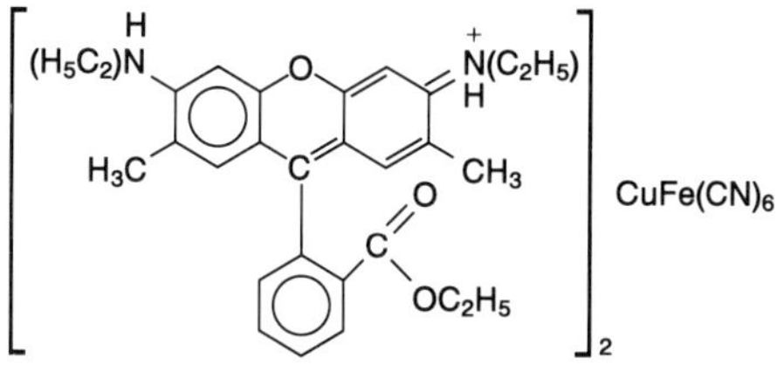

Rhodamine 6G (CFA)

C.I. Pigment red 170
C.I. 12475

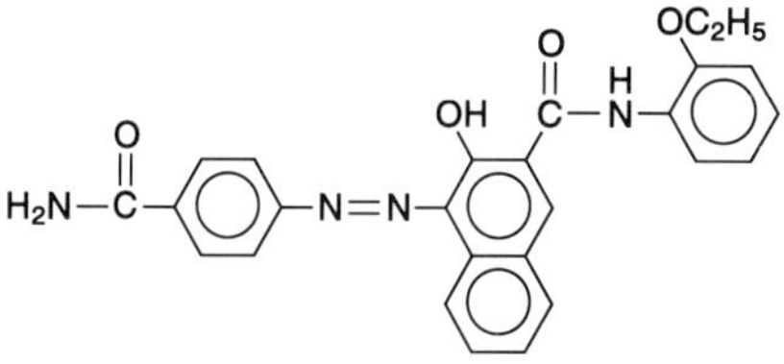

Naphthol red F5RK

C.I. Pigment red 184
C.I. Unassigned

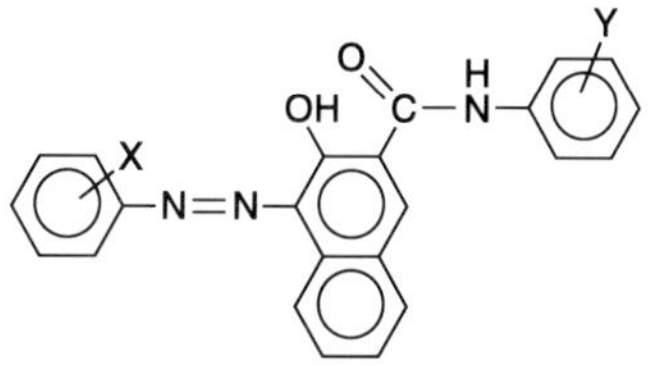

Naphthol rubine F6B

C.I. Pigment red 210
C.I. Unassigned

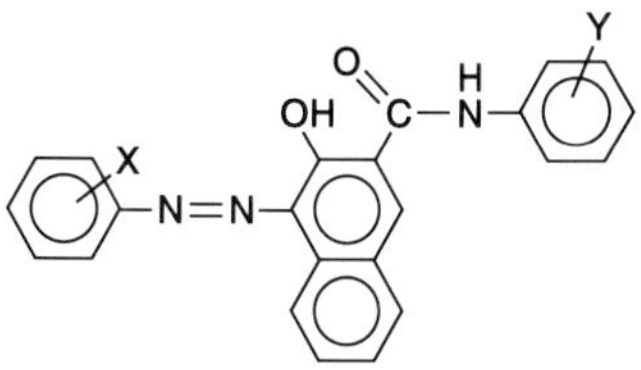

Naphthol red F6RK

C.I. Pigment red 238
C.I.

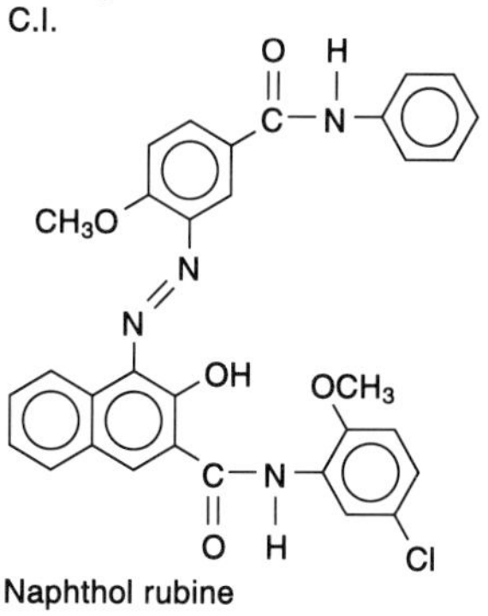

Naphthol rubine

Figure 4.14 *Continued*

C.I. Pigment violet 1
C.I. 45170:2

$(H_6C_2)_2N$ ... $\overset{+}{N}(C_2H_5)_2$ ... $]_6$ $O_3 \cdot P_2O_5 \cdot xWO_3 \cdot yMoO_3$

x = 12 − 23
y = 1 − 12

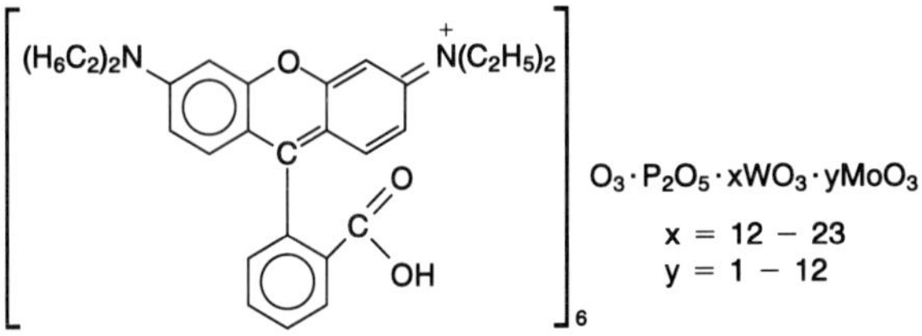

Rhodamine B (PTMA)

C.I. Pigment violet 3
C.I. 42535:2

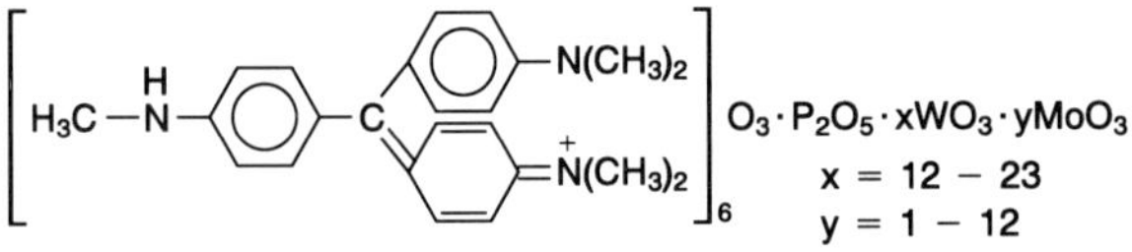

H_3C-NH- ... $N(CH_3)_2$ ... $=\overset{+}{N}(CH_3)_2$ $]_6$ $O_3 \cdot P_2O_5 \cdot xWO_3 \cdot yMoO_3$

x = 12 − 23
y = 1 − 12

Methyl violet (PTMA)

C.I. Pigment red 210
C.I. Unassigned

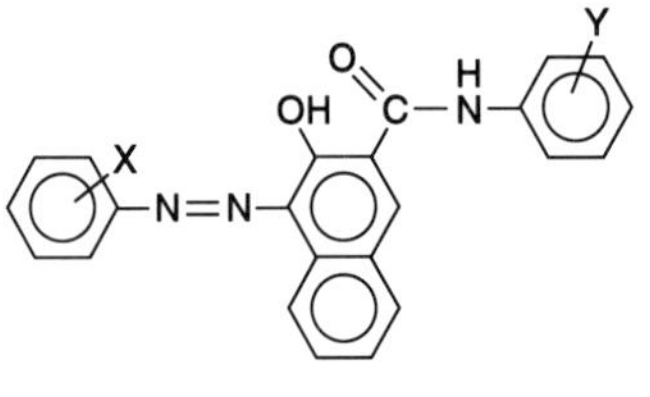

Naphthol red F6RK

C.I. Pigment red 238
C.I.

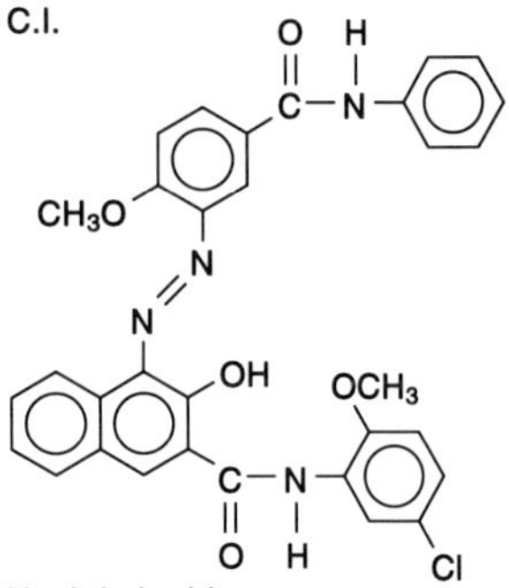

CH_3O, OCH_3, Cl

Naphthol rubine

C.I. Pigment violet 1
C.I. 45170:2

$(H_6C_2)_2N$ ... $\overset{+}{N}(C_2H_5)_2$ ... $]_6$ $O_3 \cdot P_2O_5 \cdot xWO_3 \cdot yMoO_3$

x = 12 − 23
y = 1 − 12

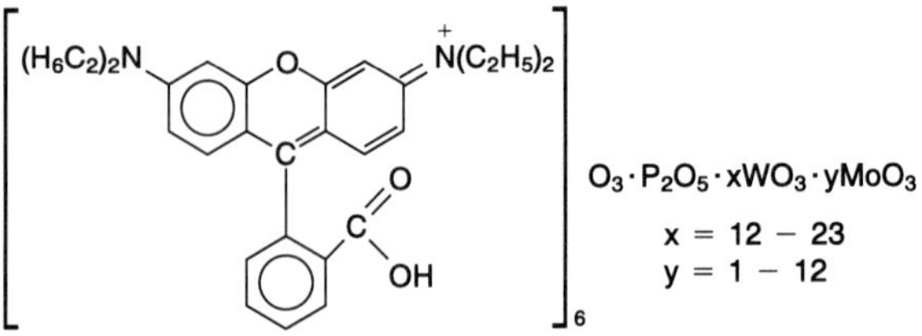

Rhodamine B (PMTA)

Figure 4.14 *Continued*

C.I. Pigment violet 3
C.I. 42535:2

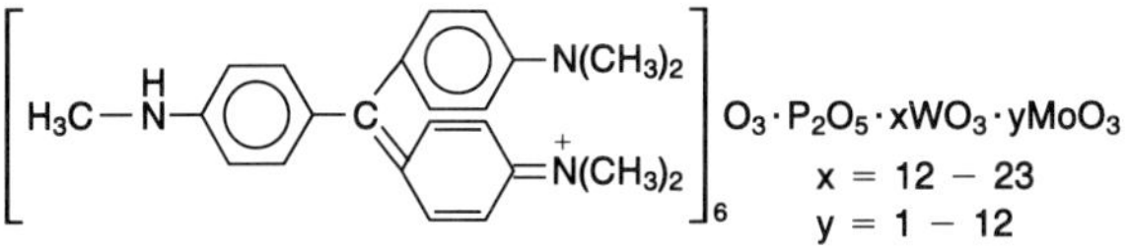

$[\ldots]_6 O_3 \cdot P_2O_5 \cdot xWO_3 \cdot yMoO_3$

x = 12 − 23
y = 1 − 12

Methyl violet (PMTA)

C.I. Pigment violet 19
C.I. 73900[a]

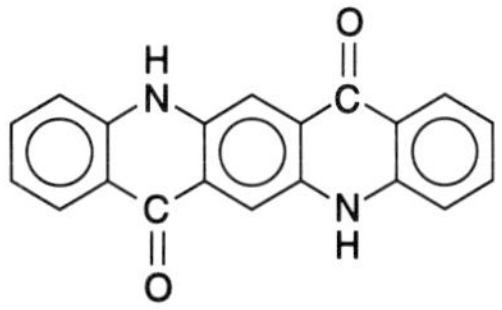

Quinacridone violet,
Quinacridone red

C.I. Pigment violet 23
C.I. 51319

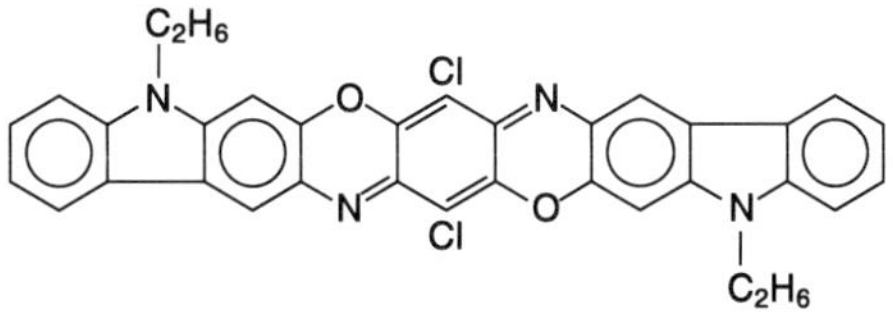

Carbazole violet

C.I. Pigment violet 27
C.I. 42535:3

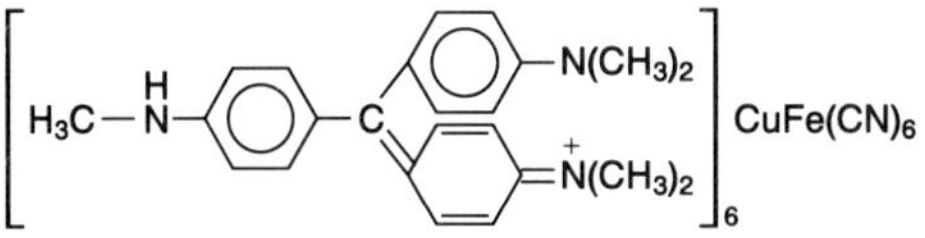

$[\ldots]_6 CuFe(CN)_6$

Crystal violet (CFA)

C.I. Pigment yellow 3
C.I. 11710

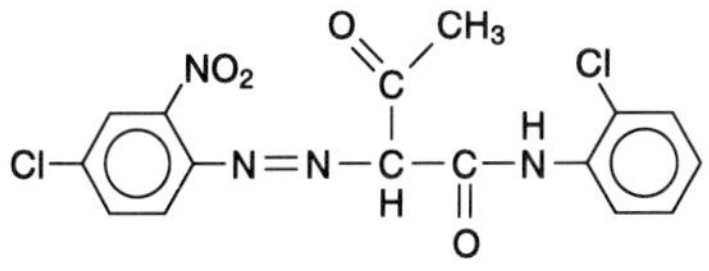

keto

Diarylide yellow 10 G

C.I. Pigment yellow 12
C.I. 21090

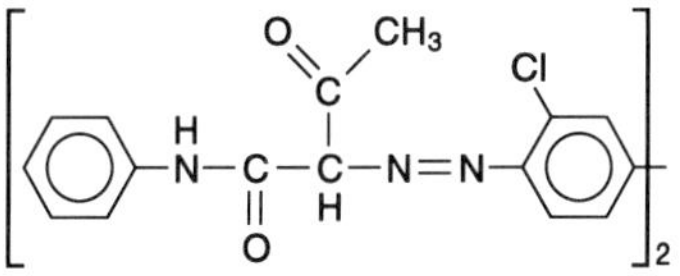

keto

Diarylide yellow AAA

Figure 4.14 *Continued*

C.I. Pigment yellow 13
C.I. 21100

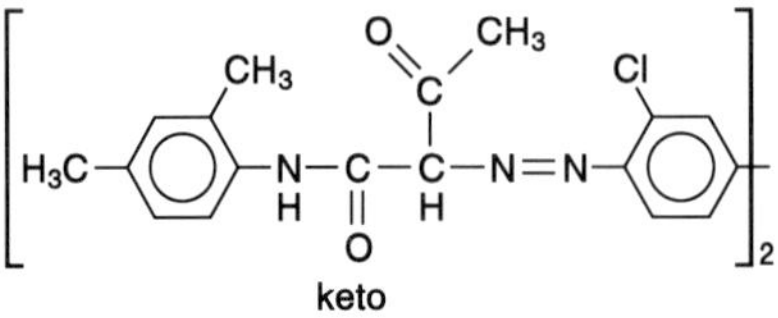

keto

Diarylide yellow AAMX

C.I. Pigment yellow 14
C.I. 21095

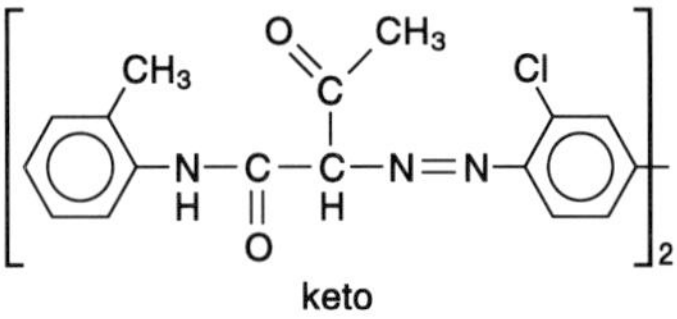

keto

Diarylide yellow AAOT

C.I. Pigment yellow 17
C.I. 21105

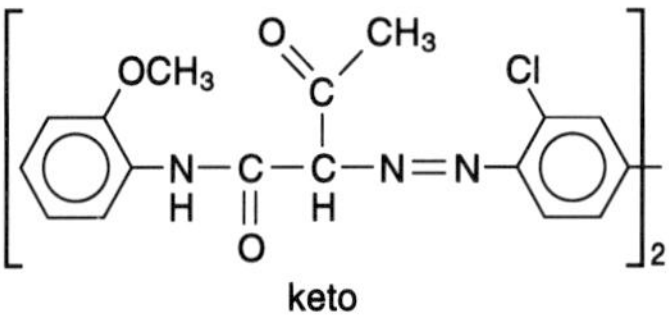

keto

Diarylide yellow AAOA

C.I. Pigment yellow 55
C.I. 21096

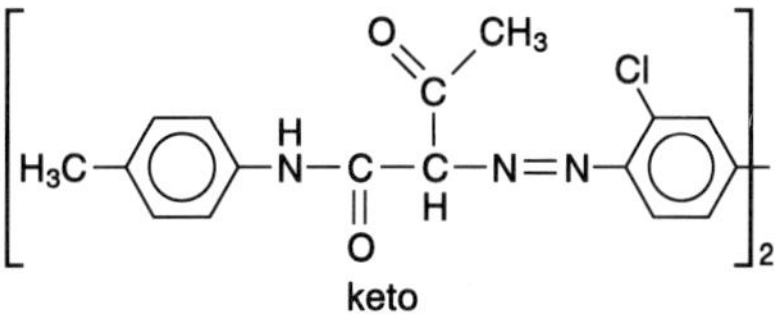

keto

Diarylide AAPT

C.I. Pigment yellow 65
C.I. 11740

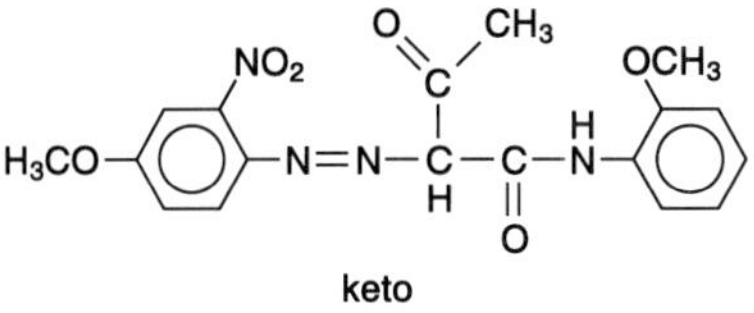

keto

Arylide yellow

C.I. Pigment yellow 74
C.I. 11741

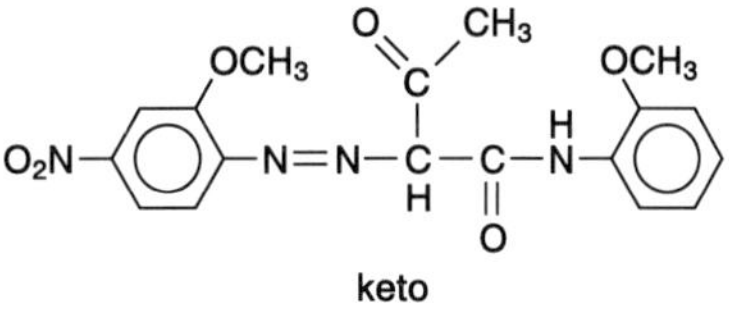

keto

Arylide yellow

C.I. Pigment yellow 83
C.I. 21108

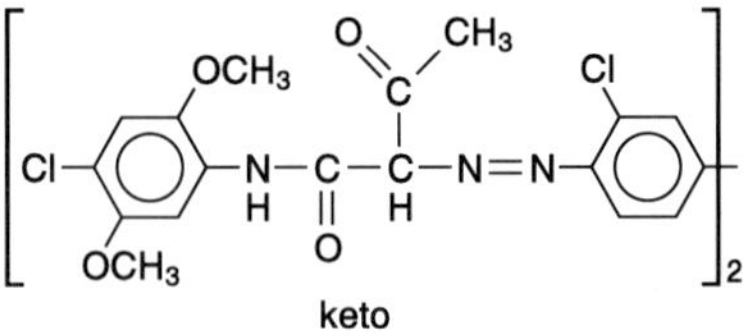

keto

Diarylide yellow HR

Figure 4.14 *Continued*

5 Carbon black

W. SCHUMACHER

5.1 Classification and types of carbon black

Carbon blacks are one of the most important items used in industry. Worldwide production figures are higher than 6 million metric tons per year. Various manufacturing methods are used, the most important ones will be covered in Section 5.3. By far the leading process is the manufacture of furnace black, which accounts for over 98% of total carbon black production [1].

Using this process the raw material (oil) is only partially burned at high temperatures in a closed furnace, a preferably large part is split through thermal oxidative decomposition into carbon (carbon black) and hydrogen. Also using the thermal oxidative decomposition process, but in a totally different form, carbon blacks are made using the Degussa gas black process or the lamp black process. These carbon blacks are used in speciality applications showing a certain advantage compared with the furnace black process. In contrast, acetylene blacks or thermal blacks are made through a thermal composition process only using acetylene or natural gas.

Of the previously mentioned 6 million tons of carbon black more than 90% are used in the rubber industry. Here it is used as a reinforcing filler in such items as tires and mechanical rubber goods. Using carbon blacks in those items improves the mechanical properties such as elasticity and reduced wear of tires on roads. In addition it also acts to prevent destructive UV rays and, at reasonably high concentrations of the rubber mixture, helps to impart the right thermal and electrical properties.

Traditionally, some of the rubber blacks are used in water based inks. These are standardized per ASTM D 1765. Typical items in that area for instance are: N-110, N-220, N-330, N-351, N-550 and N-762. Approximately 40 carbon blacks that fall under that ASTM standard are in existence. The N in these instances means 'normal curing'; the first of the three numbers typifies the particle size (the smaller the number the smaller the particle size). The other two numbers are more or less added for analytical purposes. A technical expert would recognize these in reference to particular user technical properties.

As previously mentioned, more than 90% of the carbon blacks are used in the rubber industry. The major users for the remaining amounts are the plastic, printing ink and paint industry where carbon blacks are used as black pigments, UV protectors, to improve rheological properties or as conductive pigments. In

Translated from original German text by Bert Auer.

addition, there are numerous other uses where carbon black is used as black pigment in toners, textiles, fibers, paper and concrete; also carbon brushes, graphite forms, metal carbides, refractories, foam glass and metal castings.

In printing inks, apart from organic pigments, pigment black is the largest component; approximately 150 000 tons per year are used worldwide. This is understandable, considering that newspapers and books, in particular, are today usually printed in black and white, and even where colors are printed black still plays a major role. Black also has an influence in four color process printing, where its use is growing.

To differentiate between rubber blacks and carbon black used in printing we use the term 'pigment black' or 'speciality blacks' when referring to printing inks, since in this application carbon black is used almost entirely to print black.

Pigment blacks are not standardized in the same way as are rubber blacks. They are grouped according to jetness. For historical and technical reasons gas blacks are grouped with channel blacks, which in today's world barely play any roles. As a reference for color depth the primary particle size can be considered since, as will be explained later, it is the major influence for jetness. We differentiate therefore between high color channel (HCC) and furnace blacks (HFC) with particle sizes of 10–15 nm, medium color channel (MCC) and furnace blacks (MCF) with particle sizes of 16–24 nm, regular color channel (RCC) and furnace blacks (RCF), particle size 25–35 nm and low color furnace blacks (LCF). Post-treated, that is, oxidized blacks may contain the number (0) as a reference. Speciality blacks for conductivity are usually designated as conductive (CB) blacks or conductive furnace blacks (CF). The words 'long flow impigment black' (LFI) for channel blacks with especially good rheology properties in printing inks are no longer in use today. Although black pigments are made up primarily of carbon they are considered inorganic pigments. The color index number 77266 pigment black 7, (exception-lamp black-pigment black 6), the CAS reg. number 1333-86-4 (lamp black, 133-86-4), the MITI section/class reference number 10/5-3328 (lamp black 10/5-5222), EINECS number 2156099, is uniform for all carbon blacks.

For the use of carbon blacks in toys, cosmetics and textiles which come in contact with food many countries use various standards, but these are too numerous to mention here. Details may be found in [2].

Although pigment blacks contain the smallest primary particle sizes of all pigments, they are remarkably heat and light stable [3] and resistant to chemicals. This can be explained when the chemical structure of carbon black is considered.

5.2 Composition of carbon black

The atomic structure of carbon black is similar to that of microcrystalline graphite which, according to the types of carbon black can be categorized into

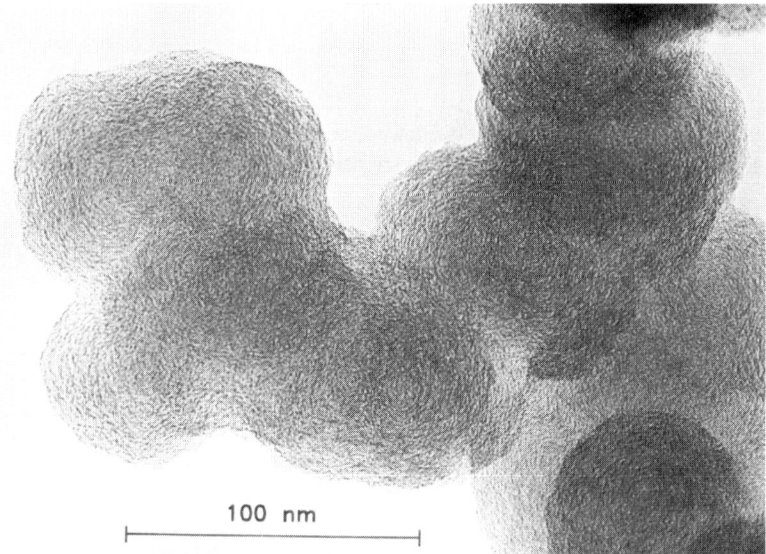

Figure 5.1 Structure of carbon black as determined by using transmission electron microscopy.

different forms [4]. Because of high definition electronmicroscope recordings (TEM) as in Figure 5.1 and X-ray examinations it seems less certain, as previously assumed, that these are crystalline layer parcels, as in Figure 5.2 (left), but rather larger disorganized areas, which are interconnected with more particles, Figure 5.2 (right). As carbon black has a structure similar to that of graphite, it is an electrical conductor and can absorb a very wide range of

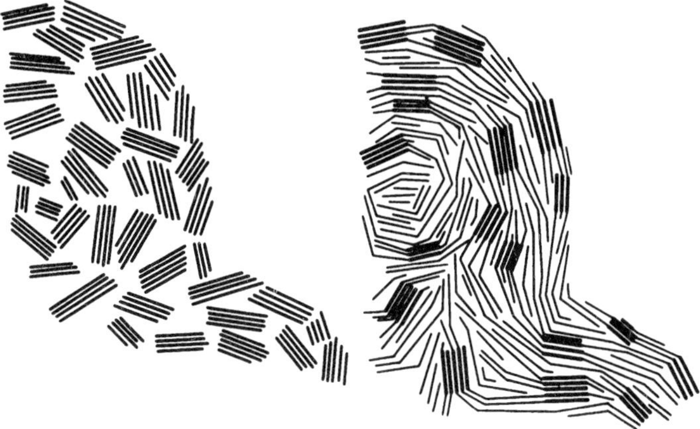

Figure 5.2 Schematic representation of the structure of carbon black as crystalline layer parcels (left diagram) and a more disorganized structure (right diagram).

wavelengths. Carbon black is not only black at visible wavelengths, but is also black in the IR and UV spectra.

Depending on the manufacturing process and raw material used carbon black contains more or less oxygen, sulphur, nitrogen, and smaller amounts of trace metals. More than anything else, oxygen on the surface of a carbon black particle, have considerable influence on the technical use of this material [5]. Oxygen in the form of chemical groups on carbon atoms are securely bound to the carbon black surface. The types and amounts can be determined by titration with different bases and acids [4]. It is known that, depending on the type of carbon black, and particularly in larger amounts on oxidized carbon blacks, not only acidic but also alkaline groups are present. The more important ones are illustrated in Figure 5.3.

5.3 Manufacture of carbon black

The printing ink industry mainly uses furnace blacks, but in specialized areas gas blacks are still used. The manufacturing processes will be described here. As a comparison older production processes, such as lamp black and channel black, are mentioned. The most important feed stocks are oils for the majority of carbon blacks. Oils with preferably a high content of polycyclic aromatic hydrocarbons are used. A higher concentration of these compounds and a larger

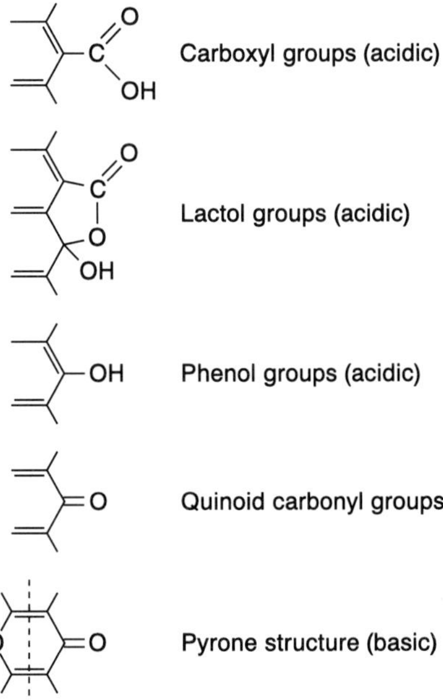

Figure 5.3 Some chemical groups present on the surface of carbon black.

amount of aromatic rings in these molecules increase the ratio of carbon versus hydrogen and therefore the yield of carbon black in the manufacturing process. Residual oils from catalytic or thermal crack processes are mostly used, but heavy distillates from coal tar are also processed.

The lamp black process is probably the oldest manufacturing process still in use. The feed stock is contained in a cast iron pan with a diameter of up to 1.5 m. Separated by a small space above the vessel is a fume hood which is lined with refractory bricks (Figure 5.4). With air entering through the above-mentioned space part of the oil is burnt off, heat from the fume hood helps to evaporate the oil. By heating at a high temperature the rest of the oil is converted to carbon black. The gas–solid mixture is then cooled and filtered to separate the carbon black. By changing the space between vessel and fume hood in small increments and maintaining the system at low pressure the amount of air controlling the flame can be changed, thus regulating the quality of the product. The process is fairly simple and it produces carbon black with large–medium sized primary particles which, because of their low jetness and tinting strength make them hardly suitable for printing inks. Characteristically these carbon blacks contain a very wide particle size range which cannot be duplicated using other manufacturing processes. Lamp blacks are used in special rubber mixtures, for toning in paints in the manufacture of graphite forms and in many other applications where they cannot easily be replaced with blacks from other manufacturing processes.

The channel black process was developed in the US during the middle of the last century. It used natural gas as the only raw material. The gas was partially

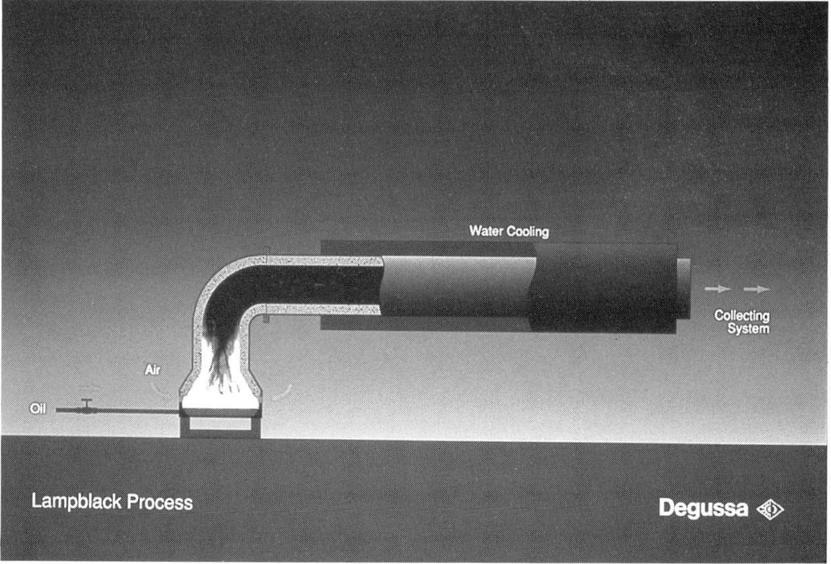

Figure 5.4 Lamp black process.

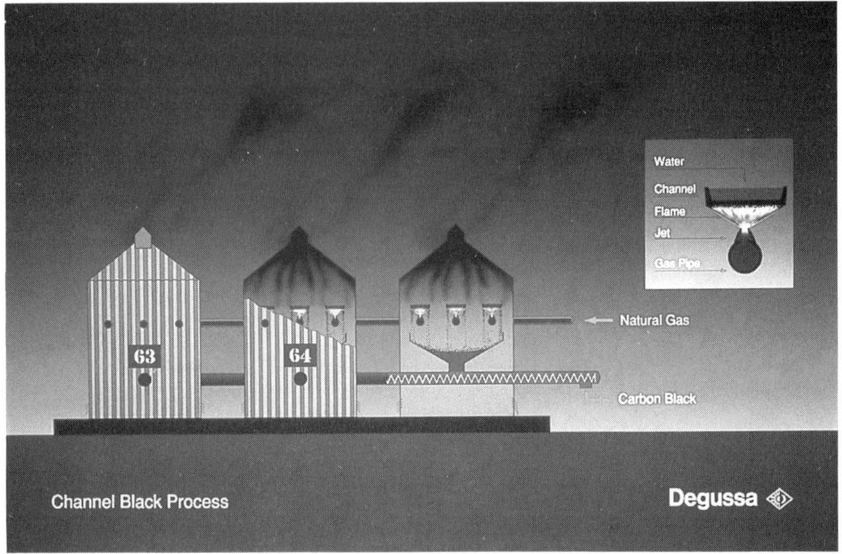

Figure 5.5 Channel black process.

burnt away in a large number of small flames, above which were a number of water cooled iron channels (Figure 5.5). The small amount of carbon on the channels was then removed mechanically [6]. By varying the flames, the distance of the channels and the gas–air relationship, carbon blacks with very different particle sizes could be produced; which, when used in the final product, were easily distinguishable.

For decades channel blacks were the most important reinforcing fillers for rubber, while pigment blacks were used for printing inks and paints. The development of the furnace black process in the middle of this century, price increases in natural gas, the low yield versus feed-stock and environmental concerns caused the demise of the channel black process. In the US these producers closed their manufacturing plants in the 1970s.

The Degussa gas black process was developed in Germany in the mid-1930s. It is very similar to the channel black process but uses coal tar distillates for feed-stock. As one can see in Figure 5.6; schematically presented, the feed-stock is first heated and evaporated, and together with a carrier gas containing a higher amount of hydrogen is conveyed to the burners. Here too a part of the evaporated oil along with the carrier gas is burnt away using numerous small flames, toward a rotating water cooled cylinder. Part of the carbon black collects on the rotating cylinder and is separated using scraper blades. The remaining carbon black contained in the after-gasses is then separated in filters and added to the scraped off material. Several vessels producing the same material are connected in groups which form the actual production units.

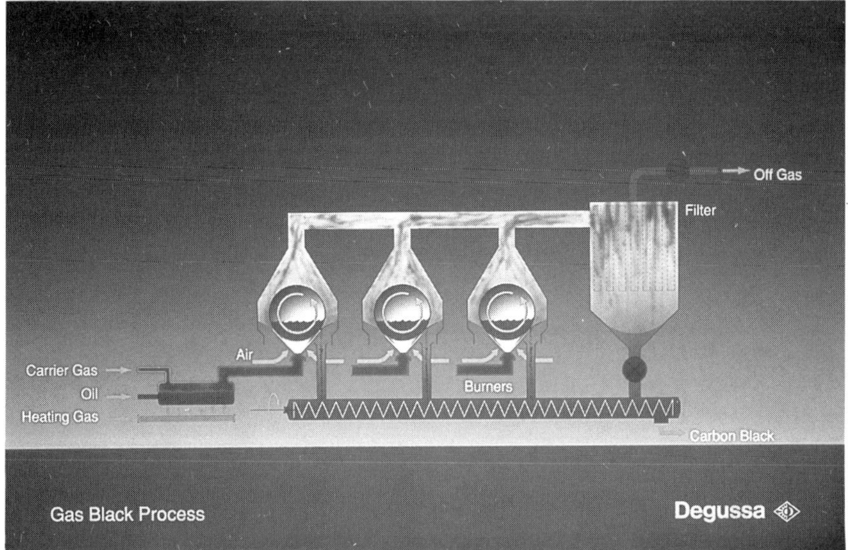

Figure 5.6 Gas black process.

The gas black process also offers the opportunity to manufacture carbon blacks with considerable variations in particle sizes, which have similar application characteristics as channel blacks. The product yield when compared to the channel black process is much higher because of the feed-stock. Also, the processing units are mostly contained, therefore no pigment is introduced into the environment. This process is still in use today and is expected to be so in the future.

Gas blacks, because of the process, are always acidic, which means they contain a larger amount of oxygen containing groups on the surface. For this reason these blacks are easy to wet out and are easily dispersible in various binders.

By oxidizing carbon blacks with nitric oxides, nitric acid or ozone the number of acidic surface groups can be further increased. Such after-treated blacks are used primarily in high-quality paints, and also in printing inks.

The most recent development is the furnace black process, which, as mentioned earlier, is today's most important manufacturing process for almost all applications. Referring to Figure 5.7, a horizontal furnace is used as a reactor, preheated air in combination with natural gas is burnt, creating a very hot (up to 1900°C) flame [1, 7]. (The furnaces therefore are lined with fire resistant ceramic material.) A very fine oil spray (the feed-stock) is introduced into this flame, which evaporates the oil, carbon black is built with high yield. To avoid secondary reactions the gas carbon black mixture is then sprayed and cooled in heat exchangers immediately after the reactor. There, the air is again pre-heated. Filters then separate the carbon black from the so-called 'tail gas',

Figure 5.7 Furnace black process.

the combustible part containing carbon monoxide, etc. The tail-gas is burnt off, the heat created is partially used to dry wet beaded carbon black. In numerous plants in Europe for instance, additional steam is created which in turn runs electrical generators creating enough energy to supply the carbon black plant. The electric energy may be more than is needed, the remainder is sold to the public utility company.

In contrast to gas blacks, furnace black processes create very few surface acid groups or none at all, therefore furnace blacks are generally alkaline. A small amount is oxidized, post-treated for special applications such as printing inks and paints.

By varying the gas–air–oil ratios and with different configurations of the furnace, carbon blacks with very different primary particle sizes may be created; the useful range is even greater than those of the gas black process.

There is another way to influence the quality of carbon black. Normally primary particles will attract each other in the flame, therefore partially growing together and cannot be separated when trying to make a dispersion. These are considered aggregates [8]. In addition, these aggregates form loose agglomerates. Both forms are termed 'carbon black structure'.

The furnace black process is the only one where the latitude of the structure can be influenced to a very large degree. Small amounts of alkali metal complexes, mostly potassium, are injected into the flame. The complexes will evaporate at high temperatures and are present as ions in the gaseous form. The positive potassium ions will be present in the carbon black particles, which will then also have a positive charge. Electric repulsion prohibits the formation of aggregates. By using different amounts of alkali metal complexes the structure of the carbon black can be controlled. As will be explained later, structure has a considerable influence on the technical use of a carbon black. The only mention here is to say that carbon blacks with high and low structures are used for different applications in the printing ink industry.

Carbon black, after manufacture, is a very 'fluffy' powder and could not be used in this form as it may be substantially contaminated. These contaminants can be eliminated by the use of grit separators and the apparent density can be increased with the help of special equipment. Even in its higher density form carbon black is very dusty and not free flowing. This is the reason why this material is packaged in bags only containing 10 to 20 kg (25 to 50 lbs).

The majority of carbon black is manufactured in beaded form, which is mostly free of dust. It can be shipped in containers up to rail cars in large quantities and stored in silos where automated equipment is used for end use processing. There are two different processes that play a major role in the manufacture of black pellets. By far the largest part is processed using a wet beading method. Pin mixers are used to form wet beads with water and carbon black, which are then dried in heated drums. Wet beaded black is normally very hard, a high amount of energy is required to disperse these products. This is the reason why most pigment blacks are dry beaded using rotating drums in which powder black is densitized into small, soft pelletss. Dry beaded black when used in dispersions is approximately half way between wet beaded and dry powder in energy requirement. It can also be shipped in large containers and can be processed dust free using automated equipment. Internal end use handling requires special precautions to avoid premature breakage and dusting. Carbon black manufacturers can advise on the use of the proper systems.

5.4 Properties and testing methods of carbon black

5.4.1 *Particle size and surface area*

The most important fundamental measure of a carbon black is its average primary particle size, which influences to a large extent its technical applications. It is determined with the use of transmission electron micrographs (TEM) which are enlarged on transparent films through measurements and counting of approximately 4000 to 6000 single particles and the calculation of their average size [9]. This method is very time consuming and is therefore impractical for routine or processing use. As primary particle size becomes smaller, the specific surface increases, measured in m^2/g. Using absorption measurements with nitrogen or argon at very low temperatures according to Brunauer, Emmet and Teller [10], the nitrogen or BET surface can be computed. Theoretically, the particle size can be determined this way, but often leads to other results than the measurement of TEM pictures since the surface of many carbon blacks is porous. The surface of the pores is included using the BET method, but not when measuring particles with the help of TEM photos. In these cases the values of the specific surfaces lead to smaller particle sizes. An opposite effect is noticed with highly structured blacks, since a certain amount of the surface, because of contact areas between individual primary particles, is not available for gas absorption. In addition, many of the primary particles of various carbon blacks are not round but have irregular forms. Therefore the computing of particle sizes is harder.

The BET method for routine measurements is time consuming, therefore the iodine absorption method according to ISO 1304 (ASTM D-2414, DIN 53 582) is used. Since the acidic groups interfere in the absorption, this precludes using this method with oxidized furnace or gas blacks. Through absorption of cetyl-trimethylammonium bromide the CTAB-surface is measured according to ISO 6810 (ASTM D-3765). Only the outer geometric surface is considered since the molecule cannot penetrate the small surface pores because of its size. The CTAB method is influenced less by the surface chemistry than the iodine absorption.

The 'jetness' is primarily dependent on the average particle size. In the past this determination used to be made by comparing standard carbon blacks of the same type visually in a dispersion. Even the Nigrometer index is based on visual comparisons of a carbon black dispersion with a surface of variable lightness [11]. Basic works with modern color measuring devices [12–16] show very high accuracy even with very dark black samples, independent of the human eye. Through comparisons of visual findings of a large number of black samples which were determined by numerous test persons, a formula was developed which is linear with the visual arrangement. This was adopted by the German Institute for Standardization in DIN 55 979. The blackness value My was used to determine the jetness, which logarithmically, according to color norm value Y has the relationship:

$$My + 100 \log \frac{Y_n}{Y} = 100(2 - \log Y)$$

There by definition $Y_n = 100$ (reflection of the white standard surface).

During these investigations, gloss of the samples (probes) had to be discounted since the addition of gloss may create a much higher reading than the actual measurement ($Y = 0.04 - 1$ in black samples). The measurements were done on a very smooth paint surface or with paste dispersions through colorless glasses of equal thickness. For black measurements equipment is recommended that can reproduce color values of at least 0.001. Because of the strong influence that gloss has on the accuracy of color measurements, equipment with diffused sample illumination (for instance d/8° geometry) gloss has to be effectively excluded. Measuring equipment with 45°/0° or 0°/45° effectively will supress gloss to a large degree. Densitometers used in the printing industry operate normally with a geometry of 45°/0° and are therefore suitable for measurements of the density of black.

The visual density (D_{vis}) also has a logarithmic connection with the reflectance (R) according to the equation:

$$D_{vis} = (2 - \log R)$$

R and Y may be considered as similar when small differences in lighting and filtering are considered. Therefore, $My = 100\ D_{vis}$ for the comparison of black value and visual density. Note that the accuracy of most densitometers is only guaranteed up to $D_{vis} = 2.5$. Pigment blacks for printing inks do not normally exceed that value.

With the use of good color measuring equipment, color depth (density) and undertone of pigment blacks can be determined. Therefore the use of a^*, b^* values according to CIELAB, or their polar coordinates C^* and h_{ab}

$$C^* = \sqrt{a^{*2} + b^{*2}} \qquad h_{ab} = \text{arc tg}\,\frac{b^*}{a^*}$$

This measuring technology can also be applied for black prints. With regard to low gloss sample prints scattered light can influence the value of the measurements. Because of this only samples with comparative gloss should be measured.

Carbon blacks with small particle sizes, when measuring solid coverage will show higher density values than blacks with coarser particles. This is shown in Figure 5.8 where the relationship between My values and particle size is given (measured in linseed oil dispersion). Manufacturing techniques, surface chemistry and structure of pigment blacks also influence the density. The differences

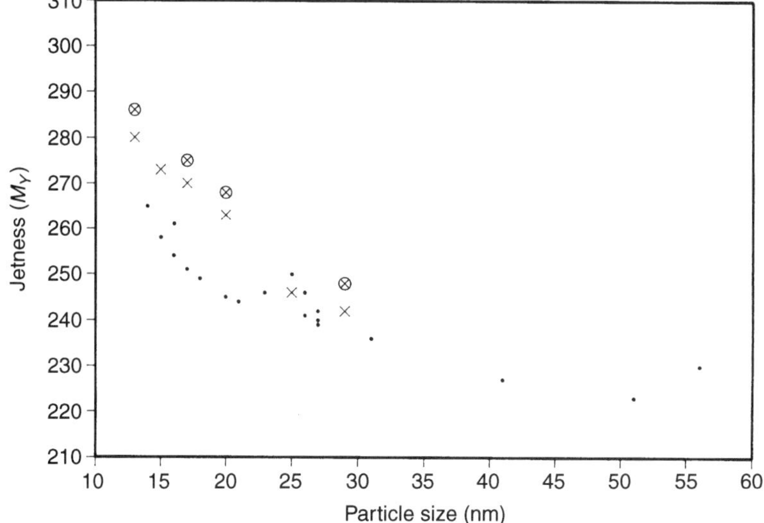

Figure 5.8 Relationship between jetness (My) and particle size. × gas blacks, ⊗ oxydized gas blacks, • furnace blacks with different structure.

in furnace blacks are caused mainly because of the wide range of structure sizes.

The largest part of incident light is absorbed by carbon black, only a small amount is scattered, short waves more than long waves. Therefore, if one looks through a sample (e.g. thin plastic film, normal print, or mixture of carbon black with white pigment), all pigment blacks look a little brownish due to the higher portion of non-scattered long wave (red, yellow) light passing the film [5, 17–20]. With small-particle size pigment blacks, a greater proportion of scattered light is in the short-wave region; therefore when looking through a print it shows a browner undertone. Since scattered light sources also show up on aggregates, agglomerates and flocculates, higher structure, poor dispersed pigments or flocculations show a bluer undertone.

In total covered prints (screen printing) or full coverage black paints where only scattered light sources are reflected into the aperture of the measuring apparatus, conditions are completely reversed [5, 15]; fine particle size pigment blacks here show a blue tone. Therefore it can be shown that the concentration of pigment black influences not only density but also undertone. The more concentrated, or thicker the coverage, the bluer the print.

The tinting strength of pigment blacks according to ISO 787/16, 787/24 (ASTM D-387) is increased to a point by smaller particle sizes, but with very small particle size blacks it decreases. For use in printing inks, tinting strength, with the exception of inks for color mixing systems, is not evident since it does not correlate with print strength.

The primary particle size has a very strong influence on oil absorption of pigment blacks which, for instance, can be determined according to ISO 787/5. Small particle size materials absorb more binders due to their higher specific surface. Rising oil absorption also increases the viscosity of the system. Because of this, printing inks which normally contain 10 to 20% pigment blacks, in some cases up to 25% contain only pigment blacks with particle sizes larger than 20 nm. Finer blacks at those concentrations would have too high a viscosity.

It stands to reason that fine particle size blacks have a higher energy demand when dispersing. A larger surface has to be wetted out. In addition a larger amount of particles has to be separated, which, because of physical attraction, are interconnected. The influence of primary particle sizes on important application properties of pigment blacks is shown schematically in Table 5.1.

5.4.2 *Structure*

The structure mentioned earlier, which can be varied only using the furnace black process, is an extremely important factor in the behavior of pigment blacks in printing inks. It is measured as DBP absorption in ml/100 g black pigment, according to ISO 4656 (ASTM D-244, DIN 53 601). In this test, an automatic burette adds the dibutyl phthalate (DBP) into a standardized chamber where it is mixed with a specific amount of carbon black. Initially, the void

Table 5.1 Application properties of pigment blacks as a function of primary particle size

Coarser	⟵	Particle size	⟶	Finer
Easier	⟵	Dispersion	⟶	Harder
Lower	⟵	Binder demand	⟶	Higher
Lower	⟵	Viscosity	⟶	Higher
Lower	⟵	Jetness	⟶	Higher
Browner	⟵	Mass tone	⟶	Bluer
Bluer	⟵	Print tone	⟶	Browner
Lower	⟵	Tinting strength	⟶	Higher

volume between the black particles are filled, the final point is reached when the torque of the apparatus increases. These measurements are done automatically and are therefore much more accurate than if done manually. DBP absorption depends not only on the specific structure, the determined value is also to a small degree influenced by the size of a wetted-out surface, and therefore also by the particle size. This means that of two pigment blacks with the same DBP absorption, the larger has the higher structure.

Viscosity and structure viscosity or yield value of printing inks are to a large extent dependent on the structure of the carbon black [5]. Highly structured blacks, in general, when used with a good binder, will increase viscosity, whereas in poor wetting binders the yield value increases; that means in printing inks, the ink is 'short'.

A hypothesis concerning rheological behavior of pigment blacks is possible only when the wetting properties of the binder are known and when pigment blacks with approximately similar primary particle sizes are considered. As mentioned above the particle size has a considerable influence on viscosity, but relatively little on the DBP absorption.

Jetness and tinting strength can be improved by lowering the structure of prime particle sizes, since the number of particles in a dispersed environment can be considerably higher. At the same time the diameter of the aggregates is reduced; therefore greater amounts of scattered light at shorter wavelengths are observed. In other words, at equal concentrations, lower structured blacks in printing inks are browner than high structured blacks of equal primary particle sizes.

Many, particularly high quality, printing inks require high gloss, which again is dependent on the structure of the pigment black. For these inks, carbon blacks with a low structure and a DBP absorption between 40 and 70 ml/100 g should be used exclusively. The aggregates of high structure blacks lead to roughing of the surface in the sub-micron area. The incident illumination is no longer correctly reflected but more or less separated, therefore the print is flat.

Low structured (LS) pigment blacks have a higher apparent density when compared to high structured (HS) blacks; this is true for powders as well as beads. This is immediately noticed when comparing the bags of LS blacks which are smaller, or contain larger amounts of material. The reason for this is the difference in the void volume which is caused by the locked structure elements. With powder blacks the apparent density is between 50 and 250 g/l, with beads 300–600 g/l.

The dispersibility gets also worse with lower structure since it is harder for the binder to penetrate the empty spaces, the wetting speed is reduced. Because of the low viscosity with LS blacks less shear energy is usually used, but high shear is necessary to give good dispersion. It is therefore recommended that higher LS pigment concentrations be considered, when compared to HS blacks. This is especially true when LS blacks with larger particle sizes are used. In this way the dispersability can at least be partially improved, and at the same time productivity will increase.

Because of the low binder demand it is possible to reduce the resin content, calculated on the amount of pigment black when using LS pigments. This is valid for the dispersion process and when the dispersion is converted into a finished product. In this way the pigment concentration is raised. The results in a finished print are higher density as well as stronger, bluer undertone.

The rub resistance in a finished print is also largely dependent on the structure of the pigment black used. LS pigments in this case surpass HS pigments. The same phenomenon applies here as was discussed concerning the development of gloss concerning the influence of structure. Larger aggregates sit on the surface of the very thin layer of a print and therefore when HS blacks are used rub problems may develop. Low structures give smoother surfaces in prints. During rub tests no black particles will come off, even the finger of the reader stays clean. Gloss readings usually give indications of rub resistance, or their lack of it.

A negative aspect of LS blacks should be included here. As mentioned earlier, flow properties of LS blacks in inks are superior to HS blacks. When printing on absorbent paper inks with LS blacks may penetrate too much, and in extreme cases could cause strike-through. A reduction in optical density is caused. With uncoated paper containing high amounts of ground wood penetration will occur where paper fibers are present, but not where small wood particles are on the surface. This causes irregular gloss. Increasing the amount of resin in the ink will lessen the effect but it makes more sense to use pigment blacks with higher structures. Gloss is usually not an important factor anyway when printing on uncoated substrates. Table 5.2 shows the influence of structure on the most important attributes of pigment blacks.

5.4.3 *Surface chemistry*

A simple method to understand the surface chemistry of pigment blacks is to determine the pH value. According to ISO 787/9 or ASTM D-1512 the pH value is measured from a mixture of pigment black and deionized water. Additional information can be taken from volatile matter which is determined according to DIN 53 552. The pigment black is heated to 950°C for 7 min, when all the

Table 5.2 Application properties of pigment blacks as a function of structure

Higher	⟵	Structure	⟶	Lower
Lower	⟵	Apparent density	⟶	Higher
Easier	⟵	Dispersion	⟶	Harder
Higher	⟵	Binder demand	⟶	Lower
Higher	⟵	Viscosity	⟶	Lower
Lower	⟵	Jetness	⟶	Higher
Bluer	⟵	Print tone	⟶	Browner
Lower	⟵	Tinting strength	⟶	Higher
Lower	⟵	Gloss	⟶	Higher
Less	⟵	Strike through	⟶	More
Worse	⟵	Rub resistance	⟶	Better

oxygen containing groups are eliminated from the surface. Normal furnace blacks contain 0.5 to 1.5% volatile matter, gas blacks and oxidized furnace blacks up to 6%, oxidized gas blacks up to 20%. With the rising content of volatiles, the pH decreases.

The surface chemistry also influences the technical applications of pigment blacks. Normally acidic pigment blacks are wetted better with polaric resins. They are therefore easier to disperse, have lower viscosities and often show higher densities. In water based inks in particlar, some disadvantages are observed. Resins used in these systems overwhelmingly have a low polarity, therefore wetting of the acidic surface groups is not as good. In addition, low molecular weight amines, which are used to make the acidic resins water soluble, are absorbed. This absorption takes time, therefore the pH value of the printing ink is lowered, resulting in instability in water, gelling or flocculation. Therefore the use of acidic pigment blacks in water is applicable only in a few speciality areas. Table 5.3 shows in schematic form the influence of the surface chemistry in determining the properties of pigment blacks.

5.4.4 *Further testing methods*

Contaminations in pigment blacks can have a detrimental effect. For that reason, for instance, sieve residue tests are performed on a regular basis. The determination is made according to ISO 787/18, ASTM D-1514 where pigment black with water is rinsed through a fine sieve (for instance 325 mesh). The residue in the

Table 5.3 Application properties of pigment blacks as a function of volatiles

Lower	⟵	Volatiles	⟶	Higher
Higher	⟵	pH value	⟶	Lower
Harder	⟵	Dispersion	⟶	Easier
Higher	⟵	Viscosity	⟶	Lower
Lower	⟵	Jetness	⟶	Higher

sieve is then dried and weighed. Normally it is less than 250 ppm. In a few instances the maximum allowable amount may be higher or lower.

Depending on the manufacturing process and the structure, pigment blacks contain some inorganic material which is determined after burning off the pigment black and checking the ash content according to ISO 1125 (ASTM 1506, DIN 53 552). Gas blacks, where the feedstock is evaporated before the actual manufacture, normally contain less than 0.02% ash. Furnace blacks, where the entire feedstock is sprayed into the flame, contain, with exceptions, more inorganic material. Through the addition of potassium compounds the ash content may rise to 1%. Normally, by this means, quality is not influenced.

In practice, pigment blacks do not contain water immediately after manufacture. As with all other fine particle size materials with large specific surface areas, pigment blacks are susceptible to adsorbing moisture from the atmosphere. How much depends on the pigment black surface area, the number of acid groups and storage conditions. Therefore moisture content cannot be specified by the manufacturer. However, it can be determined by using a drying oven, according to ISO 787/2 (ASTM D-1509). Large particle size furnace blacks may acquire up to 2% moisture content in a moist climate and extended storage. With fine particle size, highly acidic blacks the moisture content can rise to 10% in extreme cases. As the humidity drops so does moisture content in pigment blacks. For that reason the moisture content is normally lower than the previously mentioned figures. In water based printing inks moisture content in the black pigment does not matter.

The apparent density is measured according to ISO 787/11 (ASTM D-1513). Depending on structure and densitizing procedures variations can be very different. With powder blacks, density influences dispersibility, it is therefore kept low when possible by the manufacturer. There are limits, however, because transport and storage conditions induce movement and pressure which further densitize the product. This is especially true for pigment blacks with low structure.

The pellet hardness of pigment black beads can be measured according to ASTM D-3313. It is an important factor in the behavior of pigment blacks in in-plant transport operations such as pneumatic apparatus and also in the behavior in dispersions. Dry pearled blacks are softer than wet pearled and therefore easier to disperse, but are damaged much easier in pneumatic transport apparatus. Because of this it is recommended that equipment is used that transports pigment blacks at a fairly slow rate [21].

Table 5.4 summarizes the most important standard test methods.

5.5 Applications of carbon black

Section 5.4 explains the influence of the most important base properties of pigment blacks on their technical behavior in printing inks. This should ease the

Table 5.4 Methods to determine analytical pigment black data

Designation	ISO	ASTM	DIN
Nitrogen surface area	4652	D-3037/4820	66132
Iodine absorption	1304	D-1510	53582
CTAB number	6810	D-3765	
DBP absorption	4656	D-2414	53601
Oil absorption	787/5		*
Jetness			55979
Tinting strength	787/16 787/24	D-387	*
Volatile matter		D-1620	53552
Ash	1125	D-1506	53586
Moisture	787/2	D-1509	*
Sieve residue	787/18	D-1514	*
pH value	787/9	D-1512	*
Apparent density	787/11	D-1513	*
Pellet hardness		D-3313	

* DIN–ISO method.

choice for the printing ink manufacturer in the development process for a particular application. In inks that contain no water it is relatively simple, but water based inks cause larger problems. The reason is that water as a solvent has an extremely high polarity and therefore is able to be absorbed in large quantities during wetting and dispersing because of the constantly active centers of the pigment black surfaces which, for instance contain acid groups. Surface areas that are coated with water molecules can no longer absorb resin. But absorbed resin molecules are needed to keep the particles at the same configuration that was achieved through the dispersion process. If the pigment surface is not properly coated with resin molecules, the dispersion will not be stable. In other words, the separated particles are attracted to each other and flocculate. This effect can very easily be observed with a 200–300 power microscope. A good, stable black dispersion can be identified by the fact that practically no single particles are visible. The ink film on the glass has to be very thin to allow enough light going through and it shows more or less a brown undertone, depending on which pigment black is used. In a poor dispersion many black particles are visible markedly separated from the background. Flocculation is noticeable because the particles form a loose structure, like tiny balls. Between these flocculates colorless areas are visible which form irregular lines over the whole surface.

The microscope is an important tool for that reason; it shows the quality of an ink and is used in developing new formulas. However, care should be taken and it should be understood that the sample may be warmed under the microscope, therefore causing additional flocculation of the pigment black if the water had not been evaporated before viewing. For that reason the sample should be dried at room temperature before it is observed by using a microscope.

On the other hand, a wet sample that has been protected by a second glass from drying is easily followed by using microscopy to see if the pigment flocculates when the temperature is raised. This type of test leads to conclusions as to what can happen on a printing press after hot air is applied.

Flocculation is also noticeable by the fact that viscosity will increase during storage and it may be so severe that the product will gel. At the same time, due to continued flocculation, gloss is sharply reduced.

These problems are noticed fairly often in water based systems due to poor wetting properties of the resins used. Another reason for poor results has been discussed in Section 5.4; pigment blacks, especially those with acidic surface areas can absorb low molecular bases (used to solubilize resins), to such a degree that the pH may drop far enough to make the whole system insoluble (kickout). Under the microscope it is very similar, the viscosity increases, and gloss is lowered.

When developing new formulas, or the use of an alternate material (replacement), pH and viscosity should be checked during extended storage. In finished inks which regularly have low viscosities the aforementioned problems can show up within hours or days. Carbon black concentrates need to be monitored over several weeks to determine whether the dispersions are stable.

The chemical combinations of the resins play a determining role in the wettability of pigment blacks. Styrenated acrylic resins can be absorbed rather extensively by furnace blacks, particularly the untreated types with high pH values. As mentioned in Section 5.2 pigment black particles contain microcrystalline areas, similar to graphite, that are parallel to the particle surface. These crystalites contain a large amount of π electrons which are also responsible for the electrical conductivity of carbon blacks. This leads to an aromatic character, but at the same time, insolubility, due to their size.

It is known that π electrons of aromatic molecules may interact together. This is the reason why aromatic compounds are often layered in crystals. Aromatic groups of certain resins react similarly in that they attach themselves to the surface of pigment blacks, therefore solidifying the combination between the solids surface and resin molecule. For that reason, untreated furnace blacks are normally stable when dispersed, using styrenated acrylic resins. However, the styrene part of the resin plays a role.

It should be mentioned here that wetting agents with aromatic ring systems are suited much better in dispersions using alkaline black pigments rather than those with aliphatic molecules. With acidic pigment blacks the aromatic surface systems are more or less covered by the oxygen containing groups. The aromatic parts of the resin are therefore not able to contact the surface area of the black pigment and are not absorbed as much. This leads to poorer wetting and a less stable dispersion; but resins exist that contain basic reactive components such as amino or amide groups. These resins are especially suited for wetting and dispersing acidic pigment blacks. Of special interest here are resins produced with amino acids, protein resins. Acidic pigment blacks, particularly gas blacks, are especially suited for this purpose. Protein resins are used in printing inks where temperature and wet rub stability (non-rewetting) are a prime requirement. These properties are needed when printing papers that are laminated at high temperature and pressure.

Pure acrylic resins, containing neither aromatic nor basic groups usually are poor wetters and are therefore not generally suited for dispersing pigment blacks. Since a weak interaction can be expected between the carboxylated ions and oxygen-containing groups on the pigment black surface, acidic pigment blacks are best suited. Resins based on rosin usually work well with different pigment blacks which means the selection of pigment black depends on other criteria.

In comparison, it is much harder to stabilize or disperse pigment blacks in resin emulsions. The discrete particles of non-solubilized resin are not able to surround the pigment black which would stop flocculation. This is especially true in the high pigment concentrations needed in the manufacture of printing inks. Part of the resin should be soluble. Another possibility would be to use a combination of solubilized resin and emulsion. There is an advantage, however, when using emulsions, it is much easier to separate this type of ink in a de-inking process when the paper is recycled. In Europe, the norm is to combine the paper for recycling, surfactants and blaching agents like hydrogen peroxide and at a high pH the paper is repulped. In this process the 'classic' printing inks which are hydrophobic are reduced to a size of 10–100 μm. These relatively large particles attach themselves with the surfactants on air bubbles and float to the surface in the form of foam. This is called the 'flotation de-inking process'. The high pH is needed to better separate the ink from the paper fibers and to use the bleaching effect of the peroxide to its maximum. During this process with water based inks the acidic resins are resolubilized, the pigment black is again as in the original ink, made up of small hydrophilic particles. These can no longer be separated from the paper pulp through flotation but only through a wash process; this, however, reduces the fiber yield. In addition this water has to be purified, whereas the water from the flotation process is re-introduced into the pulp mixture.

Therefore the resin systems which do not resolubilize (resin–emulsions) perform much better in the flotation de-inking process. For that reason much research is being done to develop a product that also improves the pigment wetting while the above-mentioned benefits are retained, using emulsions.

The selection of a pigment black depends largely (as mentioned in Section 5.4) on the substrate to be printed and on what final specifications are required. For newspapers or low-cost wrapping paper very absorbent substrates are used where gloss is not a prerequisite. For that reason printing inks containing pigment blacks with high structure (DBP absorption $\geq$100 ml/100 g) are used. If higher gloss (for instance in corrugated or packaging paper) is required a clear coat can be applied, but requires additional work. By using pigment blacks of different particle sizes the undertone can be specified, the usable particle size is between 20 and 60 nm, normally blacks between 25 and 30 nm are used. These data are available from brochures published by the carbon black manufacturers.

High quality, glossy prints can only be achieved with the use of low structured pigment blacks in printing inks (DBP absorption 40–70 ml/100 g). Carbon black producers market blacks with particle sizes (20–60 nm), the same as high structure blacks. Since LS blacks achieve a slightly browner tone, blacks with relatively large primary particle sizes (30–60 nm) and very low structure (DBP absorption 40–55 ml/100 g) are recommended for water based gloss inks. Coarser-size low structured pigment blacks can also be used in finished inks in higher concentrations without loss of gloss. This way, prints with high density and a bluer undertone are produced. An additional toning with blue pigments is not necessary in many instances. However, a prerequisite is that the substrate has a low penetration level. Good results are achieved with coated or highly calendered papers or synthetic foils. This is true for publication gravure inks or high quality packaging inks, where additional clear coats are not required.

These recommendations, presented again in Table 5.5, should only be interpreted as an average since they do not take into account the interaction between pigment blacks and resin components. The exact chemical structure of a resin is usually not available to the user, therefore when developing new formulas trials need to be performed where aging stability should be carefully monitored.

Ink jet inks require an extremely high aging stability, where viscosities are very low. The pigments must be well dispersed, so that even during long storage periods no sediment occurs that can possibly plug the nozzles of the printing equipment. Untreated gas blacks are especially recommended for these and similar applications.

Resin systems containing amino and amido groups are suited for use with gas blacks; in this case, also, the aftertreated very acidic types. Protein resins are examples which are used in printing inks for high pressure laminates.

The selection of black pigment for use in UV inks is especially difficult. Pigment blacks absorb light not only in the visible spectrum but also in the UV and IR ranges. Therefore, it seems impossible to use pigment blacks for this application, but as explained earlier fine particle size pigment blacks scatter light

Table 5.5 Recommendations for the use of pigment blacks in water based inks

Ink system	Pigment black	Particle size (nm)	DBP absorption (ml/100 g)	Volatiles (%)
Low gloss	Furnace	20–60 25–30*	100–130	0.5–1.0
High gloss	Furnace	20–60 30–60*	40–70 40–55*	0.5–1.5
Polar binders	Gas	25–30	n.a.	5–15
Ink jet	Gas	10–30	n.a.	~5
UV curing	Furnace oxidized	30–60	40–50	2–5

* Preferably.

in the short-wave spectrum relatively more than do larger particle size blacks. For that reason, there is a difference in color tone. In this way, short-wave light, when larger particle size black is used penetrates deeper into the ink film. This process continues also in the UV spectrum. UV printing inks with larger particle size blacks harden faster, but on the other hand, increased particle size reduces color depth (density). The optimal particle size is 30–60 nm. It has also been noted that pigments with very low structure (DBP absorption 40–50 ml/100 g) influence viscosity and gloss least of all. Another factor to be considered is that alkaline, that means non-oxidized, furnace blacks – and only furnace blacks can be manufactured with the necessary particle size and structure – contain single electrons which may start a polymerization during storage of the ink. Inks will gel in the container in a relatively short time. For that reason oxidized, acid pigment blacks are more suited for this specialized application.

The dispersing process of pigment blacks is performed by using the same equipment as is used for other pigments. It has to be taken into account, however, that the primary particle sizes of pigment blacks are smaller than most inorganic and organic pigments. For that reason the energy requirement is normally higher, but pigment blacks are so thermally stable that even at high temperatures during the dispersing process color changes do not occur, which as is known, can happen with some organic pigments. It is certainly possible, however, that with the use of certain binders at higher temperatures polymerization or instability may develop, for instance with UV drying resins or monomers. In this case the temperature during the milling process must be controlled and the processing time extended.

High-speed dispersers are well suited for predispersing pigment blacks, but by themselves are not sufficient. Shot mills (media mills) are usually needed to continue the dispersion process. The highest through-put is achieved using media with high specific density such as steel or zirconium oxide in closed vertical media mills. It is important in this process to use as high a concentration of pigment as possible. Only then will optimum results in grind, through-put and minimal wear of the media be realized. This is especially true when using carbon black beads, since the beads first need to be destroyed before they can contribute to the viscosity of the system.

In closed, vertical media mills, low structured pigment blacks, usually between 40 and 50% pigment, can be dispersed. For high structured blacks the percentage is 25–35%. If open or horizontal media mills are used then the pigment concentration must be slightly lower. Ball mills, which are also well suited for dispersing carbon black beads normally run at lower viscosities and therefore lower pigment concentrations.

Rotor–stator dispersing apparatus can tolerate high pigment concentrations but the rise in temperature is very fast. In the manufacture of solid resin black concentrates, kneaders are used where the pigment black is slowly worked into the molten resin and dispersed. An especially high quality material can be made by after-dispersing on a heated two-roller mill. This method is used in manufacturing pigment black chips, which can be used by resolubilizing the resin, with the addition of suitable vehicles to make a finished printing ink. Three-roller mills still play a major role for the manufacture of high-quality offset inks but are normally not suited for the dispersing of carbon black in water based inks.

References

1. Kühner, G. (1992) *What is Carbon Black?* Degussa publication.
2. Degussa (1990) *Technical Bulletin Pigments No. 25*, 2nd edn.
3. Ferch, H., Wagner, H. and Seibold, K. (1982) *XVIth FATIPEC Congress Book*, Vol. 3, p. 235.
4. Koberstein, E., Lakatos, E. and Voll, M. (1971) *Ber. d. Bunsen-Gesellschaft f. phys. Chem.*. Vol. 75, p. 1106.
5. Schumacher, W. (1974) *Farbe and Lack*, **80**, 290.
6. *Encyclopedia of Chemical Technology*, Vol. 3, p. 39 (1949).
7. *Ullmann's Encyclopedia of Industrial Chemistry*, Vol. A5, p. 144 (1986).
8. Ferch, H. and Seibold, K. (1984) *Farbe and Lack*, **90**, 88.
9. Endter, F. and Gebauer, H. (1959) *Optik*, **13**, 97.
10. Brunauer, S., Emmet, P.H. and Teller, E. (1938) *J. Amer. Chem. Soc.*, **60**, 309.
11. US patent 1.780.231 (1928).
12. Schumacher, W. (1976) *XIIIth FATIPEC Congress Book*, p. 582.
13. Koth, D., Bode, R. and Schumacher, W. (1977) *Farbe and Lack*, **83**, 490.
14. Bode, R., Ferch, H., Koth, D. and Schumacher, W. (1979) *Farbe and Lack*, **85**, 7.
15. Ferch, H., Eisenmenger, E. and Schäfer, H. (1981) *Farbe and Lack*, **87**, 88.
16. Lippok-Lohmer, K. (1986) *Farbe and Lack*, **92**, 1024.
17. Schumacher, W. (1984) *Farbe and Lack*, **90**, 454.
18. Fiorenza, A. (1956) *Rubber Age*, **80**, 69.
19. Fiorenza, A. (1959) *Rubber Age*, **84**, 945.
20. Ravey, J.C. and Prelimat, J. (1970) *J. Chim. Phys. Physicochim.*, **67**, 1913.
21. Degussa (1993) *Technical Bulletin Pigments No. 45.*

6 Solvents

T. HOLSHUE and H. GAINES

6.1 Introduction

The surface-coating industry is an ancient one. The origin of paints dates back to prehistoric times when the inhabitants of the earth recorded some of their activities in colors on the walls of their caves. These crude coatings consisted probably of colored earths or clays suspended in water. The Egyptians developed the art of painting and by 1500 BC had a wide number and variety of colors. Around 1000 BC they discovered the forerunner of our present day varnishes, using naturally occurring resins or beeswax for their film-forming ingredient. The importance of the role of solvents is brought out most clearly by the fact that many substances exhibit their greatest usefulness when in solution. The modern solvent industry is comparatively new. It was only yesterday, so to speak, that the only industrial solvents were methyl and ethyl alcohols, acetone, glycerol and aliphatic and aromatic hydrocarbons.

In World War I the urgent need for acetone was supplied by the development of a new fermentation process. Butyl alcohol was produced as a by-product, but very little use could be found for it at that time. As the supply increased, activity among research workers was greatly accelerated, with the ultimate employment of butyl esters in flavors, lacquers and plasticizers. The petroleum industry was impelled by the same fever of activity, with the subsequent conversion of ethylene, propylene, butylene, and ethyl, isopropyl, butyl and amyl alcohols.

The classification of solvents can be made in many different ways; some people advocate a grouping according to evaporation rates while others think dilution ratios are more useful. None gives a complete picture comprising purity, density, stability, evaporation rates, dilution ratios, etc. . . .

In the mid to late 1980s, the United States Government applied pressure to remove solvents from the printing industry. The Clean Air Act of 1990 indicated that the government was willing to prescribe some strong medicine to printers who were using alcohol. The Texas Air Control Board were doing their best to shut down printers, due to pressure from the Texas Water Commission who did not want alcohol in the sewer.

The main concern is ozone, that enemy of clean air found in the lower atmosphere. It had reached proportions that are of concern to many nations, and most scientific minds blame VOCs—volatile organic compounds. To the printing industry, this means solvents, isopropanol, and the oils used in the printing inks.

At this time safety became a large concern with the National Institute for Occupational Safety and Health (NIOSH). They found indications that at high levels of exposure, over prolonged periods, isopropyl alcohol may lead to respiratory problems, so new thresholds were established. Along with the possible exposure problems, a greater threat was addressed—the fire hazard. Operators become too comfortable with the use of alcohol, and proper containers and transfer procedures were not used, too often leading to a flash fire.

The use of isopropyl alcohol, as an example, was to demonstrate the process of all solvents used in the printing process, from ink manufacturing, to washes, to wash-ups at all levels of printing. This is where the trend to remove solvent from the process starts, but this is no easy task. So how does a firm get solvents out of the process? Because solvents have played such an important role in the chemical interaction of printing, their removal or reduction can have repercussions on the ink train as well as plate, ink, and fountain solution chemistry. It now requires manufacturers to get involved and become a partner in the whole process.

A major impact in the change of products, comes from the new Clean Air Act which was amended in 1990. These amendments represent the fifth major effort by Congress to address clean air legislation. The first Clean Air Act, passed in 1967, provided authority to establish air quality standards. The Clean Air Act of 1970 was far stronger and more comprehensive, laying the foundation for regulator efforts during the past 20 years. In 1974, the Energy Supply and Environmental Coordination Act modified certain air quality requirements to balance energy needs following the Arab oil embargo of October 1973. In 1977, Congress enacted additional revisions. Following completion of the report of the National Commission on Air Quality in 1981, Congress engaged in an intensive debate over further amendments to the statute; these debates blocked further amendments until 1990.

Until this time, the entire national strategy to improve air quality had been centered around a basic set of National Ambient Air Quality Standards. The basic idea behind ambient air quality standards is that they are to be based on scientific determinations of the threshold levels of air pollution below which no adverse effect will be experienced by humans or the environment. The Environment Protection Agency (EPA) promulgated standards establishing numerical criteria to be applied uniformly throughout the entire country (Table 6.1).

The biggest push to move from old standard products such as toluene, methyl ethyl ketone, and xylenes, came with the EPA's list of good and bad HAPs (hazardous air pollutants), and their famous 33/50 Program. The 33/50 Program is EPA's voluntary pollution prevention initiative to reduce national pollution releases and off-site transfers by 50% in 1995 (with an interim goal of 33% in 1992) there were 17 target chemicals which were chosen on the basis of three criteria: (1) they pose environmental and health concerns; (2) they are high-volume industrial chemicals; and (3) they can be reduced through pollution prevention. The 17 target chemicals are: benzene, cadmium and its compounds, carbon tetrachloride, chloroform, chromium and its compounds, cyanides, lead and its compounds, mercury and its compounds, methyl ethyl ketone, methyl isobutyl ketone, methylene chloride, nickel and its compounds, tetrachloroethylene, toluene, trichloroethane, tichloroethylene, and xylenes.

In an effort to comply with the new air regulations companies use the other products that appear on the 'good' HAP list. This stop gap approach has seen the decline in use of hydrocarbons and aliphatics and the increase in use of esters to fill the printing needs. Even though some ketones appear on the 'bad list' use is still increasing in these areas due to the strong solvency and varied evaporation rates (Figure 6.1). Two ketones used in ink manufacture are methyl ethyl ketone and isophorone. Methyl ethyl ketone is used because of its strong solvency and fast evaporation rate, which makes it an important component of gravure printing inks, particularly in Type C and Type V inks. It also finds some applications in silk screen printing. Another strong factor is its ability to form an azeotrope with water (Table 6.2). Isophorone is used in the preparation of high-quality vinyl letterpress inks and ink pastes. Its high solvency and slow evaporation rate improve the flow properties of these speciality inks.

Table 6.1 National ambient air quality standards (as of November 15, 1990)

Pollutant	Primary standards (protective of health)*
Ozone	0.120 ppm (235 $\mu g/m^3$) (1-h average)
Carbon monoxide	9 ppm (10 mg^3) (8-h average) 35 ppm (40 mg^3) (1-h average)
Particulate matter (PM-10)	150 $\mu g/m^3$ (24-h average) 50 $\mu g/m^3$ (annual arithmetic mean)
Sulfur dioxide	0.140 ppm (365 $\mu g/m^3$) (24-h average) 0.03 ppm (80 $\mu g/m^3$) (annual arithmetic mean)
Nitrogen dioxide	0.053 ppm (100 $\mu g/m^3$) (annual arithmetic mean)
Lead	1.5 $\mu g/m^3$ (arithmetic mean averaged quarterly)

* See 40 C.F.R. Part 50. The Clean Air Act also requires that EPA establish secondary standards, which protect against adverse effects on the environment. Secondary standards have been established for most of the listed pollutants, and in most cases the levels are lower than the primary standards.

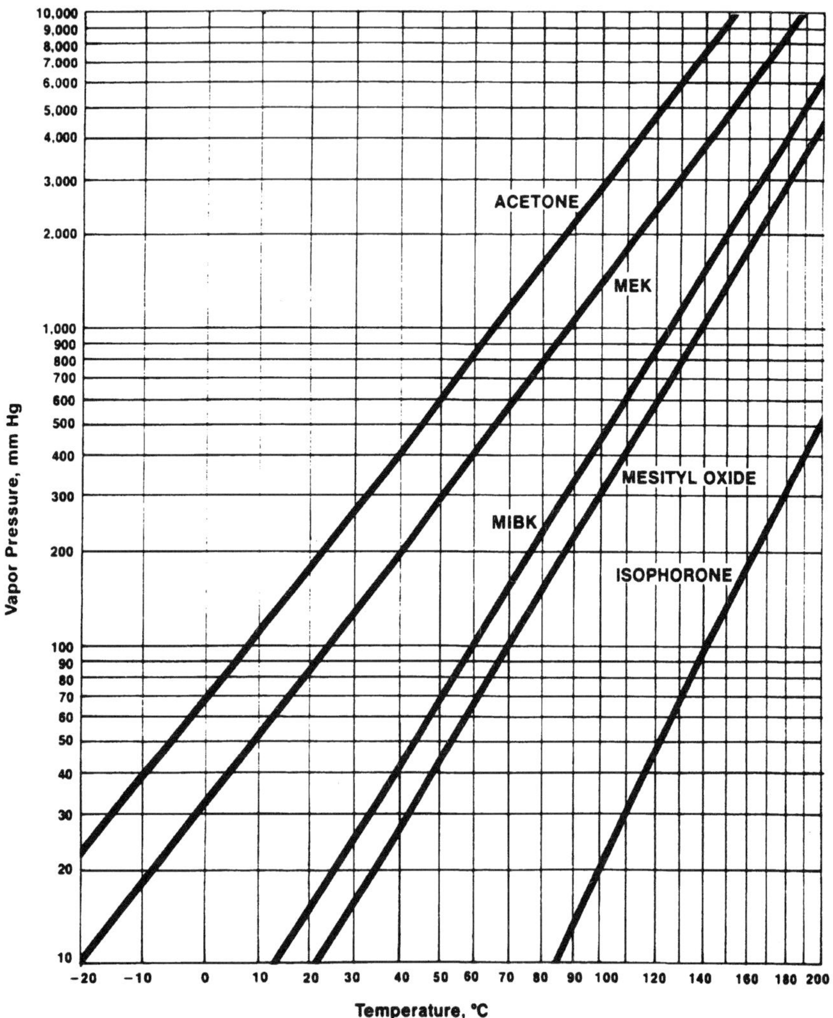

Figure 6.1 Relationship between vapor pressures of ketones and temperature.

6.2 Glycol ether solvents for water based ink applications

6.2.1 *Properties of glycol ethers*

Glycol ethers, one of the most versatile classes of organic solvents, are generally produced by reacting alcohols with ethylene oxide. The resulting glycol ether contains an ether and an alcohol functional group in the same molecule, thus providing unique solvency characteristics for a wide diversity of applications. Union Carbide's glycol ethers are produced by reacting one mole of an alcohol

Table 6.2 Physical properties of various solvents used in ink manufacture

Properties of the pure compounds* (all values are at 20°C, except where noted)	Solvent				
	Acetone	MEK	MIBK	Mesityl oxide	Isophorone
Autoignition temperature, vapor, °F	1000	960	854	—	860
Azeotrope with water, BP, °C	—	73.4	87.9	91.8	99.5
wt% ketone in vapor	—	88.7	76.0	65.2	16.1
Boiling point °F	133.0	175.3	241.2	265.6	419.4
°C	56.1	79.6	116.2	129.8	215.2
Coefficient of cubic expansion, per °C	0.00143	0.00119	0.00116	0.00109	0.00085
Critical pressure, atm.	47	43	32	—	35
Critical temperature, °C	235	260	298	—	441
Density, g/ml	0.7907	0.8037	0.8006	0.853	0.920
Dielectric constant	21.45	18.51	13.11	15.4	—
Dilution ratios, toluene	4.5	4.3	3.6	3.7	6.2
aliphatic naphtha	0.7	0.9	1.1	0.9	1.0
Dipole moment, Debye units	2.72 (25°C)	2.74	2.7	2.7	4.0
Electrical conductivity, $\times 10^{-8}$ mho	5.5	5.0	—	—	—
Explosive limits, in air, vol. %					
upper	12.8	10.0	7.5	8.8	3.8
lower	2.6	1.8	1.4	1.3	0.8
Flash point, °F					
Tag open cup	15	34	74	98	205
Tag closed cup	1	28	60	83	200
Freezing point °C	−94.7	−86.3	−80.2	−53	−8.1
°F	−138.5	−123.3	−112.4	−63	17.4
Gallons per 100 pounds	15.15	14.90	14.97	14.04	13.02
Heat of combustion, BTU/lb	13 260	14 540	15 980	—	15 630
Heat of fusion, cal/g	23.4	24.7	—	—	—
Heat of vaporization, BTU/lb at BP	219.76	190.69	155.97	157	139
Molecular weight	58.087	72.104	100.156	98.14	138.2
Pounds per gallon	6.60	6.71	6.68	7.12	7.68
Refractive index n_D	1.3592	1.3791	1.3956	1.4456	1.4781
Relative evaporation rate (*n*-butyl acetate = 100)	1160	572	165	90	3
Solubility, wt%					
—of ketone in water	Infinite	26.3	1.7	3.1	1.2
—of water in ketone	Infinite	11.8	1.9	3.4	4.3
Solubility parameter	10.0	9.3	8.4	9.2	9.7
Specific gravity	0.7911	0.8051	0.8042	0.854	0.922
Specific heat, BTU/lb, °F	0.528	0.549	0.459	0.520	0.426
Surface tension, dynes/cm	23.32	24.6	23.64	22.9	32.3
Thermal conductivity, cal/(cm^2)(s)(°C) $\times 10^6$	428	358	—	—	—
Vapor pressure, mmHg	185.95	80.21	14.96	7.9	0.2
Viscosity, centipoise	0.32	0.43	0.58	0.6	2.6

* These are properties of pure chemical compounds, and should not be taken as specifications for commercial products.

Table 6.3 Union Carbide's glycol ethers

Ethylene oxide, moles	Alcohol				
	Methanol	Ethanol	Propanol	Butanol	Hexanol
1	Methyl cellosolve solvent (Ethylene glycol monomethyl ether) CAS* 109-86-4	cellosolve solvent (Ethylene glycol monoethyl ether) CAS 110-80-5	Propyl cellosolve solvent (Ethylene glycol monopropyl ether) CAS 2807-30-9	Butyl cellosolve solvent (Ethylene glycol monobutyl ether) CAS 111-76-2	Hexyl cellosolve solvent (Ethylene glycol monohexyl ether) CAS 112-25-4
2	Methyl carbitol solvent (Diethylene glycol monomethyl ether) CAS 111-77-3	Carbitol solvent (Diethyl glycol monoethyl ether) CAS 111-90-0		Butyl carbitol solvent (Diethylene glycol monobutyl ether) CAS 112-34-5	Hexyl carbitol solvent (Diethylene glycol monohexyl ether) CAS 112-59-4
3	Methoxytriglycol (Triethylene glycol monomethyl ether) CAS 112-35-6	Ethoxytriglycol (Triethylene glycol monoethyl ether) CAS 112-50-5		Butoxytriglycol (Triethylene glycol monobutyl ether) CAS 143-22-6	

* Chemical Abstract Service (CAS) number.
Glycol ethers having longer hydrocarbon-like alkoxide groups display solubility more characteristic of hydrocarbons. Thus glycol ethers produced from higher molecular weight alcohols, such as hexyl cellosolve® solvent, have limited water solubility. The ether groups introduce additional sites for hydrogen bonding with improved hydrophilic solubility performance.

with one, two, or three moles of ethylene oxide (Table 6.3). The synthesis routes for some typical glycol ethers are illustrated in Figure 6.2.

Glycol ethers with longer hydrocarbon-like alkoxide groups display solubility more characteristic of hydrocarbons. Thus glycol ethers produced from higher molecular weight alcohols, such as hexyl cellosolve solvent, have limited water solubility. The ether groups introduce additional sites for hydrogen bonding with improved hydrophilic solubility performance.

The glycol ethers are characterized by excellent solvency, chemical stability, and compatibility with water. Their properties are summarized in Tables 6.4 and

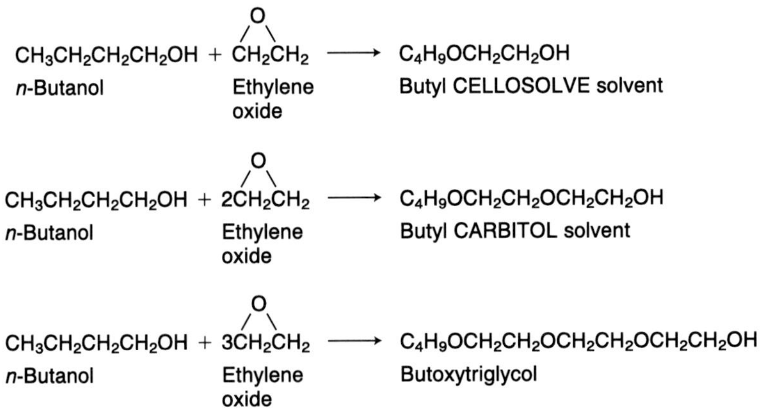

$$CH_3CH_2CH_2CH_2OH + CH_2CH_2(O) \longrightarrow C_4H_9OCH_2CH_2OH$$

n-Butanol — Ethylene oxide — Butyl CELLOSOLVE solvent

$$CH_3CH_2CH_2CH_2OH + 2CH_2CH_2(O) \longrightarrow C_4H_9OCH_2CH_2OCH_2CH_2OH$$

n-Butanol — Ethylene oxide — Butyl CARBITOL solvent

$$CH_3CH_2CH_2CH_2OH + 3CH_2CH_2(O) \longrightarrow C_4H_9OCH_2CH_2OCH_2CH_2OCH_2CH_2OH$$

n-Butanol — Ethylene oxide — Butoxytriglycol

Figure 6.2 Synthesis routes for some typical glycol ethers.

6.5. The dual functionality present in cellosolve and carbitol solvents, and alkoxytriglycol glycol ethers accounts for their unique solvency properties. These glycol ethers are:

1. Miscible with a wide range of polar and non-polar organic solvents
2. Mild-odored solvents for many resins, oils, waxes, fats, and dyestuffs
3. Coupling agents for many water/organic systems
4. Miscible with water in most cases

This strong solvency has led to the selection of cellosolve and carbitol solvents, and alkoxytriglycol glycol ethers for a broad array of end uses:

1. Dye solvents in the textile, leather, and printing industries
2. Solvents for grease and grime in industrial cleaning and speciality formulations
3. Solvents for insecticides and herbicides for agricultural applications
4. Couplers and mutual solvents for soluble oils, hard-surface cleaners, and other soap–hydrocarbon systems
5. Solvents and co-solvents for conventional solvent-based lacquer enamel, and wood stain industrial coating systems
6. Co-solvents for waterborne industrial coating systems
7. Jet fuel additives
8. Printed circuit board laminating formulations
9. Freeze–thaw agents in latex emulsions
10. Chemical reaction solvents

The uses are summarized in Table 6.6.

Cellosolve and carbitol solvents are also useful chemical intermediates. These glycol ethers will undergo many of the same reactions as alcohols because they contain the hydroxyl (—OH) functional group. Some typical examples are:

1. Reaction with carboxylic acids, carboxylic acid chlorides, anhydrides, and inorganic acids to produce esters
2. Reaction with organic halides to produce ethers, such as glymes
3. Reaction with alkenes and alkynes to produce ethers
4. Reaction with halogenating agents to produce alkoxy alkyl halides
5. Reaction with epoxides to produce polyether alcohols
6. Reaction with aldehydes and ketones to produce hemiacetals and acetals

Some physical properties of glycol ethers are shown graphically in Figures 6.3 (A to C) and 6.4.

6.2.2 *Butyl cellosolve solvent (ethylene glycol monobutyl ether)*

Butyl cellosolve solvent is an excellent solvent widely used in coatings and cleaner applications, including consumer products. It offers superior performance in many hard-surface cleaners and related cleaning formations. Mineral

Table 6.4 Typical physical properties for 12 glycol ethers produced by the Union Carbide. Products are arranged by parent-alcohol group in the upper part of the table which impacts on solvency characteristics. In the lower part of the table, products are arranged by boiling point

Solvent	Formula molecular weight	Boiling point, °C	Freezing point, °C	Flash point, °F[a]	Vapor pressure, mmHg at 20°C	Specific gravity, 20/20°C	Pounds per gallon	Coefficient of expansion at 20°C	Solubility at 20°C, % by wt		Relative evaporation rate (nBuAc = 100)	Surface tension at 25°C, dynes/cm	
									In water	Water in		Neat product	25% Aq. solution[b]
Product family order													
Methyl cellosolve solvent	76.1	124.5	−85	103	6.2	0.966	8.04	0.00094	100	100	62	32.1	54.3
Methyl carbitol solvent	120.2	194.0	−85	188[d]	0.1	1.023	8.51	0.00086	100	100	1.5	35.9	54.3
Methoxytriglycol	164.2	249.0	−44	238[d]	<0.01	1.050	8.74	0.00084	100	100	0.04	34.7	48.4
Cellosolve solvent	90.1	134.9	−90	108	4.1	0.931	7.74	0.00097	100	100	41	29.4	47.1
Carbitol solvent	134.2	201.6	−78[c]	182[d]	0.08	0.991	8.25	0.00090	100	100	1.3	35.2	49.6
Ethoxytriglycol	178.2	255.9	−19	255	<0.01	1.025	8.53	0.00086	100	100	0.04	32.2	45.7
Propyl cellosolve solvent	104.2	150.1	−90	135[d]	1.6	0.913	7.60	0.00095	100	100	21	26.3	32.3
Butyl cellosolve solvent	118.2	171.2	−70	160[d]	0.6	0.902	7.50	0.00092	100	100	7.8	28.6	28.9
Butyl carbitol solvent	162.2	230.6	−68	214	0.01	0.954	7.94	0.00088	100	100	0.24	31.0	33.2
Butoxytriglycol	206.3	279.8[e]	−48	276[d]	<0.01	0.989	8.19	0.00085	100	100	<0.1	30.0	32.2
Hexyl cellosolve solvent	146.2	208.1	−50	179	0.05	0.889	7.40	0.00086	1.00	18.80	0.82	30.3	28.5[g]
Hexyl carbitol solvent	190.3	259.1	−40	271[d]	<0.01	0.935	7.78	0.00084	3	56.30	0.03	29.2[f]	—
Boiling point order													
Methyl cellosolve solvent	76.1	124.5	−85	103	6.2	0.966	8.04	0.00094	100	100	62	32.1	54.3
Cellosolve solvent	90.1	134.9	−90	108	4.1	0.931	7.74	0.00097	100	10	41	29.4	47.1
Propyl cellosolve solvent	104.2	150.1	−90	135[d]	1.6	0.913	7.60	0.00095	100	100	21	26.3	32.3
Butyl cellosolve solvent	118.2	171.2	−70	160[d]	0.6	0.902	7.50	0.00092	100	100	7.8	28.6	28.9
Methyl carbitol solvent	120.2	194.0	−85	188[d]	0.1	1.023	8.51	0.00086	100	100	1.5	35.9	54.3
Carbitol solvent	134.2	201.6	−78[c]	182[d]	0.08	0.991	8.25	0.00090	100	100	1.3	35.2	49.6
Hexyl cellosolve solvent	146.2	208.1	−50	179	0.05	0.889	7.40	0.00086	1.00	18.80	0.82	30.3	28.5[g]
Butyl carbitol solvent	162.2	230.6	−68	214	0.01	0.954	7.94	0.00088	100	100	0.24	31.0	33.2
Methoxytriglycol	164.2	249.0	−44	238[d]	<0.01	1.050	8.74	0.00084	100	100	0.04	34.7	48.4
Ethoxytriglycol	178.2	255.9	−19	255	<0.01	1.025	8.53	0.00086	100	100	0.04	32.2	45.7
Hexyl carbitol solvent	190.3	259.1	−40	271[d]	<0.01	0.935	7.78	0.00084	3	56.30	0.03	29.2[f]	—
Butoxytriglycol	206.3	279.8[e]	−48	276[d]	<0.01	0.989	8.19	0.00085	100	100	<0.1	30.0	32.2

(a) Tag closed cup unless otherwise noted
(b) All solutions are percent by volume
(c) Sets to glass below this temperature
(d) Pensky–Martens closed cup
(e) Decomposes at 760 mmHg, boiling point extrapolated
(f) at 2°C
(g) 1% solution

Table 6.5 Constant boiling azeotropic mixtures of glycol ethers and other solvents

	Components		Azeotrope					
				Composition, % by wt, at 20°C				
Solvent	Specific gravity at 20/20°C	Boiling point at 760 mmHg, °C	Boiling point at 760 mmHg, °C	in azeotrope	in upper layer	in lower layer	Relative volume of layers at 20°C	Specific gravity at 20/20°C of azeotrope layer
Methyl cellosolve solvent	0.966	124.5		25	—	—	—	
Toluene	0.868	10.6	105.9	75	—	—	—	0.887
Methyl cellosolve solvent	0.966	124.5		15	—	—	—	
Water	1.000	100.0	99.9	85	—	—	—	—
Methyl carbitol solvent	1.023	194.0		70	—	—	—	
Ethylene glycol	1.115	197.6	192	30	—	—	—	1.051
Cellosolve solvent	0.931	135.6		35.7	—	—	—	
Butyl acetate	0.88	126.0	125.8	64.3	—	—	—	0.896
Cellosolve solvent	0.931	135.6		10.0	—	—	—	
Toluene	0.868	110.6	110.0	90.0	—	—	—	0.874
Cellosolve solvent	0.931	135.6		28.8	—	—	—	
Water	1.000	100.0	99.4	71.2	—	—	—	1.003
Carbitol solvent	0.991	202.7		54.5	—	—	—	
Ethylene glycol	1.115	197.6	192	45.5	—	—	—	—
Propyl cellosolve solvent	0.913	150.1		30	—	—	—	
Water	1.000	100.0	98.8	70	—	—	—	—
Butyl cellosolve solvent	0.902	171.2		20.8	57	10	—	
Water	1.000	100.0	98.8[a]	79.2	43	90	—	0.989[b]
Butyl carbitol solvent	0.954	230.6		27.5	—	—	—	
Ethylene glycol	1.115	197.6	196.2	72.5	—	—	—	1.074
Hexyl cellosolve solvent	0.889	208.1		9	81.2	1.0	U 11	U 0.915
Water	1.000	100.0	99.7	91	18.8	99.0	L 89	L 1.000
Hexyl carbitol solvent	0.935	259.1		2	43.7	1.7	U 0.5	U 0.982
Water	1.000	100.0	100.0	98	56.3	98.3	L 99.5	1.000

(a) Heterogeneous at this boiling point (b) Homogeneous at 20°C

Table 6.6 End-use areas for glycol ethers

	Glycol ether											
Industry	Methyl cellosolve solvent	Methyl carbitol solvent	Methoxytriglycol	Cellosolve solvent	Carbitol solvent	Ethoxytriglycol	Propyl cellosolve solvent	Butyl cellosolve solvent	Butyl carbitol solvent	Butoxytriglycol	Hexyl cellosolve solvent	Hexyl carbitol solvent
Agriculture	●	●		●	●			●	●			
Chemical			●	●		●			●	●		
Coatings		●	●		●	●	●	●	●	●	●	●
Electronics	●								●			
Food		●			●				●			
Household/institutional					●			●	●		●	●
Printing									●			●
Textile		●	●		●	●		●	●	●		
Transportation	●	●	●		●	●		●	●	●		

oils, water, and soaps are miscible with butyl cellosolve solvent. It also functions as a coupling agent to stabilize immiscible ingredients in industrial metal cleaners and vapor degreasers, in the clarification of oil–water dispersions, and in the formulation of concentrated liquid cleaners. This solvent is an effective coupling agent for the addition of water to dry-cleaning soap. Butyl cellosolve solvent also assists in removing water-soluble stains, aids in brightening the colors of garments, and improves penetrating and wetting action of phosphoric acid-type rust removers.

Butyl cellosolve solvent exhibits strong solvency for alkyd, phenolic, nitrocellulose, and maleic-modified resins. It is one of the best retarders for nitrocellulose lacquers, strengthens blush resistance, heightens gloss, improves flow-out and prevents orange peel formation. Hot-spray lacquers may include up to 10% butyl cellosolve solvent by weight. Related applications are in special coatings for polystyrene, in epoxy coatings, in high–low thinners for reduction of the viscosity of alkyd baking enamels, and as a solvent for insecticides and herbicides. In addition, butyl cellosolve solvent is the most widely used cosolvent for industrial waterborne coatings because of its excellent balance of properties.

6.2.3 *Inverse solubility*

Some glycol ethers exhibit an inverse solubility in water at elevated temperatures. For example, butyl cellosolve® solvent becomes less soluble and separates from solution as the solution temperature is increased (Figure 6.5). In some cases, this property can be utilized to remove butyl cellosolve solvent from

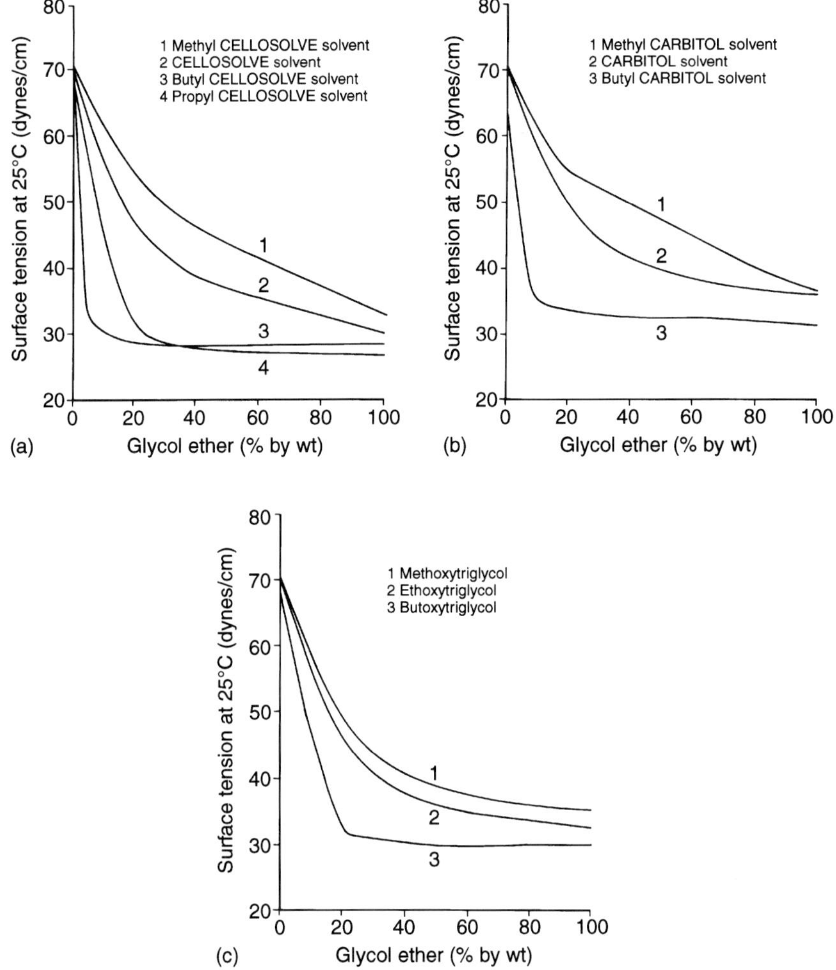

Figure 6.3 Surface tension of aqueous solutions of glycol ethers. (a) Cellosolve solvents. (b) Carbitol solvents. (c) Triglycols.

an aqueous waste stream. At temperatures above 50°C, it may be possible to decant an aqueous layer from an upper organic layer, thus reducing the organic content of the aqueous waste stream.

6.2.4 *Hexyl cellosolve solvent (ethylene glycol monohexyl ether)*

Hexyl cellosolve solvent is a high-boiling solvent that is useful as a coalescing aid for dispersion paints. The greater hydrocarbon-like chargers of hexyl cellosolve solvent makes it a superior solvent for greases and waxes in cleaning applications.

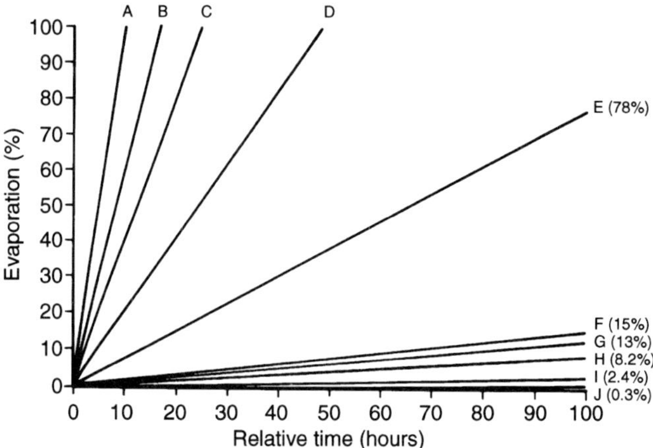

Figure 6.4 Relative evaporation of glycol ethers. (A) Butyl acetate, (B) methyl cellosolve solvent, (C) cellosolve solvent, (D) propyl cellosolve solvent, (E) butyl cellosolve solvent, (F) methyl carbitol solvent, (G) carbitol solvent, (H) hexyl cellosolve solvent, (I) butyl carbitol solvent, (J) hexyl carbitol solvent.

6.2.5 *Methyl carbitol solvent (diethylene glycol monomethyl ether)*

Methyl carbitol solvent has the solvency and low volatility necessary to produce excellent results in brushing lacquers. Its mild odor and ability to improve dye penetration make it a superior solvent for non-grain-raising wood stains, stamp pad and stencil inks, spirit-type dyes, and textile dye pastes.

Methyl carbitol solvent is a suitable solvent for dyes, oils, fats, waxes, and resins. It can also be used as a coalescing aid for polyvinyl acetate latex paints. Methyl carbitol solvent-fuel additive grade is used by the aviation industry to prevent icing in aircraft jet fuel systems.

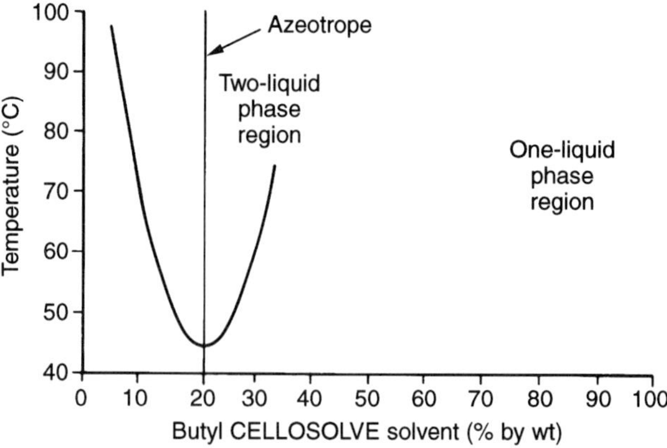

Figure 6.5 Mutual solubility of butyl cellosolve® solvent/water versus temperature.

6.2.6 *Carbitol solvent (diethylene glycol monoethyl ether)*

Carbitol solvent is the principal solvent in most non-grain-raising wood stains. It is widely used in non-aqueous wood stains, and serves as a component of many industrial cleaners. Latex paint formulations frequently include carbitol solvent to improve film coalescence, particularly at low temperatures.

Textile printing and dyeing processes utilize carbitol solvent to facilitate deep penetration into fibers and to promote the production of intense, bright shades. Its high boiling point, low vapor pressure, and ability to penetrate textile fibers make it useful for twist setting, as well as yarn and cloth conditioning.

Mineral oil–soap and mineral oil–sulfonated oil mixtures require fast-acting mutual solvents. Small quantities of carbitol solvent in these mixtures form translucent-to-clear dispersions, in contrast to the usual turbid dispersion that indicates incomplete solubility.

6.2.7 *Butyl carbitol solvent (diethylene glycol monobutyl ether)*

Butyl carbitol solvent is useful in lacquers, dopes, and stamp pad and printing inks that require a solvent with an extremely low rate of evaporation. This glycol ether is incorporated in high-baked enamels to contribute desirable flow and gloss characteristics. Butyl carbitol solvent is a dye solvent widely used to promote rapid and uniform ink penetration for printing box board and similar materials.

The solvent serves as a mutual solvent for soaps, oils, and water, as a component of liquid cleaners, 'soluble' oils, and textile oils. It has found increased use in cleaning formulations because of its good solvency for greasy soils and waxes.

Butyl carbitol solvent is used as a coalescing aid for latex paints and as a dispersant for vinyl chloride resins used in organosols. It is a diluent for hydraulic brake fluids and an intermediate in the manufacture of plasticizers.

6.2.8 *Hexyl carbitol solvent (diethylene glycol monohexyl ether)*

Hexyl carbitol solvent, like hexyl cellosolve solvent, displays a stronger hydrocarbon-type solvency than other glycol ethers. The increased ether functionality of hexyl carbitol solvent provides greater solubility with water than is found with hexyl cellosolve solvent. Hexyl carbitol solvent has shown excellent performance for the removal of greasy soils in evaluations of cleaning formulations. It is also used as a retarder solvent for inks and as a coalescing solvent in waterborne coatings.

6.2.9 *Alkoxytriglycols*

The alkoxytriglycols, methoxytriglycol (triethylene glycol monomethyl ether), ethoxytriglycol (triethylene glycol monoether ether) and butoxytriglycol (triethylene glycol monobutyl ether), are used as low-volatility components in paint stripping formulations, as dye carriers for textile dye processes, as chemical process solvents, and as intermediates for making esters of interest as solvents, surfactants, and plasticizers. The alkoxytriglycols are also useful as components of high-boiling hydraulic brake fluids.

6.2.10 *Storage of glycol ethers*

Glycol ethers can be stored in carbon steel for most applications. Stainless steel or high-baked, phenolic-lined tanks could be considered for application where trace iron contamination or slight discoloration must be avoided.

Methyl cellosolve solvent, cellosolve solvent, or propyl cellosolve solvent should not be stored or used in aluminum tanks, pumps, or lines. Aluminum rank trucks can be used for shipment purposes only. Copper or galvanized iron can cause discoloration of glycol ethers. Hexyl carbitol solvent should not be stored in galvanized steel or zinc-lined tanks.

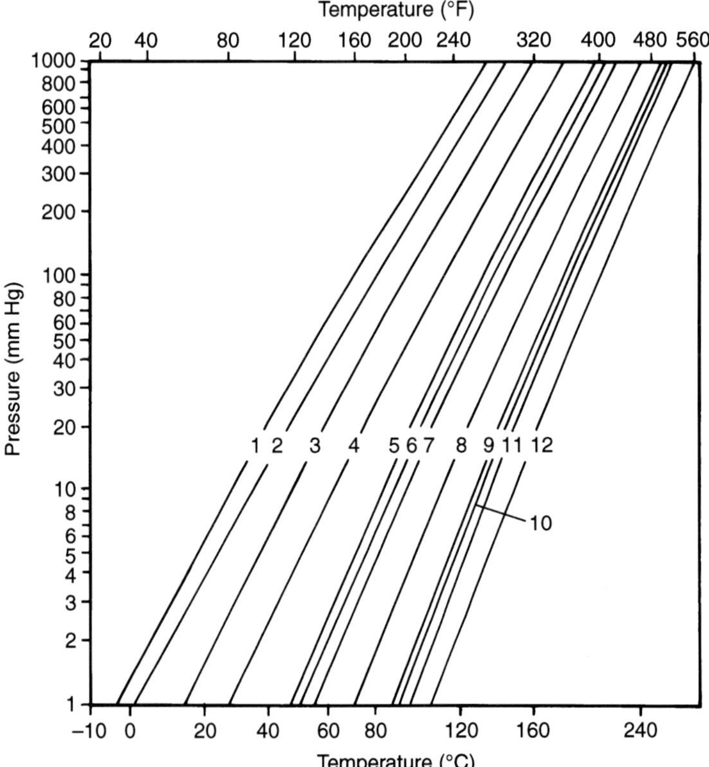

Figure 6.6 Vapor pressures of glycol ethers. (1) Methyl cellosolve solvent, (2) cellosolve solvent, (3) propyl cellosolve solvent, (4) butyl cellosolve solvent, (5) methyl carbitol solvent, (6) carbitol solvent, (7) hexyl cellosolve solvent, (8) butyl carbitol solvent, (9) methoxytriglycol, (10) ethyoxytriglycol, (11) hexyl carbitol solvent, (12) butoxytriglycol.

Glycol ether quality is best preserved in long-term storage by nitrogen-blanketing the storage tanks. The use of emission-control domes or desiccant chambers on tank vents will protect the solvent from atmospheric moisture if low water levels are important. Most glycol ethers do not present a significant flammability hazard at normal storage temperature. They have relatively low vapor pressures (Figure 6.6). The most volatile produce is methyl cellosolve solvent, which as a flash point of 105°F (41°C).

Glycol ethers have relatively low viscosities and freezing points. Heated storage is not normally required. Ethoxytriglycol is a possible exception, since it freezes at −2°F (−19°C).

Piping can be made of the same material as the storage tank. The use of PVC pipe is not recommended for glycol ether service. A centrifugal pump is suitable for transfer service. Butyl rubber or EPDM can be used for gaskets and packing. 'Viton', neoprene, and natural rubber are not recommended for use with glycol ethers.

6.2.11 *Hazards of glycol ethers*

Union Carbide has carried out a series of laboratory studies on the biodegradation and ecological effects of its glycol ethers. Biodegradation was measured by 20-day biochemical oxygen demands values. Most of the glycol ethers evaluated in this study showed moderate to rapid biodegradation with unacclimatized microorganisms. Although several products only achieve about 30 to 45% biooxidation over the 20-day test period, this rate indicates that the chemical structure would not persist in the environment. Natural acclimatization in a waste-water treatment plant or the general environment would be expected to increase the rate and extent of biodegradation.

Bacterial inhibition tests indicate that none of these glycol ethers should be inhibitory to conventional waste-water treatment processes at reasonable discharge levels (500 to 1000 mg/l in compliance with applicable regulations).

Aquatic toxicity measurements showed only moderate toxicity to the test species, which included fathead minnows and *Daphnia magna*. Only two of the 12 glycol ethers tested indicated LC_{50} values below 1000 mg/l. These tests were conducted by procedures which follow recent EPA (1)/ASTM (2)/Standard Method (3) techniques.

7 Emulsion and solution polymers

J.J. HARTSCHUH, G.H. CABIRO,
M.A. DALTON, R.P. GRANDKE,
T. SMITH and G. OLIVER

7.1 Designing innovative water based polymers for graphic arts

Acrylic polymerization techniques have evolved over the past several decades to encompass a wide array of procedures and composition combinations that impart very specialized performance attributes to water based inks and coatings. Although some experts feel that acrylics have reached their maximum technical capability in water based systems, they still dominate the industry. As emission regulations force printers into converting the most demanding applications to water based systems, improvements in resin technology are essential. Acrylic hybrids such as polyester-acrylic and polyurethane acrylic are likely to extend the technical capability for the next several years.

New polymer research is aided by more sophisticated analytical tools and intensive formulation, which the polymer chemist utilizes in developing products that provide additional improvements in performance. This chapter offers several ideas regarding new uses for current acrylic technology and illustrates how new ideas in formulation and design enhancement are supported by analytical technology such as particle size analysis, shear rheometry, and dynamic surface tension.

The market for water based inks and coatings is still growing rapidly and there is strong demand for high performance in applications previously satisfied by solvent based systems. These include high-speed gravure, flexible film inks, and detergent and chemical resistance coatings. Water based polymer compositions and processing methods that were initially developed for early flexographic corrugated printing inks have had to be upgraded to meet the demands of today's flexo presses and the diverse functional and high quality end use requirements.

For a number of years it has been common practice to design polymers around a solid, styrene acrylic resin solution in a medium molecular weight range. This solution fraction supports subsequent emulsion polymerization *in situ* and in effect eliminates the need for high levels of surfactants or emulsifiers. The solution fraction of a polymer offers such properties as pigment wetting, transfer and rewettability to inks while the emulsion fraction can provide functional properties such as gloss, dry speed, adhesion and water resistance. We have learned that not only the amount of solution resin used in support but

Table 7.1 Supported polymers

A. Standard level SAA resin	Polymers in which the first stage uses a standard amount of a neutralized, mid-range molecular weight styrene/acrylic resin solution and the second fraction is a 'hard', high Tg emulsion.
B. Combination of molecular weights	A 'hard' polymer in which the first stage incorporates a combination of molecular weights of styrene/acrylic resin solutions.
C. High % levels SAA resin	A 'hard' polymer in which the first stage utilizes very high levels (as a percentage of total solids) of styrene/acrylic resin solution.

also the type, including non-acrylic chemistries, has a significant influence on the performance properties of the total polymer. Some of the possible 'supported' combinations of polymer chemistry are given in Table 7.1.

One possible application for these hard acrylic polymers is in the area of water based gravure inks. When designing polymers for use in high-speed gravure, end use properties such as rewettability, dry speed, printability and high sheer rheology must be considered. Today's presses are equipped with high screen count and shallow etched cylinders for maximum image reproduction and ink transfer. If the ink and/or polymer contains large or hard abrasion particles, cylinder wear and plugging may occur. We have been able to improve polymer design and address this concern with the help of particle size analysis.

Figure 7.1 demonstrates the differences in particle size obtained from polymer compositions and processes as described above. The standard styrene-acrylic resin supported latex (A) results in a product that has a wide particle size distribution and which tends to have a greater amount of larger particles.

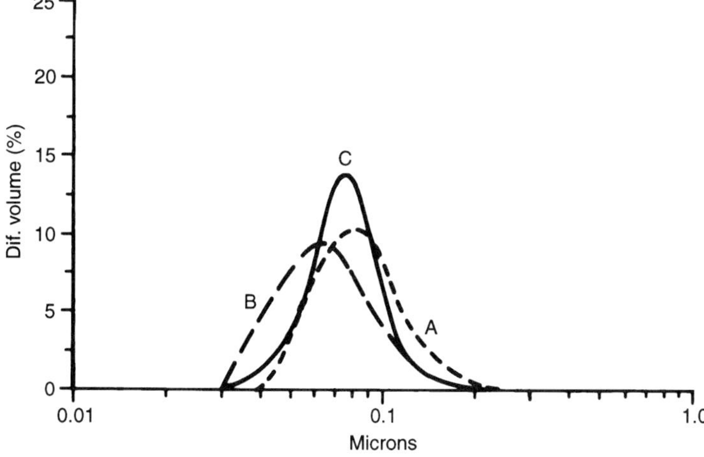

Figure 7.1 Differences in particle size obtained from polymer compositions outlined in Table 7.1.

Product (B), which utilizes a combination of styrene-acrylic molecular weight resins has a lower overall range. Lowering the particle size and narrowing the range can also be accomplished by increasing the total percent of supporting styrene-acrylic in the first fraction (C). A smaller particle size held within a narrow range will show improved performance in gravure transfer and rewettability.

Another important performance criterion in water based flexographic and gravure latex systems is the ability to remain stable when combined with other polymers and ink ingredients. A lower particle size indicates a lower molecular weight polymer. Low molecular weight correlates with better compatibility including viscosity stability, with a wide range of polymers.

Investigation into non-acrylic chemistries has shown that a substantial impact can be made on particle size when urethane or polyester is used in the backbone of a styrene-acrylic polymer. Figure 7.2 illustrates the net decrease in particle size of a polyester (PE)-containing polymer over the styrene-acrylic supported system B and C as previously described.

A water based polyurethane (PU) reacted within a styrene-acrylic latex system (Figure 7.3) also shows an overall decrease in particle size with the additional benefit of a very narrow range. Both the polyester-acrylic and polyurethane acrylic show excellent compatibility and stability with conventional systems, thus giving the ink formulator increased options. The polyester-acrylic improves running speeds and rewettability in gravure applications while the polyurethane-acrylic improves detergent and chemical resistance.

Another aspect of modern day printing that influences potential polymer design is the high shear forces generated by press speeds of 1000 feet per min and higher. At these speeds it is imperative that the ink be able to flow out of the

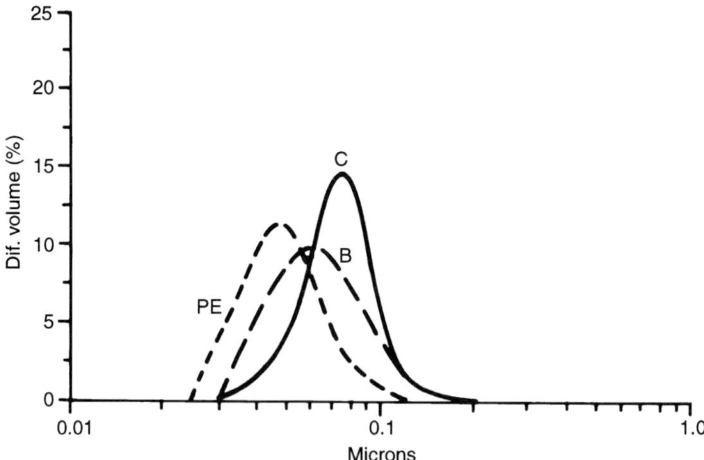

Figure 7.2 Net decrease in size of a polyester-containing polymer over the styrene-acrylic supported system B and C.

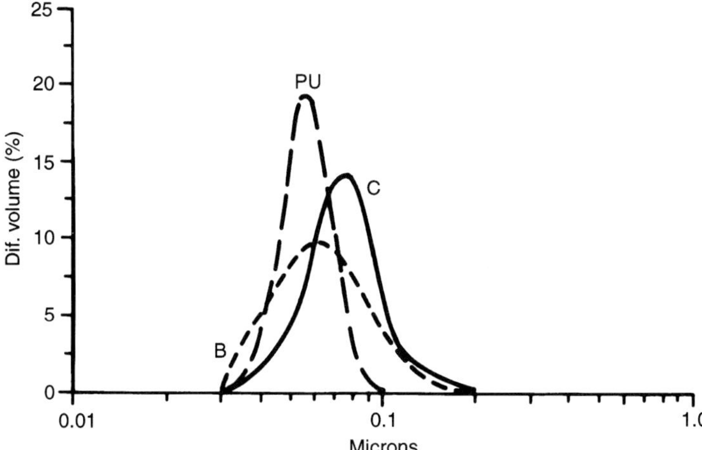

Figure 7.3 The overall decrease in particle size when a water based polyurethane reacts within a styrene-acrylic latex system.

cells and level quickly on the substrate. Mottle or poor lay can result from an ink exhibiting inadequate rheology characteristics. Using a rotational dynamic mechanical analyzer, a comparison of shear rate to viscosity enables us to predict which polymer designs will perform the best under high shear conditions.

A fluid is said to be Newtonian if the viscosity does not vary with the shear rate. Viscoelastic polymers are non-Newtonian in behavior because their viscosity will either decrease or increase with changing shear rate. Figure 7.4 shows viscosity as measured in units of Pascals versus shear rate as presented in units of reciprocal seconds. The shear rate scale plotted logarithmically is divided into regions in which certain physical properties are directly related. In the low shear rate range of 10^0 the properties of flow and leveling prevail. The mid shear rate range of 10^1 to 10^2 relates to the properties of consistency and appearance. If the low shear viscosity is too high, the material may show drag during application and not level out properly. Figure 7.4 shows that a polymer designed with an increased percentage of resin supports (C) shows close to Newtonian behavior at low shear rates. Products with an average level of styrene-acrylic resin support (A) tend to have an increased low shear viscosity and may show poorer flow and leveling.

Experimenting with the possible application of polyester chemistry for the gravure market, we can see a good correlation to desired Newtonian behavior. We can expect that these products would show improved gravure performance over standard styrene-acrylic emulsions. The ability to maintain viscosity under high shear is also an indication of gloss (Figure 7.5). The improved surface leveling characteristics exhibited by polyester and polyurethane modified polymers are demonstrated by gloss readings in the 80° range depending on substrate

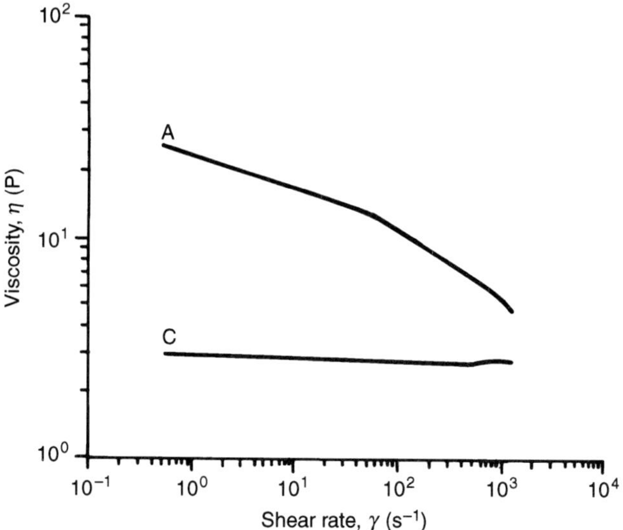

Figure 7.4 Relationship between viscosity and shear rate for a product with an average level of styrene-acrylic support (A), and a product with an increased amount of resin support (C).

and application coat weight. The high percentage resin supported products also exhibit better gloss than standard acrylics.

The use of styrene-acrylic solution chemistry in support of 'soft', film forming polymers is also well documented (Table 7.2). These types of products

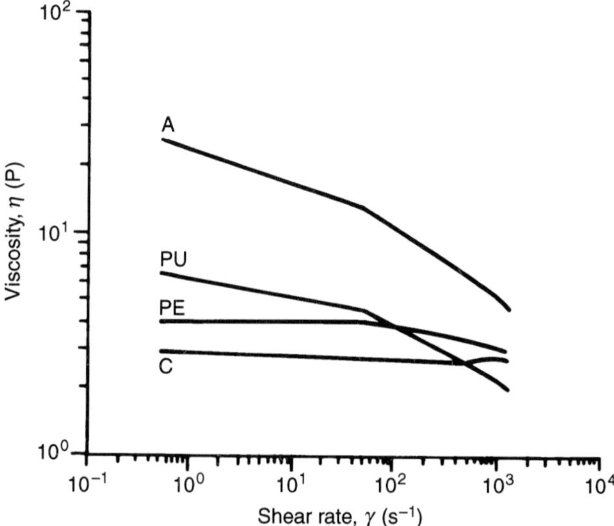

Figure 7.5 Relationship between viscosity and shear rate for polyester (PE) and polyurethane (PU) products compared with products A and C referred to in the caption to Figure 7.4.

Table 7.2 Supported soft polymers

D. Low level SAA resin	The first stage contains low levels of styrene-acrylic resin support.
E. Standard level SAA resin	The first stage contains a standard level of mid-range molecular weight styrene-acrylic resin support.
F. High and/or combinations of molecular weights	A high level and/or combination of molecular weight resins are used in the polymer design.

typically contain butyl acrylate or 2-ethyl hexyl acrylate in the second fraction are designed to have low Tg values in the range of 10 to 40°C. As with a hard polymer different compositions can impact final performance.

Applications for these polymers include flexible films and foils that pose some unique design challenges. We have been able to utilize particle size and dynamic surface tension analysis in improving properties such as adhesion and clarity.

The use of low levels of styrene-acrylic solution resin as a support in a film forming polymer is desirable for improved resistance properties and dry speed on non-absorbent substrates. This decrease in solution fraction (Figure 7.6), however, impacts on the formation of polymer particles, which results in a higher average particle size latex (D) than a system supported with a standard amount of solution resin (E). Performance advantages of a low particle size product such as (E) include excellent clarity and gloss for transparent inks of films and foil.

The relationship of ink surface tension to press performance has been investigated to a considerable degree by many in the industry. The water based polymer manufacturer plays a major role in providing the chemistry that

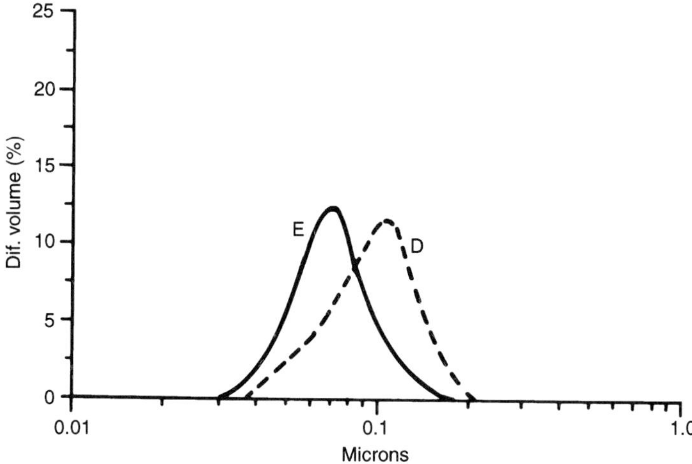

Figure 7.6 Relationship between diffusion volume and particle size when styrene-acrylic solution is used as a support in film-forming polymers.

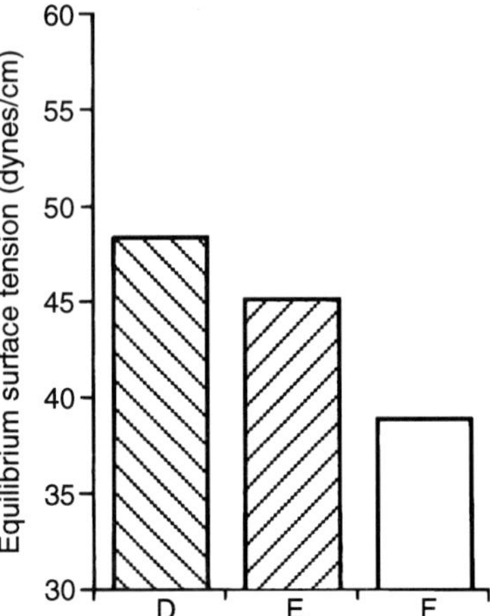

Figure 7.7 Effect of level of resin support on equilibrium surface tension.

addresses the desired characteristics of low surface energy. The use of dynamic surface tension analysis can help predict how a polymer will perform under shear conditions. Surface tension modifiers such as surfactants or alcohols can easily be added to an ink or polymer by the intrinsic surface tension value contributed by the polymer design offers guidelines to further formulation. Figure 7.7 shows that polymer with low levels of resin support (D) has higher equilibrium surface tensions than a polymer that utilized wither standard (E) or high levels of resin support (F). When the stress/shear is increased on these same polymers, the dynamic surface tension drops significantly for the higher resin supported systems (Figure 7.8).

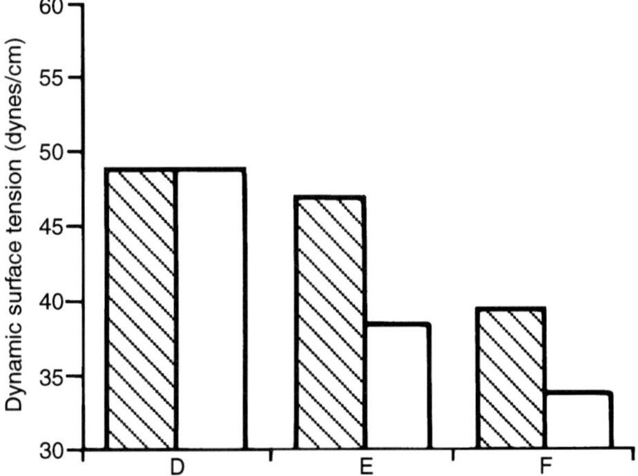

Figure 7.8 Effect of increased shear/stress on dynamic surface tension.

These products have better adhesion and wetting properties on most films. Evaluation of surface tension properties under similar conditions with polyurethane and polyester modified polymers designed for paper substrates indicates no notable improvements over conventional acrylics, but further characteristics for systems designed for films.

We have seen how variations in styrene-acrylic supported polymers and non-acrylic supported systems can impact particle size, rheology characteristics, and surface tension. Learning more about the relationships between these attributes and press performance will help guide the chemist in designing future polymers.

As we head toward the next century we will see the evolution of polymer design rapidly embracing a whole new array of chemistries. Printers will be converting their most demanding applications from solvent to water and the technical improvements necessary in ink and coating polymers will be achieved with novel designs assisted by information gathered from new analytical equipment.

7.2 Reduction of volatile organic compounds through aqueous acrylic polymers

The ink and overprint varnish industries' progress in conversion to water based technology has led to a dramatic source reduction in volatile organic compounds (VOCs). These producers should be proud of their progress, which aims to improve the quality of life we all enjoy and has allowed converters and printers to meet federal and state air quality regulations. The move to water based technology has shown that industry and the environment can co-exist and that source reduction can be mutually beneficial.

Water based inks and overprint varnishes (OPVs) can now rival traditional solvent systems for print quality and ease of use for many applications. This VOC reduction has been made possible through the expertise of skilled formulators and the increasingly broad variety of polymer systems available for water based use. In this section we will review these systems and provide guidelines on their selection and use in high quality inks and OPVs.

7.2.1 *Comparison of water based and solvent based ink requirements*

Apart from the obvious differences in evaporation rates between water and hydrocarbons as solvents, the greatest differences between water and solvent systems are the way in which resistance properties and resolubility are controlled (Figure 7.9).

Resolubility is the ability of an ink or OPV to dissolve its own dry film. In other words, any film formed on the printing play during a press stop must be readily soluble in the liquid ink to prevent print defects when the press restarts.

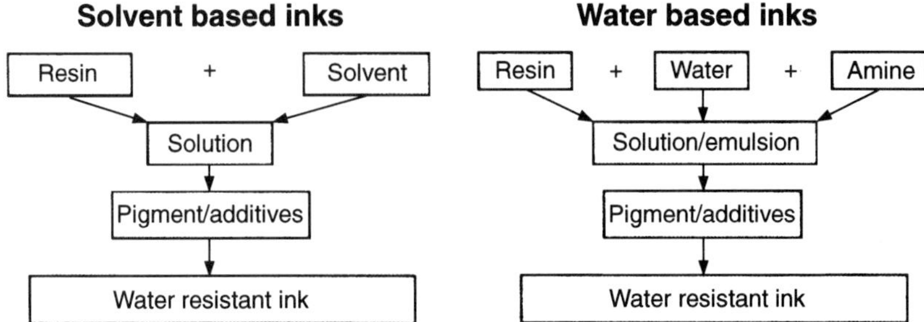

Figure 7.9 Comparison of solvent based and water based inks.

This is easy for organic solvent based ink systems where the dry ink film is readily soluble in the liquid ink. However, in water systems the formulator must learn to build in water resolubility on the press without sacrificing the water resistance on the substrate when the ink has fully cured. The balance between resistance properties and press performance is accomplished through experience and the careful selection of polymer systems.

Acrylic polymers are the most successful of all waterborne technologies available and can be classified into four categories:

1. Alkali soluble resins
2. Acrylic emulsions
3. Colloidal emulsions and solutions
4. Acrylic speciality polymers

7.2.2 *Alkali soluble resins*

Alkali soluble resins are widely used in waterborne inks and OPVs. These resins, typically styrene acrylic, are used in a broad variety of applications for the properties they impart:

1. In pigment dispersions to improve stability, color strength, gloss, and transparency
2. In inks to increase gloss and add resolubility on the press
3. In OVPs to increase gloss and resolubility

Acrylic resins are widely utilized in grinding and dispersion of pigments since they are highly efficient wetting and grinding agents that also help to stabilize the dispersion. There are new acrylic resins in the marketplace which, through careful design, have been able to improve significantly the color development, gloss, and transparency of pigment dispersions. With the potential to reduce organic pigment usage, normally the most expensive component of an ink, acrylic resin selection becomes a key formulating consideration. Lower pigment usage also benefits the environment through source reduction.

In addition to dispersion of organic and inorganic pigments, acrylic resins are also used in inks and OVPs for both gloss and resolubility. In their acidic form,

Table 7.3 Amine selection

Amine	Boiling point (°C)
Ammonia	−33
Morpholine	128
Diethylaminoethanol	163
Monoethanolamine (MEA)	170
Diethanolamine (DEA)	268
Triethanolamine (TEA)	335

resins are insoluble in water, but once neutralized, the resins and press resolubility for high quality printing. Then as the amine evaporates, the dried ink film develops the water resistance necessary for most end uses.

Ammonia is the most common neutralizing agent due to its low cost and ease of evaporation (Table 7.3). Monoethanolamine is also used in many applications. Occasionally, other amines are used, because they help improve resolubility. However, the amines are more expensive, require more energy to evaporate, and add to the VOC content. Just as with inorganic alkalis, such as sodium or potassium hydroxide, less volatile amines will reduce water resistance due to the formation of nearly permanent salts.

7.2.3 *Selection of alkali soluble resins*

The most important selection criteria for the use of alkali soluble resins in inks and OPVs are molecular weight (Mw) and acid value (AV). Figure 7.10 compares the molecular weight distribution of conventional resins to the tightly controlled molecular weight of resins produced with a coninuous process. At Johnson we utilize a patented continuous process called SGO(TM). The area on the left represents the low molecular weight volatile oligomeric fractions present in conventional resins. These species are the cause of unpleasant odor as well as

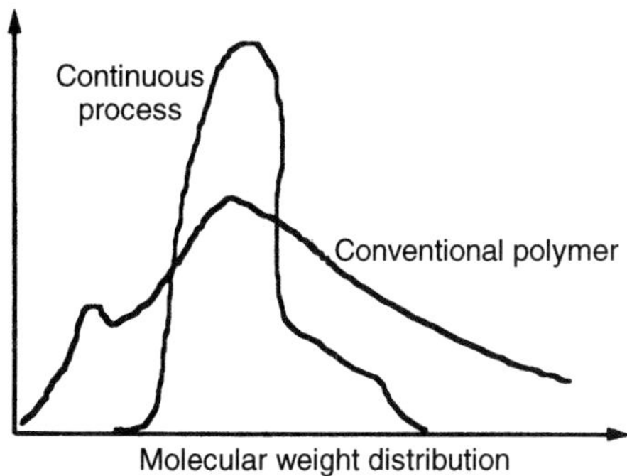

Figure 7.10 Molecular weight distribution of conventional resins and those produced by a continuous process.

an often unexpected source of VOCs. The area on the right portion of the curve represents high molecular weight species or super-polymer. The presence of these fractions produces extremely high and unpredictable viscosities in the final formulation.

In selecting an alkali soluble resin it is important to consider that higher molecular weight resins will have a higher viscosity for a fixed solids level. Additionally, they will disperse pigment better. Conversely, low molecular weight resins will allow for much higher solids levels without reaching high viscosities. The higher solids levels attainable with low molecular weight resins allows the formulator to maximize the gloss of inks and OPVs. Additionally, higher solids content in an ink or varnish reduces the amount of water that has to be evaporated, therefore, increasing the drying speed. These effects are seen in Figure 7.11.

High acid number resins result in higher resolubility and improved pigment wetting but reduce wet rub resistance. In contrast, a low acid number relates to greater alkali and detergent resistance, but at the expense of pigment dispersion and resolubility properties (Figure 7.12).

7.2.4 *Acrylic emulsions*

Acrylic emulsions are used to give inks and OPVs their key application properties. The preferred acrylic emulsions are small particle size lattices engineered to maintain many of the desirable application properties of solution resins. For example, on our company, the Joncryl (R) rheology controlled emulsions exhibit near-Newtonian rheology. Emulsions such as these maintain their viscosity stability even under the high shear conditions of the printing process. This important property is known as rheology control (Figure 7.13).

Other key properties of rheology controlled emulsions include their lower polarity, which contributes to faster water release than with acrylic solutions and improves the drying speed of the inks. Also, in contrast to true solutions, they

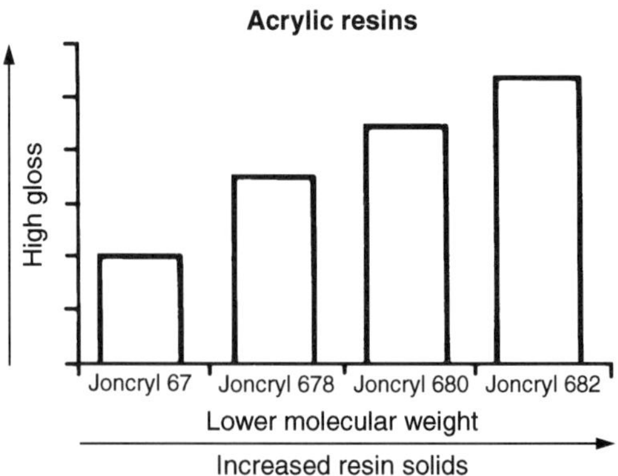

Figure 7.11 Effects of increasing the amount of solids in a resin.

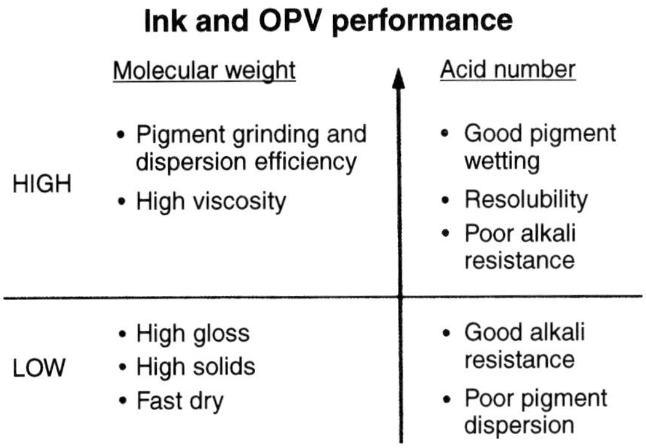

Ink and OPV performance
Molecular weight
Acid number
HIGH
• Pigment grinding and dispersion efficiency
• High viscosity
• Good pigment wetting
• Resolubility
• Poor alkali resistance
LOW
• High gloss
• High solids
• Fast dry
• Good alkali resistance
• Poor pigment dispersion

Figure 7.12 Comparison of ink and OPV performance.

become completely insoluble upon drying, which greatly improves the inks' and OPVs' water resistance. The small particle size of these emulsions, around 800 Å (0.08 μm), provides the transparency and high gloss required for inks and OPVs. Finally the chemistry of these emulsions tends to reduce foaming in comparison to conventional emulsions.

7.2.5 *Selection of acrylic emulsions*

In selecting the right blend of emulsions, the formulator must look at the balance of properties required and choose a polymer with the appropriate glass transition temperature (Tg). Tg is the temperature at which a polymer changes from an amorphous, elastic stage to a brittle stage in which the molecular mobility that originates film formation has stopped.

Minimum filming temperature (MFT) is closely related to Tg. MFT is the lowest temperature at which a continuous film will be formed upon drying of the polymer. High Tg polymers will not form a film at room temperature. Conversely, low Tg values will result in better film formation (Figure 7.14).

Near Newtonian rheology
Conventional emulsion
Viscosity
Rheology controlled emulsion
Shear rate

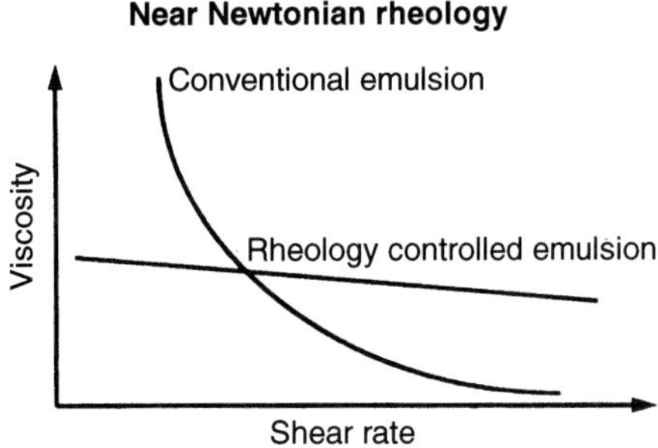

Figure 7.13 Relationship between shear rate and viscosity.

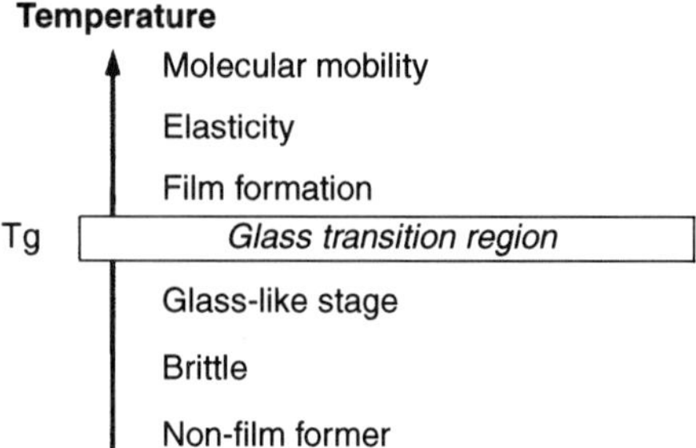

Figure 7.14 Relationship between Tg values and film formation.

Rheology controlled emulsions are utilized as letdown vehicles to provide final properties to the ink or OPV. In selecting the appropriate emulsion, careful consideration must be given to the desired properties. For example, with higher Tg polymers, hardness, gloss, dry speed and heat resistance will be increased. Low Tg emulsions are excellent film formers and will add to film properties such as water and grease resistance, flexibility and adhesion to non-porous substrates like polyethylene, polypropylene, polyester, and foil.

There are several ways to balance the desirable properties of high Tg polymers with film formation.

(a) *Heat.* Through heating, molecular mobility is imparted to the polymer allowing latex to coalescence and film formation to occur. The film will remain stable even after cooling.

(b) *Plasticizers.* Plasticizers work as internal lubricants allowing film formation. The disadvantage is that plasticizers permanently soften the film and reduce heat resistance among other physical and chemical resistance properties.

(c) *Coalescing agents.* Coalescing agents are solvents that work like plasticizers, but once the film is dry they evaporate, restoring the original hardness and resistance properties to the system. One disadvantage is that the evaporated coalescent is a VOC.

(d) *Blends of polymers.* The addition of soft polymers can promote film formation in hard and brittle systems. This is the best way to control film formation, such blending of hard and soft emulsions is the key to obtaining the right balance of properties. Also, the use of soft polymers does not add to VOC.

7.2.6 *Colloidal emulsions and solutions*

Colloidal emulsions are used extensively in low-cost inks for porous substrates, such as corrugated board and multi-wall kraft bags, where high gloss or adhesion to film is not required. Frequently, colloids are utilized as letdown vehicles in organic colored inks. However, their high molecular weight (Mw) (between 25 000 and 100 000) makes them very popular as grinding or dispersing vehicles for carbon black pigment. Colloidal emulsions are also utilized as additives in different types of inks and OPVs in order to reduce cost, improve press transfer properties, and to adjust viscosity.

Colloidal emulsions consistent with an acidic polymer are dispersed in water and stabilized with a surfactant. Colloidal emulsions should be neutralized to form a colloidal solution before they are used in the ink formation. As the carboxyl groups are neutralized, the colloidal particle uncoils and the solution goes through a tremendous viscosity increase and flattens out to a stable viscosity within the range of use. For convenience, colloids are available as neutralized solutions, but where possible the unneutralized emulsion offers formulating latitude. Figure 7.15 shows how a 20% solids emulsion increases in viscosity from 40 cps to 4000 cps with the addition of only 3% of ammonia.

It is precisely this high viscosity efficiency that makes colloidal emulsions ideal for the formulation of low cost inks (Figure 7.16). Very stable ink systems can be formulated with just the minimum amount of polymer to wet the pigment, maintain a suitable printing viscosity, and maximize the water content of the ink.

Recent advances in colloidal emulsion technology have led to products that offer excellent ink formulation economics, organic pigment dispersion capability and excellent print quality.

7.2.7 *Acrylic speciality polymers*

There are also specific acrylic emulsions designed to provide properties in very high performance applications such as:

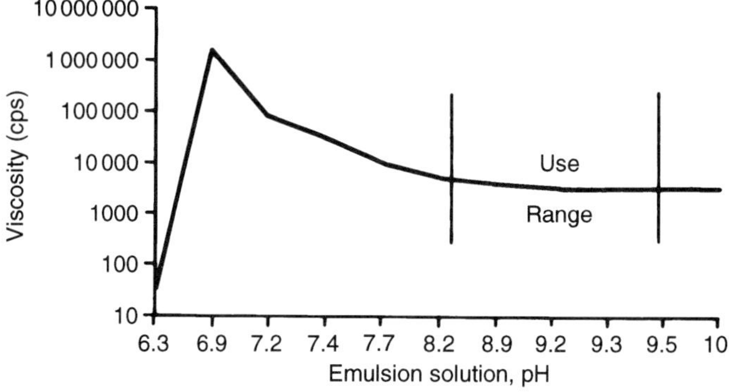

Figure 7.15 The increase in viscosity of a 20% solid emulsion upon the addition of 3% ammonia.

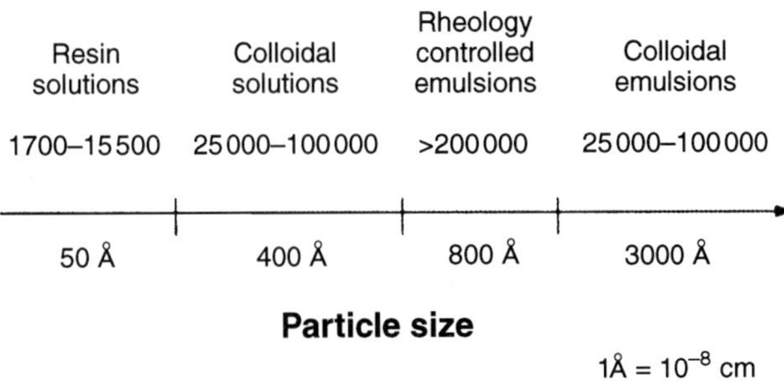

Figure 7.16 Molecular weight and particle size of acrylic polymers.

1. Alkali resistance
2. Gasoline resistance
3. Metallic pigment stability
4. High slip properties
5. Hot mar resistance

Other polymers are often added at small levels to modify ink and OPV properties such as slide angle and rub or scuff properties.

In addition to acrylic polymers, other additives can be used for formulators to improve the performance of a specific ink or OPV formulation. Typical additives include:

1. Wax emulsions to decrease slide angle of coated board
2. Wax emulsions to add rub or scuff resistance
3. Wax emulsions to provide a matte finish
4. Zinc oxide solution add film toughness and heat resistance through crosslinking
5. Wax emulsions for water beading

Up to this point, the discussion has shown how to select and use each of the major aqueous polymer categories. However, these systems are rarely used alone. The skilled formulator will use a combination of any or all of these types of polymers along with the additives and specialities discussed above to produce waterborne inks and OPVs to meet a broad range of demanding requirements.

7.2.8 *Trends for the future*

Section 7.2 has covered the basics on use and selection of today's resins for water based inks and OPVs. However, far from being satisfied, the polymer industry continues to push the envelope of waterborne performance as this sampling of recent advances shows:

1. Polymers for pre-printed corrugated, which require no zinc or zinconium to reach 450°F hot mar resistance
2. Dispersion resins, which develop chip quality gloss and transparency in conventional media milling equipment
3. Corrugated vehicles, which require half the resin solids of conventional systems allowing for very low cost inks
4. Film printing polymers, which wet out olefins without the addition of any VOC contributing alcohols
5. News flexo polymers, which virtually eliminate second impression set off
6. Polymer systems for waterborne publication gravure

7.3 The waterborne way

SC Johnson Polymer expect new polymer chemistry to be the springboard for polymers that can be used for waterborne inks for high quality coated magazine stock. While it is true that solvent-borne inks are presently preferred for publication printing, using the gravure process, times are changing. In the United States, with the adoption of the 1990 Clean Air Act Amendment, there is a need for printers continually to reduce the volatile organic compounds (VOC) that find their way into the environment. Even with high efficiency solvent recovery equipment, it is estimated that from 2 to 7% of the solvent contained in the inks finds its way into the environment.

In addition to air pollution concerns, toluene, the primary solvent used in publication gravure printing, is considered undesirable in the press room. The reduction of air pollutants and worker exposure are not just problems in the United States. These situations present a global problem that must be dealt with; one obvious choice is the use of waterborne inks. This choice has been largely successful in paper and paperboard package printing using both flexographic and gravure printing methods.

Press trials to evaluate waterborne ink for publication printing are not new. Until recently, waterborne inks just did not have the necessary attributes to be a viable replacement for solvent inks. Print quality, gloss, excess VOC, poor dot formation and an inability to print well in both the mid-tone range and the solid areas have all been contributing factors.

In 1991 a dedicated research and development programme was initiated to resolve the problems associated with waterborne inks for publication printing. The objective was to develop polymers, from which inks could be made and applied to calendered or super-calendered stock. This included newspaper advertising supplements, catalogs, television program guides and other booklets.

The result was the development of two polymers designed to be used in combination that eliminates many of the problems and substantially improves ink and print quality. SCX-686 is a solid alkali soluble resin specifically designed to perform on two-side high-speed web gravure, with excellent carbon black dispersion properties. The companion polymer, SCX-2160, is a non-film forming rheology controlled emulsion that provides excellent print characteristics in aqueous publication gravure inks.

Press trials were run on a Cerutti press at Western Michigan University. The SC Johnson Polymer R & D group evaluated a carbon black ink system on super-calendered stock. Initial trials were used to correlate print quality with polymer structure. Table 7.4 shows the ink formula evaluated.

The results of the printing trials were presented in a series of photomicrographs taken from print samples. Printing on the 50% midtone range showed a typical styrenated acrylic ink system. Carbon black dispersions made with the resin were very 'short' and contributed to a mottled print. An ink utilizing a polymer structure change showed directional improvement, but was not considered acceptable. Prints based on the SCX-686 and SCX-2160 polymers showed vast improvement in print quality. The inks did not contain additional additives typically used to improve flow and levelling. Yet the experimental ink had a Newtonian reheology with excellent flow and levelling. The result is a

Table 7.4 Starting point formulations using SCX-686 and SCX-2160

Compound	Percent
Resin solution, SCX-686	
SCX-686	43.0
Ammonia, 28%	14.0
Water	43.0
	100.0
(pH = 9.2, viscosity* = 1200 cps)	
Pigment dispersion	
Carbon black	23.0
SCX-686	48.0
Water	28.8
Defoamer	0.2
	100.0
(Predisperse pigment in high-speed disperser. Use media mill for final dispersion)	
Finished ink formula	
Pigment dispersion	42.0
SCX-2160	21.0
Water	36.7
Antifoam	0.3
	100.0
(pH = 8.5, viscosity† = 21 s)	

* Brookfield LVF #3 spindle, 12 rpm, 25°C.
† #2 shell cup.

polymer system that will allow ink makers to formulate a waterborne ink that exhibits dot spread more like a typical solvent ink and not observed with previous waterborne inks.

Results for solid print areas were similar to those noted in the previous paragraph for half-tones. We consider this quite important. Customers tell us that with other polymer systems, careful formulation techniques and the use of expensive additives, they can get acceptable print quality in either the mid-tone range or the solid print area, but not both.

The inks evaluated based on SCX-686 and SCX-2160 have been run with cylinders specifically designed for waterborne ink systems as well as for solvent inks. We have seen excellent print quality with both. We believe this continues to point to the improved flow and leveling of the ink systems and expect them to be more robust or forgiving.

In conclusion the observations include excellent pigment dispersion, very fast dry, little or no ink viscosity build, good inherent rub resistance, minimal or no doctor blade build-up and excellent gloss. In addition, dot skip, striations and snow flaking were reduced using the new polymers, and these conditions are virtually eliminated with electrostatic assist. In other works, printers can now get more solvent-like qualities from waterborne inks.

Future work and press trials will involve four-color process printing on calendered and super-calendered stock. For the present, SCX-686 and SCX-2160 offer the ink formulator the technology necessary to begin waterborne publication gravure printing.

8 Additives

A.C.D. COWLEY and T. WALSH

8.1 Hyperdispersants: the problem solvers

The ink industry is changing as never before. A combination of economic recession, growing imports and the implementation of environmental legislation is forcing major technological changes. The economic recession has put pressure on profit margins, and this in turn has created the need to reduce raw material and labor costs in order to remain economically viable. At the same time, the ink industry must also face up to the difficult task of reducing solvent emissions. These challenges are creating quite a few problems for the inkmaker.

However, it is precisely in these areas that hyperdispersants are finding major outlets. The phenomenal growth in the use of hyperdispersants is directly related to their role as problem solvers. Hyperdispersants are excellent pigment wetting and dispersion stabilizing agents and give beneficial effects on the rheology of inks. These properties make the hyperdispersants effective in meeting the challenges and solving the problems not only in the current but also in the newer technologies, such as water based and UV cured coatings.

8.1.1 *The problems*

A brief survey of ink formulators was recently carried out to ascertain the major technical problems facing the ink industry. The major concern was, not surprisingly, the effect of environmental legislation, but there was also increasing awareness of commercial factors such as raw material and processing costs. Table 8.1 shows the problems, grouped under three headings.

Environmental concern has been growing inexorably around the world ever since the impact of Rule 66, implemented in Los Angeles in 1967. That regulation restricted the concentration of specific types of solvent, but did not limit the total amounts emitted.

More recently, the Clean Air Act, the Clean Water Act, the Resource Conservation and Recovery Act, the Superfund Amendments and SARA Title III have all placed an enormous economic and technological burden on the industry. The EPA recently announced that in 1992 they sought $135 million in civil and criminal penalties. This amount will undoubtedly better focus industries' attention on the problem of pollution than will the threat of ozone holes, acid rain and smog.

Under the latest regulatory initiative, the HON rule of 29 October 1992, the

Table 8.1 Summary of the major technical problems facing the ink industry

Legal compliance	1990 Clean Air Act Amendments SARA III, Toxic Substances Release Inventory EPA Voluntary 33/50 Project Hazardous Organic National Emissions Standard for Hazardous Air Pollutants
Commercial	Raw material cost reduction ISO 9002 registration High processing costs
Technical	Need to improve rheology Storage stability of inks

EPA announced that the total amount of industrial emissions must be reduced by 75 to 90% over the next 10 years. While no one can realistically argue about the need to reduce solvent emissions, there will be a major cost associated with complying with these regulations.

During the mid-1980s, the needs and demands of customers were the paramount forces in promoting new produce development, but now it is the EPA that is determining the direction and speed of technological change. The move toward water based inks is already here. There is pressure from the voluntary EPA 33/50 project to move away from aromatic hydrocarbons, and newer technologies such as UV cured coatings are growing rapidly.

These technologies represent clear opportunities for innovative companies, but they also present new problems to be solved. Whichever technology is adopted, the inkmaker still has to produce high quality inks and supply them at competitive prices. The cost-intensive stage in ink production is the dispersion and stabilization of the pigments. This is precisely the area where hyperdispersants are of value. They are excellent pigment wetting agents, dispersion stabilizing agents and rheology modifying agents. These properties make hyperdispersants the ideal tool to help formulators solve the problems of complying with the VOC regulations.

8.1.2 *Hyperdispersants*

Before describing how these hyperdispersants may be used to achieve VOC reductions, it is useful to understand their chemistry and mode of action. The hyperdispersants are based on a two-part structure; namely, an anchoring group and a polymeric chain, and it is the particular combination of these that leads to their effectiveness.

As the nature of the surface of pigments varies widely with their chemical type, it is impossible to use a single anchoring group successfully with all pigments. A variety of functional groups have been used in order to achieve efficient adsorption on the pigment surface, such as: amines, alcohols, carboxylic acids, sulphonic acids, phosphates and blocked isocyanates. Generally speaking, more efficient anchoring is achieved with polymeric anchoring groups than with structures containing just one anchoring group. It is essential that the anchor group does not adversely interfere with the curing of the coating, e.g. in UV cured coatings.

A very successful approach to wetting out the particularly difficult pigment types, such as the non-polar, phthalocyanines and carbon blacks, is the use of so-called synergist hyperdispersants. These synergists have very strong affinity for certain pigment types and furnish the surface of the pigment with anchoring sites for a conventional polymeric hyperdispersant. It may thus be beneficial to use a combination of synergist and polymeric hyperdispersants for wetting-out and stabilizing organic pigments.

The nature of the polymeric chain is critical to the performance of the hyperdispersant. The molecular weight of the hyperdispersant needs to be sufficient to provide polymer chains of sufficient length to overcome the van der Waals forces of attraction between pigment particles. If the chains are too short, they will not provide a sufficiently thick barrier to prevent pigment particles attracting one another, which leads to an increase in viscosity and commonly to flocculation or gelling. There is generally an optimum chain leangth over and above which the effectiveness as a hyperdispersant ceases to increase, and indeed in some cases molecules with longer chains can be less effective. Ideally, chains should be free to move in the solvent/resin medium. To achieve this, chains anchored at one end only have been shown to be most effective in providing steric stabilization.

It does not matter whether the polymeric chains are provided by hyperdispersants containing single chains or up to many hundreds of chains, providing they are successfully anchored. The requirement is to cover the surface of the pigment with a sufficient density of chains to ensure the minimum pigment–pigment interaction. There is, generally speaking, an optimum amount of hyperdispersant for each pigment, dependent on its surface area in the mill-base. This is commonly known as a monomolecular layer. If too little is used, then the full benefits will not be realized. If too much is used, then not only is the final result inferior, but final film properties such as adhesion or a tendency to block may be adversely affected due to the pressure of free hyperdispersant molecules in the drying film. The optimum amount is easily determined by carrying out a ladder series of hyperdispersant dosages on a given pigment weight. Measurements of millbase viscosities of the ladder series will show a minimum at the optimum dosage. This minimum has been shown to correspond to approximately 2 mg/m^2 of the pigment surface area as measured by the BET nitrogen adsorption method (Figure 8.1).

The viscosity depression results from the particles becoming progressively encapsulated by the adsorbed hyperdispersant. As a result, a physical or steric barrier is created around the particle, preventing any influence from neighboring particles, and this reduces the thixotropy/puffiness in the millbase. The effect of the hyperdispersant is to make inks of more Newtonian rheology, so less solvent

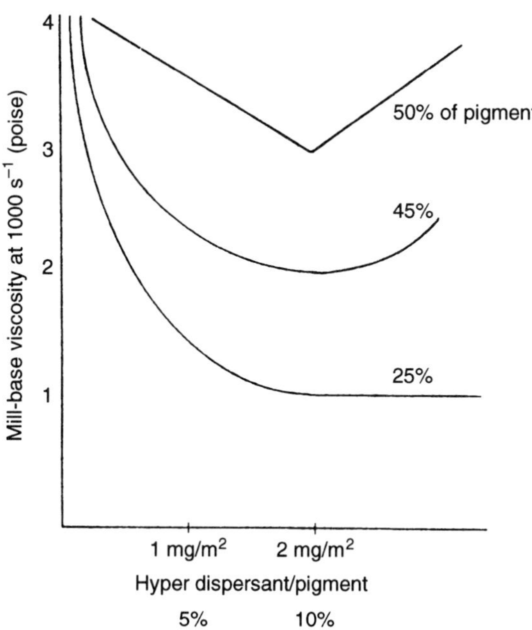

Figure 8.1 Viscosity profile of C.I. Pigment Red 57 : 1 in toluene.

is needed to reduce their viscosity. This allows the production of millbases or ink concentrates with higher solids content, where the hyperdispersants are acting as rheology modifiers.

It is essential that the polymeric chain is compatible with the solvents and resins in the ink. If the chains are not sufficiently solvated, they will collapse onto the surface pigment allowing the pigments to aggregate or flocculate. This need for compatibility extends throughout the final drying stage of the ink film. If it ceases to be compatible, flocculation might occur, leading to a loss of gloss, tinctorial strength and, with some pigments, a loss of transparency. In order to meet this demand, several different polymeric types are required to span the full range of solvents encountered, from aliphatic hydrocarbons to water. Happily, it is possible to produce hyperdispersants that work across the sufficiently wide ranges of solvents required for practical use. More detailed explanations of the mechanisms by which hyperdispersants work are available [1].

8.1.3 *Use of hyperdispersants*

When conventional ink millbases are being manufactured, it is the resin or vehicle that normally wets-out and stabilizes the pigment. If hyperdispersants are being used, there will be competition between the resin and hyper-dispersant for the pigment surface. In order to obtain the full benefits from the hyperdispersant, it is preferable to minimize this competition by using vehicles of lower resin solids than normal. For example, in water based or solvent based liquid ink, vehicles of 10% resin solids content should be used. Whereas for heat set inks, the vehicle solids in the millbase can be reduced to 40%.

In practice, millbases can be prepared by natural methods of adding the hyperdispersant to the resin and solvent, before incorporating the pigment. On pre-mixing it will be immediately obvious that the hyperdispersants have a dramatic effect in depressing the millbase viscosity. It is essential to monitor this millbase viscosity instrumentally, and raise the pigment loading (not the resin content) in order to achieve a practical viscosity.

Clearly, it is necessary to maintain the required hyperdispersant : pigment ratio, in line with the surface area of the pigment, in order to obtain the best performance from the hyperdispersants. Figure 8.2 represents a nitrocellulose based gravure ink in which a blue shade phthalocyanine green is being dispersed. By using 12% of 4 : 1 blend of polymeric : synergist hyperdispersant on weight of pigment it is possible to disperse a millbase at a pigment loading of 35%, and this millbase displays more Newtonian behavior than does the control at 20% pigment content.

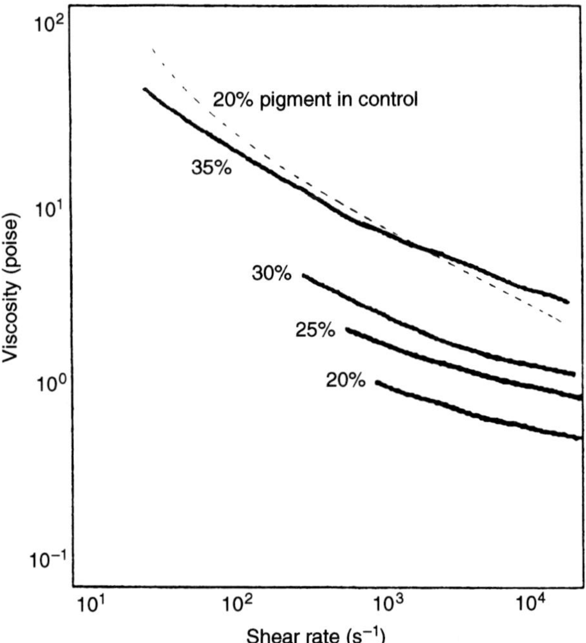

Figure 8.2 Effect of hyperdispersants on rheology: monolite green GNX in nitrocellulose. Control contains 16% nitrocellulose solution; others contain 10% nitrocellulose solution plus hyperdispersant.

8.1.4 *Solving the problems*

(a) *Raw material cost.* When you consider the raw material cost of an ink, it is usually the pigment that is the most expensive component. It is therefore economically sensible to ensure that the full color strength of the pigment will be able to be obtained on dispersion.

As the hyperdispersants were designed specifically to anchor onto pigments, they wet-out pigments much more effectively than most ink vehicles. Also, because the hyperdispersants create a physical or steric barrier around the dispersed pigment particles, there is much less tendency for flocculation or aggregation to occur, thus allowing the full color strength to be displayed.

Figure 8.3 shows the typical rate of strength development curve for pigments with and without hyperdispersants present. It is important to note that in the millbase containing hyperdispersant, a higher strength is realized, thereby allowing the possibility of using a lower pigment content with a savings in raw material costs. As a spin-off, the better state of pigment dispersion will also be manifest in terms of clearer colors with higher gloss and transparency.

(b) *ISO 9002.* An increasing number of inkmakers are seeking registration under ISO 9002. Anyone who has been through this daunting process will understand the value of good quality control. One area where poor quality control is often observed occurs in the ink production stage, where large shade and strength variations are possible from one batch of ink to the next. Figure 8.3 shows that by using hyperdispersants to wet-out pigments effectively, it is

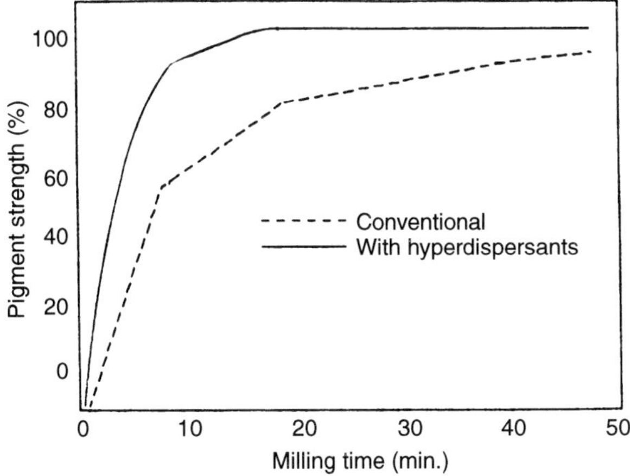

Figure 8.3 Effect of hyperdispersants on raw material costs: monolite green GNX in NC packaging ink. Faster rate of strength development.

possible to reach the flat part of the strength development curve faster than with conventional vehicles. As a result, the color strength is much less dependent on milling time and temperature, and quality control is enhanced.

(c) *Processing costs.* Although ink production costs are still relatively small compared to raw material costs, they still represent an area where cost savings can be made. By incorporating hyperdispersants into the millbase it is usually possible (indeed, often essential) to increase pigment loadings by 50% (Figure 8.2). It will also become apparent that the millbases only require half the usual milling time to obtain an off-gauge reading. The reasons for their faster rate of dispersion are threefold. First, the higher pigment loading increases the incidence of particle–particle impacts leading to breakdown of agglomerates. Second, the hyperdispersants were designed to wet-out and give strong adsorption to the pigment; and third, the polymeric chains stabilize the dispersed particles, preventing any reaggregation during the milling process. The combination of higher pigment loadings and shorter plant occupation time will result in significant savings in processing costs.

(d) *Rheology.* One of the constant quests in the ink industry is to improve ink rheology and overcome the thixotropy due to the pigments. This thixotropy or puffiness is normally associated with inter-particle forces and generally increases in line with pigment surface area. The problem can be mitigated either by using surface-treated pigments, or by dispersing the pigments with sufficient hyperdispersant to build up a monomolecular layer around each particle. For example a low pH, high surface area black can be dispersed at 24% pigment content by using 25% polymeric hyperdispersant on weight of pigment in a flexographic ink vehicle, and this millbase displays similar rheology to a conventional millbase at 15% pigment content (Figure 8.4). Hyperdispersants can be used in this way to improve rheology in all ink formulations, whether they are liquid inks, past inks or solvent-free UV cured coatings.

One interesting application for hyperdispersants is in flush color production. As offset ink manufacturing becomes more automated, demand for flush colors of lower viscosity is increasing. To achieve these low-viscosity flushes, the hyperdispersant should be added to the presscake in the flushing unit before adding the flushing vehicle. On processing, the water breaks clear rapidly. If the presscake contains pigment that has been surface treated with hyperdispersants, the water break is usually faster, and a larger amount of water breaks clear due to the more hydrophobic nature of the pigment. This leads to lower conversion costs. The flushes are much easier to pump and they mix more readily with other ink components.

(e) *Storage stability.* Although instability of inks on storage is not a common problem today, when it does occur it can be costly. Carbon black inks can gel on prolonged storage, and AAA yellows can lose strength and transparency in

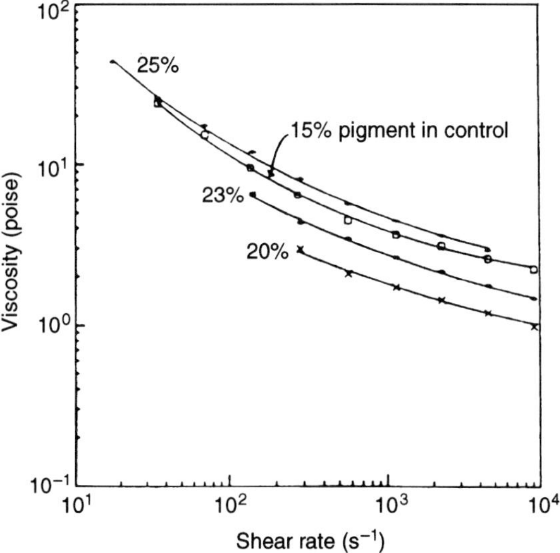

Figure 8.4 Effect of hyperdispersants on rheology: millbase viscosity of special black 4 in nitrocellulose and ethanol. Control contains 16% nitrocellulose solution; others contain 10% nitrocellulose solution plus hyperdispersant.

publication gravure inks that are stored under warm conditions. These failures are situations that can often be prevented by stabilizing the pigment dispersions more efficiently.

Carbon black pigments can be stabilized very easily, provided that the quantity of hyperdispersant used is in line with the surface area of the pigment (2 mg/m^2). For carbon blacks of, say, 50 m^2/g surface area, the use of 10% hyperdispersant on weight of pigment should give excellent viscosity stability on storage.

AAA yellows have a molecular weight of 630 and it is not uncommon for these to dissolve in toluene based publication gravure inks. When the pigment recrystallizes from solution, the particle size is much longer, resulting in water, less transparent inks. However, recent tests have shown that if the pigment is stabilized with a monomolecular layer of hyperdispersant (12 to 15% of pigment weight), excellent color and viscosity stability on storage can be achieved.

(f) *Water based inks.* Water based ink technology presents an attractive strategy for achieving lower VOCs. New hyperdispersants are now available for use in water based systems. Classical surfactants are used in lower quality water based inks to aid pigment dispersion, but as they remain in the dry film they can cause problems of water sensitivity. Surfactants may also enhance foam production during the milling stage, which can create manufacturing difficulties.

The alternative approach of using hyperdispersants to wet-out and stabilize pigments has advantages. These agents are strongly anchored to the pigment surface and are not active at the water/air interface. They are less capable of producing foam and one must think they are inherently better in terms of providing long-term storage stability if stabilizing agents are not labile. Once again, an optimum dosage of hyperdispersant on pigment (approximately 2 mg/m^2) should be determined, and it is also essential that any competing species, e.g. anti-foams and adhesion promoters, should be removed from the millbase or kept to a minimum.

There is one further benefit to using polymeric hyperdispersants in water based inks, namely their ability to act as steric stabilizing agents during the drying phase. A normal surfactant is designed to work in water, yet it is well known that, during drying, the film becomes more and more hydrophobic while still maintaining sufficient mobility for pigment–pigment interaction. Under these circumstances, a traditional surfactant is of little use if it becomes insoluble in the coalescent solvents. The physical or steric barrier surrounding the pigments may then collapse, which allows the pigments to migrate together, resulting in flocculation or loss of gloss as the film dries.

By contrast, the use of a hyperdispersant that is compatible with both water and the organic co-solvents will ensure good steric stability during the drying/curing stage. The presence of the hyperdispersant will minimize pigment–pigment interactions leading to more attractive inks of higher gloss and better viscosity stability on storage.

8.1.5 *Summary*

One of the positive aspects of the recent recession is that it has forced companies to re-examine their operating costs, seeking new ways to obtain better value from their raw materials. For this reason, hyperdispersants now have a proven track record. Not only will they generate higher strength per pound from the process colors, but they reduce processing costs and increase plant capacity, as higher pigment loadings can be dispersed in a given mill.

With regard to solvent emissions, as government rules and customer demands push formulators to reduce VOCs without sacrificing quality, new technical challenges are occurring. This will create new opportunities for using hyperdispersants to solve these problems.

8.2 Defoaming agents

A big, foamy froth is desirable in many things in life: atop a frosty mug of beer, in one's washer, or in a cappuccino cup on a cold winter's day. For these and many other applications, foam and bubbles are positive effects; but when it comes to ink, foam is the last thing a printer needs.

A number of problems can occur during printing, such as color and printing appearance faults, drying problems and problems in the reel. However, a manufacturer's trouble shooting may be most needed in the area of foaming, those tiny bubbles that can destroy an otherwise beautiful printing job. This is where defoamers are called into play.

Although defoamers are used in all types of printing, 'foaming is a condition which might happen when a monomolecular film of a surfactant, which is fairly insoluble in the bulk of the liquid is present on the surface of the liquid,' according to Reference 2. In defoaming, stable bubbles are prevented from forming so that foams cannot be created. (Prevention is, of course, better than fixing it afterwards, but where foaming does occur, it should be dealt with immediately, the Manual informs.)

Defoaming agents defoam by either being good solvents for the surfactant, causing the film to be dispersed into the bulk of the liquid, or by adding a small amount of a material that is immiscible with the system, thereby lowering the surface tension of that system, and causing the bubbles to break [2].

Defoaming an ink is by far the easiest task that can be asked of a defoamer manufacturer. 'However,' explains Dr Qun Chu, research scientist with Ultra Additives in Paterson, NJ, 'once we get into specific ink uses, then a defoamer must be formulated to perform various valued added tasks in addition to controlling foam.'

'Formulators should remember that not all defoamers work the same way. Some help to prevent foam formation while others are primarily bubble-breakers. By using two defoamers based on different chemistries, one in the grind and one in the letdown, it is possible to get better results than any single defoamer,' says M.C. Frantz, Northeast Technical Sales Manager, Daniel Products Company.

Most ink manufacturers agree that aqueous inks are the place where defoamers are needed and used most. 'Systems containing water naturally foam,' explains S. Turner, Industry Manager, Inks and Powder Coatings Division, Ashland Chemicals, 'which must be controlled to simplify manufacturing and processing, as well as insuring the quality of the final ink film.' (Surfactant is added to water based formulations, enhancing the spreading and film properties. The surface tension is in turn reduced and the tendency to form stable gas is increased.) Along with its less specialized products, Ashland has several products to help fight foaming. It has a full line of speciality defoamers for both Dow Corning, for which it produces silicone defoamers and antifoams, and Witco, which receives its Bubble Breaker non-silicone defoamers.

8.2.1 *Foam*

But let us not give water bases a bad rap. There are several advantages to water based inks. Owing to their low evaporation rate, the inks are stable on the press without drying on the stero or in the duct, giving constant viscosity. Water does not attack stero material, including photopolymers; the inks present no fire hazards and are economical as water is used as a reducer and cleaning solvent. There are also other areas where there is said to be a genuine need for defoamers. Other applications, such as oil-based and high solids and radiation curable inks often require defoamers as well. Here, their function is to help the dispersed gas to flow through a higher viscosity film and release at the film surface before the ink cures.

Surfynol (R) defoamers (Air Products and Chemicals), used across the board in water based gravures and flexographic printing inks, are being used even in lithography. 'Surfactants play a vital role in fountain solutions where low dynamic surface tension and low foam are vital to the founts' ability to wet the non-image area of the printing plate, without the use of alcohol,' says M.N. Sutovich, senior chemist with the Surfynol Surfactants Group, Air Products. The Air Products and Chemicals' product line includes classic defoamers, low/no foaming grind aids and low/no foaming surfactants for both the grind and let-down stages of water based ink formulating.

8.2.2 *New defoaming products*

Like any industry, the ink defoamer industry has seen many changes in the past few years. One such change is the ongoing effort to reduce mineral oil and VOC content. According to Ashland Chemical Company, the most dramatic change in the defoamers industry has been the steady increase in the number of water based inks themselves. 'This increases the need for defoamers in the market,' says S. Turner of Ashland.

Environmentally, this change has been for the better and companies may benefit from their taking part in this effort. When faced with a choice, a product that is socially conscious may be more appealing to a printer.

The recent upsurge of environmental concern has sparked the development of more and more uses of water based inks. However, environmental groups are calling for yet more 'friendly' agents in water based inks, in addition to the ones already in them. Although water bases may not be the way to go when dealing with publication gravure, such as magazines, they are, right now, good for other uses, such as wrapping paper. More and more research and development time and money are being devoted to water based inks. In fact, C. D'Amico of Ultra Additives believes its time has come. 'Water based defoamers are the future,' he said. 'Before the year 2000 everything will be water.' Ultra Additives has a full line of defoamers tailored for several different markets.

Current work is being done to reduce the volatile organic portion of defoamers, according to Frantz of Daniel Products. 'The ultimate goal [is to] produce effective defoamers with zero VOC.' Frantz calls for more change for the future. 'For flexo inks, efficient defoamers with long-term stability are needed that won't cause plate swelling.' DPC is currently working on both projects, along with their range of Dapro Foam Suppressors, silicone-free and silicon-modified defoamers.

Another major change is being driven by quality. Dr Chu of Ultra Additives states 'Ink makers are no longer tolerating defoamers that must be used at 1 to

2 percent of the ink formulation, only to find additional defoamers must be added press side by our customers. The ink maker today sees the advantage of cost effective defoamers that are only added once to the ink formulation. The investment up front pays dividends in shorter ink production time and maintenance free inks with superior print quality.'

8.2.3 *Future defoaming agents*

What does the future hold for defoamers? Besides the strides in progress in the past 5 years, there are still more plans in the works for the bubble-buster industry. Ultra Additives is going forward with a new series of defoamers, building on their already successful Dee FO PI-12, Dee FO PI-35 and Dee FO 3010E/50.

Air Products, Daniel Products and many other manufacturers are concentrating on the area of low VOC, water based inks. 'Observing the movement to low VOC and FDA compliance water based inks, Air Products has introduced several new defoamers and surfactants within the last 2 years, with low VOCs and with FDA compliance,' says Sutovich. Surfynol DF-70 and DF-75 defoamers and Surfynol SEF and 504 surfactants are examples. Air Products currently has a new program in the research and development stage to synthesize new and more effective defoamers and surfactants for the water based market.

Daniel's goal is also to produce effective defoamers with zero VOC. Current work is being done there to reduce the volatile organic portion of defoamers. On Daniel's future products, Frantz said, 'For flexo inks, efficient defoamers with long-term stability are needed that won't cause plate swelling.'

On the quality side of the matter, companies like Zinchem are concentrating on perfection, reducing printing defects such as pinholes and fisheyes, which are increasing with the upswing in of water based systems.

References

1. Schofield, J.D. (1992) *Handbook of Coatings Additives*, Vol. 2, p. 71.
2. Leach, R.H. (ed) (1994) *The Printing Ink Manual, 4th edition.*

9 Design and formulation of water based ink systems

P.J. LADEN and S. FINGERMAN

9.1 Factors to be considered when designing a water based ink system

This chapter on design and formulation of water based ink systems is not intended to be a complete compilation of all techniques or ideas. Rather consider it as a guide to get through some of the difficult formulation aspects that we often encountered. In retrospect, reading the many technical articles that have been introduced into the field, a study of design parameters and formulation keys has not been found. It is to be hoped that some aid to the formulator can be found in some of the observations discussed in this chapter.

To keep the multitude of variables as simple as possible in the design of water based ink systems the formulator must bear in mind that there are only four individual categories of components that make up an ink. These are polymers, pigments, volatiles (water, organic solvents) and additives (defoamers, dispersing aids).

The relationships between these components is outlined in Figure 9.1. The relationship of raw materials to the different categories of ink, such as base ink systems, heavy ink systems, standard ink systems, press ready ink systems, and varnish or overlacquer systems is outlined in Figure 9.2.

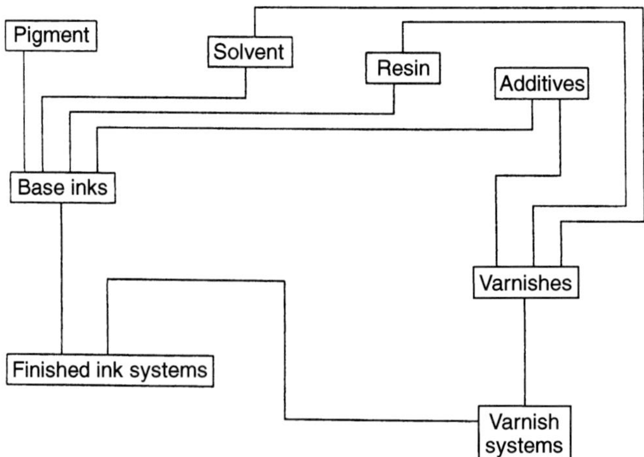

Figure 9.1 Flow chart representing the relationship between the categories of components used to make up an ink.

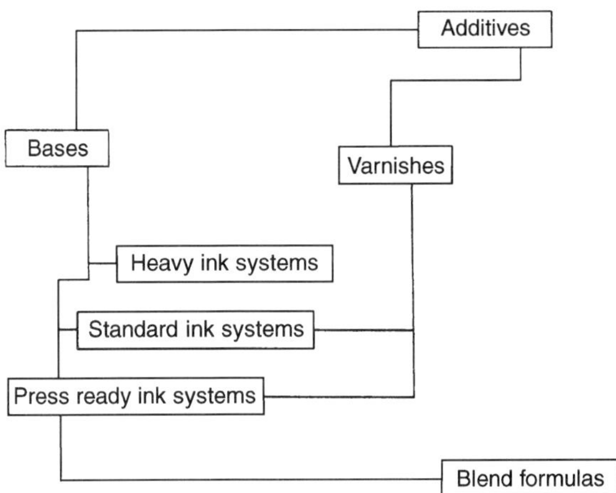

Figure 9.2 Flow chart representing the relationship between the raw materials and different categories of ink.

Virtually all the ingredients that comprise an ink system can be listed under one of the above categories. There are, fortunately, hundreds of materials to choose from for each category. This gives the formulator a very wide latitude to select from for the unique end application that is often required.

9.1.1 *Base ink systems*

A base ink system is one where the pigment binder ratio is very high. This is essentially a very unstable system because of this but it is the most economical way to manufacture large amounts of ink. This type of ink system is used by blending with various varnish systems to lend a great degree of versatility to base ink and to keep inventory levels down by limiting the varied types of polymers that may have to be utilized for different applications.

Generally speaking a base ink system is never sold out of house to customers who wish to do their own blending and is only kept for internal use. A typical base ink formulation with all physical and chemical variables is given in Table 9.A.1. All of these formulas are built up in a ridged structural hierarchy. The lowest is the Chemical Abstract System of C.A.S. file. The raw material file is built from this and then the formula file is structured from the raw material file.

9.1.2 *Extenders*

The next most important tool that the formulator has to work with is the extender. This is simply a mixture of polymer, volatile and additives that will be used to increase the pigment binder ratio which will allow better printability, better rheology and longer periods of stability. A simplistic formula for an extender with all the chemical and physical properties is given in Table 9.A.2.

9.1.3 *Heavy ink systems*

A heavy ink system is simply the base ink system with approximately 10% of extender added to it. This allows for more stability and better flow for more efficient handling or transfer in the manufacturing environment. The physical and chemical parameters are given in Table 9.A.3.

9.1.4 *Standard ink system (flexographic)*

Standard ink systems are those that are traditionally sent to customers. As is demonstrated in the computer analysis this usually is a fifty blend of both the base ink and the extender system. This mixture yields a traditional good blend for viscosity and strength. The ink formulator generally matches a color and increases the strength of the formula by 10%. This increase in strength will allow for individual differences in the presses upon which the ink will be applied. Some of the common problems that can be experienced include variations in anilox configurations, substrate differences, variations in line speed, doctor blade or nip metering, and surface tension variations.

Flexographic standard ink data are given in Table 9.A.4.

9.1.5 *Standard ink system (gravure)*

There are several important differences between gravure and flexographic ink systems. One of the most important is that gravure is direct transfer to the substrate. The ink film is split only once from pickup in the fountain to the substrate (Figure 9.3), as compared to the flexographic application method of printing where there are three transfers of ink until application to the substrate. This means that a typical water based gravure ink as applied to a non-porous substrate utilizing the cylinder configurations 175 line screen/45° screen angle/ 32 microns deep, which will deposit approximately 12 to 15 microns of dried ink on the substrate.

If we consider the flexographic water based application of 200 line screen, Dr blade, line speed, this configuration will yield approximately 5 microns of dry ink on the substrate. The end result is that there is generally 6 to 10% pigment load in a gravure ink while there is approximately 15 to 18% in a flexographic ink. The physical constants of the five types of inks are compared in Table 9.A.5.

Because of the pigment loading difference there is also a relationship regarding viscosity. In most cases the higher the pigment load the higher the viscosity. The formulator must also be aware of the morphology of the particle size distribution curve. The more fine particles the more surface area and a greater increase in viscosity will result. Gravure standard ink data are given in Table 9.A.6.

Before even the first thought of formulation is initiated several other factors have to be determined.

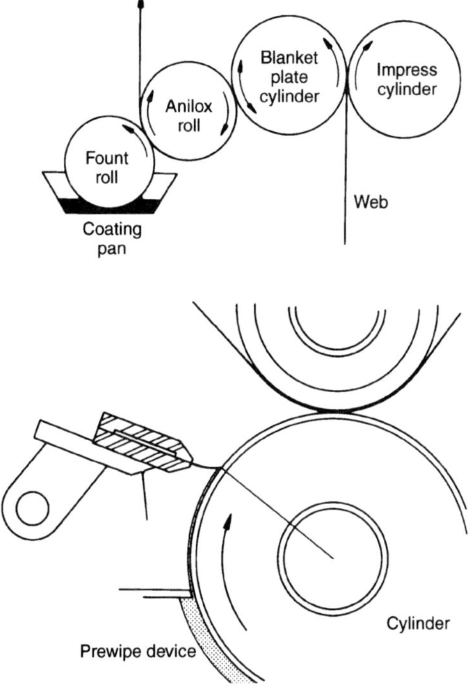

Figure 9.3 Schematic representation of a flexo press.

(a) *The printing press.* Typical factors include:

1. How fast is the job to run?
2. Gravure or flexographic?
3. Cylinder configuration
4. Drying capabilities of the press
5. Roll to roll or inline dyecut
6. Ambient variables of local weather conditions
7. How sophisticated are the press operators? A complex ink will not perform properly if there is not sufficient press-side expertise to administer it correctly

These are just a few of the many questions a formulator must pose concerning the piece of equipment on which the ink will run.

(b) *The substrate.* Typical factors include:

1. Porous or non-porous (film, foil, paper or board)
2. Surface tension. Usually anything over 38 dynes/cm^2 is appropriate
3. Chemical make-up of substrates (LDPE, HDPE, EVA)
4. Downstream specifications for printed material
5. Ultimate end use of the printed material

After these initial items of information have been determined then the preliminary steps can be initiated for the outline of an ink.

(c) *The ink.* Putting together an ink system can be divided into three distinct processes.

Design. All variables and end use applications are determined. Operating parameters are analyzed and substrates evaluated for printability. Downstream specifications taken into consideration and end use evaluated.

Formulation. The raw materials mentioned previously are selected and put together to determine suitability or fit as mandated by the variables that are not easily changed. The environmental impact of the ink must be determined at this point.

Manufacture. Small amounts of ink are made in the laboratory, evaluated, then made up on pilot plant equipment to determine if scale-up will call for further reformulation. Evaluations at press-side are mandatory and the data generated should reflect any shortcomings in the ink performance. A general form for the recording of press-side data follows.

9.1.6 *Formulation*

Having noted the most important variables that the formulator must consider, and having an understanding of the physical and chemical parameters that are indicative of the base ink, heavy ink, and extender types basic formulation can begin. The single most significant fact the formulator must always keep in mind is the 'KIS' (keep it simple) factor. Having a large number of raw materials in the formulation will do two things. First, it will be very difficult to solve a problem should one arise either in production or on the press. Quite often the addition of multiple additives (e.g. defoamer and surfactant) will solve one problem and create another. Experience has shown that the most simple inks are usually the best performers.

Secondly a formulator who utilizes a large number of components in a formula will be very unpopular with the manufacturing supervisor. If numerous raw materials are necessary for a reasonably complicated application, then that is what must be done. However, I cannot emphasize too strongly that the best formulas, both in manufacture and on press are usually the most simple ones.

Two of the most difficult problems with which a technician has to work are the divergent nature of the polymers available, and their preparation. In this context we must look at the difference between solution and emulsion polymers. Very basically an emulsion is a combination of organic polymers in an inorganic medium. The very different nature of these two types of compound do not allow 'wetting out' of pigments or substrates. There are exceptions to this but not many. Solution polymers lend themselves more readily to the 'wetting out' phenomenon.

Solution polymers are usually high in viscosity and low in solids. Emulsions are usually high in solids and low in viscosity. Solutions are low in molecular weight compared to emulsions. There are more differences but these are essential to acknowledge when formulating. Most often combinations of the two types, solution and emulsion, yield results superior to either by itself. This is what is commonly called the fortified emulsion approach. In formulating for water based applications, especially on non-porous substrates, or MG or MF paper, fortified emulsions will aid in the lay and transfer of ink to substrate. Examine the generic formula for the extender as exemplified under the master formula list. This is a workable formulation for a let down for the base or heavy ink systems.

One of the greatest difficulties to overcome in formulating for water based ink systems is inherent in the very nature of the raw materials with which the formulator has to work. This problem is called thixotropy. Emulsions are thixotropic, and with the addition of pigment and fillers the rheological problems sometimes becomes acute. Many studies have been done to determine the relationship of thixotropy to ink transfer. The summation of these reports can be indicated as, the more thixotropic the ink system the poorer the transfer capabilities. At the present time this statement may seem to be obvious to most, but at the time when water based technology was in its infancy and the entire industry was using solvent based solution technology, this was big news. This is not overly important when formulating for gravure applications because, by its very press configurations, a comparatively thick film is generated. However, it is critical when utilizing flexographic formulations. Fortunately this can be easily overcome by using several different approaches.

The first approach is to change the pigment to binder ratio if possible. At this point if an acceptable viscosity curve cannot be developed a change in pigment manufacturer would be indicated. Pigment manufacturers have not yet accepted the concept of custom particle size distribution curves for segregated markets. The author feels strongly that this will be the next logical step that pigment producers will introduce into the marketplace. Should this become available several valuable tools will then be available to the formulator that will aid in the manufacture and use of inks.

For example, air classification equipment for modification of particle size distribution is on the market now. Should a pigment manufacturer decide to invest in one the following, cuts could be made for the following markets:

1. ultrafines, 1 micron or less – emerging pigmented ink jet technology
2. fines, 1 to 10 microns – fluid ink applications
3. coarse, 10 microns and up – paint and coatings industry

This would be a natural extension of an existing product line and it would prove to be much more profitable for all. To do this would result in a win-win scenario for both the manufacturer and the user.

The second approach would be to introduce a solution polymer whenever possible. The drawback to this is that the drying time for the ink is reduced.

Depending on how efficient the dryers are at the printer, the result with too much solution polymer in the formulation may be poor drying and ultimately, blocking. Care must be exercised when utilizing this option.

A third choice that could be used would be the addition of flow agents and surfactants. Based on past experience this is usually the last choice but still a viable one. The reason for this is that in using these types of products great care must be used in manufacture so as not to exceed the amount indicated as small amounts tend to make for very large differences. If a little is good more is not usually better! Additionally many of the more sophisticated additives are sinusoidally functional, in which case a ladder type of evaluation would be indicated for its effective use.

Another area of formulation that needs to be addressed is the determination of the effect of adding solvents such as alcohols or glycolethers to the performance level of the dried ink film as compared to its solvent based counterpart. The following study regarding this question has been commissioned and preliminary results have been correlated and graphed for easy visual analysis.

As mentioned in the preface of this book the ink industry is continually changing and developing, it is to be hoped that the preceding chapter has helped in guiding formulators over some of the rough spots.

9.1.7 *A method for determining the performance of water based inks*

Water based ink systems do not have comparable performance levels to solvent based ink systems. The technology is emerging but at present performance on selected substrates for water based ink is poorer than on solvent based ink systems.

A study was carried out to determine whether increasing the amounts of solvents (VOCs) increased performance levels to those of solvent based ink systems. Two ink systems and six substrates, PET, EVA, LDPE, HDPE, cellophane, and metalized mylar were used. The testing methods included: adhesion, block test, Sutherland rub, cold water soak, ice crinkle, and wet rub.

Initially solvent based ink was run on the selected substrates to determine the solvent ink performance envelope. The solvent based run established the performance envelope for the substrates and all other variations of the water based ink systems were compared to this baseline.

(a) *Procedures.* Solvent based ink was put on the press and diluted to press viscosity with the appropriate solvent. Once press viscosity was obtained the five designated stocks were printed with the designated pattern. The water based ink was put on the press and diluted in three different ways:

1. Water
2. 75/25 Water/solvent
3. 50/50 Water/solvent

The five selected substrates were printed with the designated pattern.

Test prints from the solvent based runs and the water based dilutions at the three converters were tested. Results are presented in Figures 9.4 to 9.10. Values for the test results were assigned on the scale: 1 = poor; 2 = average; and 3 = excellent. Analysis for both inks was integrated and averaged into a spreadsheet from which the figures were constructed. This technique removed any differences that might be encountered during formulation evolution.

(b) *Summary of results*

PET as the substrate (Figure 9.4). In overall performance levels the dilution of 75/25 water/solvent has higher levels of functionality more closely approximating the baseline standards as developed for the solvent-based ink system.

EVA as the substrate (Figure 9.5). Overall functionality on this substrate is achieved by the 50/50 blend of water. On EVA while the performance levels are still below those of the solvent based envelope, this blend consistently yields better results.

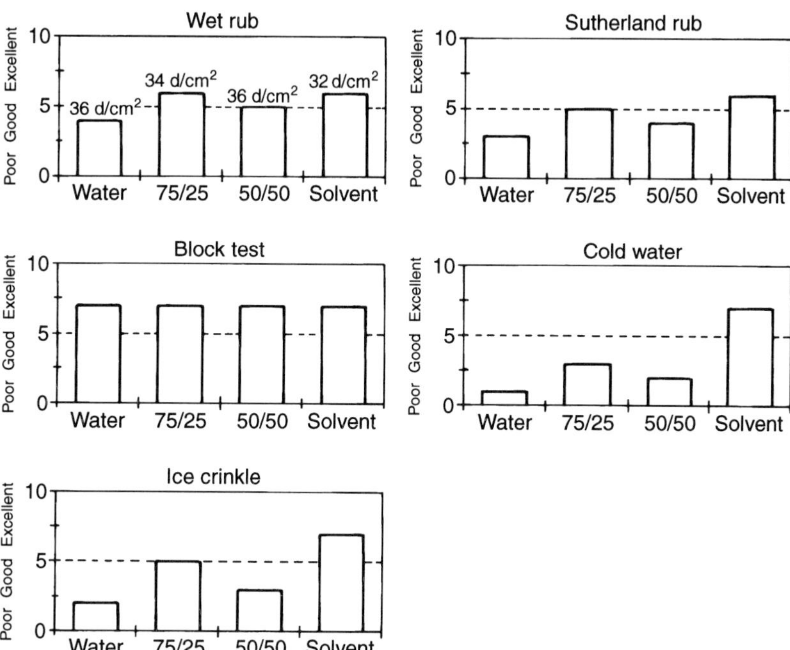

Figure 9.4 Determination of the performance of a water based ink using PET as the substrate. The various tests conducted (e.g. wet rub) are indicated for each panel. The ink type/dilution (e.g. 75/25) is given below each bar; and the surface tension of the stock (e.g. 32 dynes/cm^2) is given above each bar.

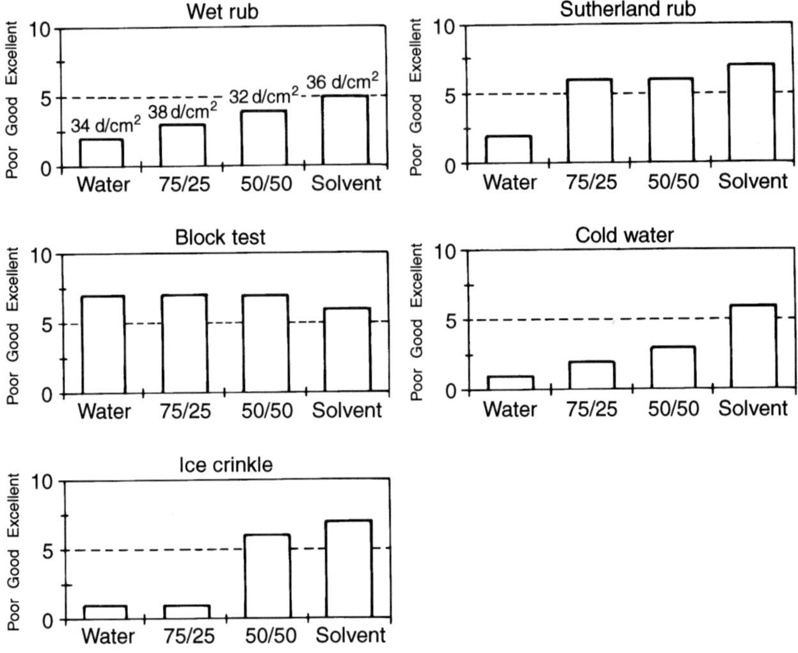

Figure 9.5 Determination of the performance of a water based ink using EVA as the substrate. Details are as given in the caption to Figure 9.4.

LDPE as the substrate (Figure 9.6). Results are inconsistent in this graph. In some tests (cold water, Sutherland rub) the 75/25 mix performs better. In the other tests the 50/50 blend has better results.

HDPE as the substrate (Figure 9.7). Overall levels indicated that the 50/50 blend yields the best results although the 75/25 blend had slightly higher levels of functionality in the wet rub test.

Cellophane as the substrate (Figure 9.8). Results were inconsistent in this area of application. Total overall performance levels were slightly in favor of the 75/25 dilution blend in most test areas.

Metalized mylar as the substrate (Figure 9.9). On this substrate the best total overall performance was developed by the 75/25 dilution blend. This ink system with 75/25 blend in some cases exceeded the solvent base line parameters.

Results for overall performance are summarized in Figure 9.10. There can be no doubt that the addition of alcohols to the water based ink systems increases the performance envelopes on almost all substrates. Interestingly enough this is not a linear function and performance increases are not uniform on all substrates

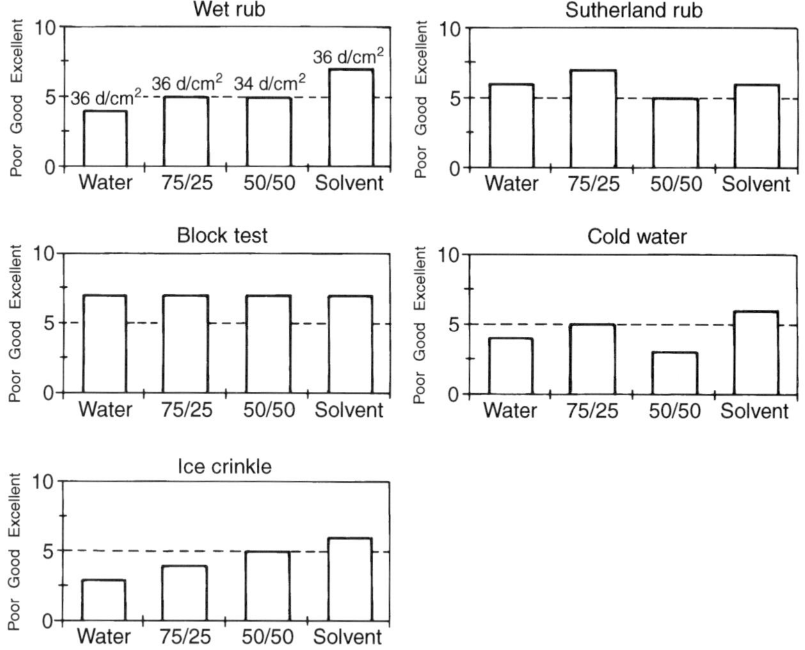

Figure 9.6 Determination of the performance of a water based ink using LDPE as the substrate. Details are as given in the caption to Figure 9.4.

tested. With this information it will be possible in some cases to formulate an ink with increased VOC levels for specific purposes and for targeted markets.

9.2 Scratch-off inks and coatings

To understand the technical properties of a good scratch-off ink and coating, one must first understand the end use characteristics that are important to the finished product. Once this is clarified, the chemist can then design the coating to work in the specific application method that will be used by the printer. Points that must be considered include:

1. The purpose of the coating
2. The method of application that the printer will use
3. The stock or substrate on which the message will be printed
4. The drying conditions of the system being used to apply the coating
5. The properties required of the finished coating

9.2.1 *Hiding properties*

A coating guarantees that the 'hidden' message is not easily identified until the person scratches the coating.

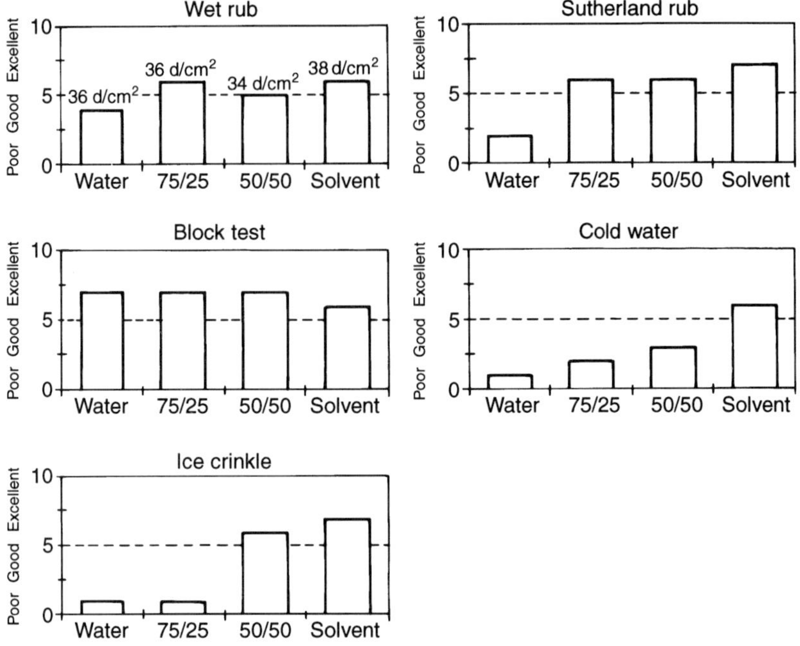

Figure 9.7 Determination of the performance of a water based ink using HDPE as the substrate. Details are as given in the caption to Figure 9.4.

To accomplish the task of building into a coating the property that will offer the best 'hiding' characteristics, we must turn to the use of fillers or clay additives. While these additives will lend a degree of hiding to the finished coating, the need to place them into the coating formulation is the challenge that confronts the chemist. The reason is that the lack of wetting out or ability to place these inorganic compounds into solution/dispersion is critical. This is best accomplished through the use of specialized wetting agents that assist in the dispersion of the fillers or clay. A brief look at a generic formulation will reveal that it takes less than $\frac{1}{2}$% to accomplish this process.

9.2.2 *Application method for scratch-off inks and coatings*

The methods may vary from flexo printing, screen printing, roll-over knife coating, etc. Each of the printing applications have their own technical questions that will impact on the use of a scratch-off coating owing to the chemical composition of the coating and its physical properties. More of this discussion will be noted when we discuss the chemistry of the product and its composition.

Because in-line finishing (flexo printing) units used to apply coatings will vary considerably both from company to company and even from press to press,

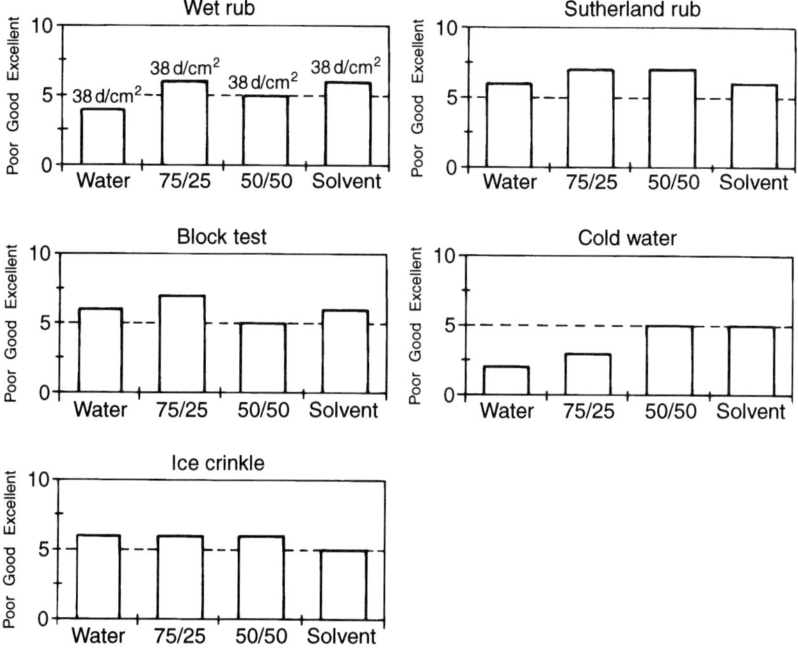

Figure 9.8 Determination of the performance of a water based ink using cellophane as the substrate. Details are as given in the caption to Figure 9.4.

the scratch-off coatings are usually supplied at a higher range of viscosity in order to allow the pressman to thin down the coating to their specific needs. This is best accomplished through experience on the part of the pressmen. For this reason, immediate thinning of the coating is sometimes required. It is a simple matter to dilute with solvents that the supplier recommends. For solvent based scratch-off coatings this can be a blend of aromatic and aliphatic solvents. A recommended composition would be a blend of 75% aliphatic and 25% aromatic. This combination offers the printer the necessary thinning or viscosity properties and the ability to dry quickly after application when going into the drying phase.

It is necessary that a printer has in place a circulation system to allow for proper mixing at all times in the ink pan. Since most printers use a Zahn cup for measurement, it is a simple matter to have the pressman add the solvent slowly (in small amounts) until the desired viscosity is achieved. For many of the solvent types of metallic scratch off coatings, the recommended viscosity using a #5 cup is between 33 and 20 s. The range is due to the ability of the pressman to put down sufficient coating to successfully hide the message, in addition to taking into consideration the paper stock in question. Over-dilution at this point will render the coating's hiding power useless and care must be taken to achieve the right balance on the press.

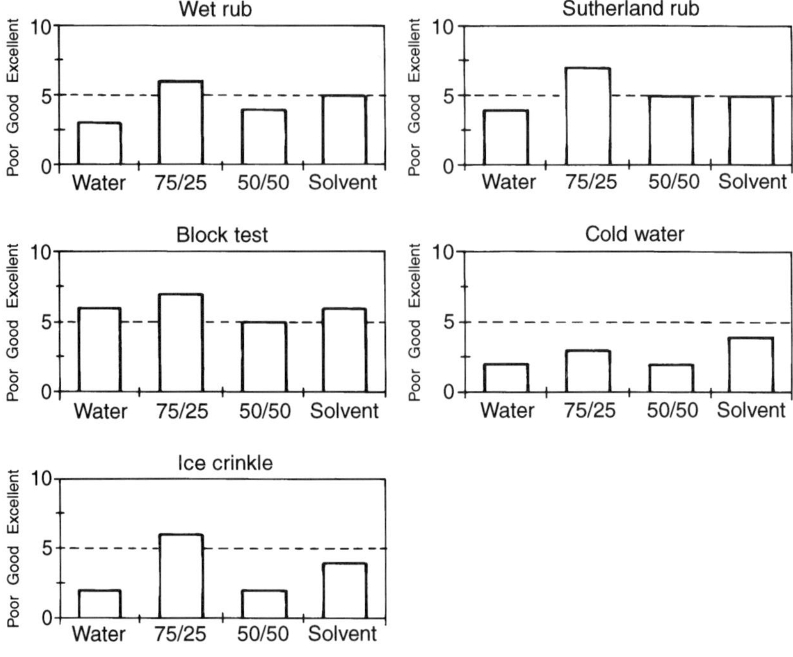

Figure 9.9 Determination of the performance of a water based ink using metalized mylar as the substrate. Details are as given in the caption to Figure 9.4.

9.2.3 *Substrate on which the message will be printed*

Depending on the stock in question, the chemist will have to vary the composition of the final coating to accommodate the different systems.

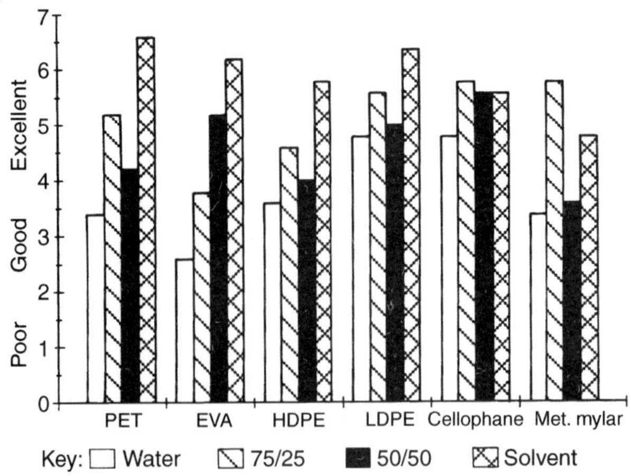

Figure 9.10 Overall performance averages for the five substrates and four ink types/dilutions.

The selection of the paper to be printed is a key variable in obtaining good results with scratch-off inks and coatings. Owing to the fact that paper will absorb the solvents in the coating thereby causing the coating to become part of the fiber of the paper, one must either use a highly coated sheet or apply some type of release coating to the surface prior to applying the scratch-off material.

The purpose of the scratch-off coating is to allow it to be removed easily once it is applied and dried properly on the stock. If it is absorbed into the stock it will defeat the product's use. Naturally the paper quality, both the screen and the color of the printed message (hidden), the thickness of the sheet and whether there is any heavy ink coverage on the reverse side, all help to determine the degree of hiding desired.

Another concern is the possible degree of stain that could occur if the paper used is of inferior quality. This is a function of the paper absorbing the solvent and some of the pigment into the top-most layer of the substrate. To avoid this, the chemist must build into the formulation sufficient film-forming properties to minimize absorption while effecting good coverage of the ink or coating to the surface. The use of coated stock is generally recommended along with the application of an undercoat of release product. In many plants, this is accomplished via a UV undercoating laid down prior to the scratch-off material.

When printing on non-absorbent stock, i.e. films, foils, etc. the key point to consider is the ability of the scratch-off coating to wet out the stock in question and print a smooth application. When and where possible, either a primed surface or a treated one is highly recommended.

9.2.4 *Drying conditions*

The need to understand the total drying characteristics of the coating and the specific system employed by the printer must be taken into consideration by the formulator.

The drying conditions of the system employed must be understood in order to ensure that the finished coating is dried properly to avoid marring, premature scratching in the converting operation, etc. Since scratch-off inks are heavily metallic pigmented inks, care must be taken to avoid residual solvent that can often occur during improper drying stages. This is the result of 'skinning' over during the drying phase. Once this happens, the coating appears to be dry to the touch; however, the residual solvent trapped in the coating will eventually find its way to the surface and cause off-setting or downstream marring, rendering the product non-usable.

The best system is one in which there is a graduated zone heating effect to remove the solvent laden air initially while passing through the drying tunnel. The key here is to have the air surface over the coating removed through proper movement and flow. The ability to impinge sufficient air on the coated surface and remove it from the area is critical to the success of any curing system. This is the same for solvent and water based systems.

9.2.5 *Properties required of the finished coating*

The finished coating must, in addition to being see-through proof, have a degree of abrasion resistance, some amount of slip to avoid marring or scratching during converting operations, and others that require building into the formulation additives to obtain these characteristics. While there are many more technical matters to consider when designing a scratch-off coating, we can discuss them as we examine the five areas noted.

As previously discussed, the need to maintain high hiding properties while considering many of the factors noted in this section, are important features for the formulator to keep in mind.

9.2.6 *A formula for a scratch-off coating*

An example of a generic formula for this type of coating is given for information:

Pigment inorganic fillers, clay, etc. 18–21% metallic pigments (bronze, aluminum type), 28–32%
Solvents aromatic and aliphatic blends, 35–40%
Latex emulsion resin emulsion binders, 7–10%
Wetting agents thickeners, defoamers, surfactants, etc., less than $\frac{1}{2}$% (as needed)

As previously noted, the key to developing a successful product containing the chemical composition known as a scratch-off coating ink lies in the ability to properly select the right combination of the listed items. The chemist must know how to blend the different additives into a finished product that will perform on press, yield satisfactory hiding characteristics, adhere to the substrate in question and be priced to allow for continued value to the end user.

If you examine the key five elements identified in this section you will see that this is no easy task and requires a knowledge of the printing industry, chemistry and performance criteria to produce a product that will offer special properties needed in a scratch-off coating ink.

Appendix 9.A: data for design and formulation

Table 9.A.1 A typical base ink formulation, with physical and chemical variables

(a) **Blue base, master formula**

Ingredient	Description	Qty (lbs)	Qty (gal)	lb/gal	TNV	PCT
PBD-65	Pigment blue 29	30.0	1.53	19.57	100	30
S-2-7	Water	53.0	6.36	8.33	0	53
S-2-2	Inksolv sol 6264, 190 proof	5.0	0.72	6.91	0	5
XA-495	Surfynol 104H surfactant	0.5	0.06	7.91	100	1
XA-518	Dee fo 1020F	1.0	0.14	7.34	100	1
XA-537	Surfynol CT-136	1.5	0.17	8.82	89	2
WRA-1-55	Joncryl 678	7.0	0.84	8.33	100	7
A-6-6	Aqua ammonia 26%	2.0	0.26	7.84	0	2

Attribute	Weight	Volume	Attribute	Value
Quantity	100.000	10.085	Density (lb/gal)	9.916
Unit cost	0.991	9.827	(VOC (lbs/gal)	0.000
Volatile (%)	60.165	84.775	PB ratio	3.050
Solids (%)	39.835	15.225	PVC (%)	99.844
Pigment (%)	30.000	15.201	Mileage-mil dry	244.202 sq. ft
Exempt solv. (%)	110.000	130.786	RWM cost/sq. ft	$0.0402 per sq. ft

Material	Quantity	Wt (%)	Vol (%)	Description	Cat.	Volatile
S-2-7	53.0	53.0	63.10	Water	Sol	53.00
PBD-65	30.0	30.0	15.20	Pigment blue 29	Pig	0.00
WRA-1-55	7.0	7.0	8.33	Joncryl 678	Res	0.00
S-2-2	5.0	5.0	7.18	Inksolve sol 6264, 190 proof	Sol	5.00
A-6-6	2.0	2.0	2.53	Aqua ammonia 26%	Add	2.00
XA-537	1.5	1.5	1.69	Surfynol CT-136	Add	0.16
XA-518	1.0	1.0	1.35	Dee fo 1020F	Add	0.00
XA-495	0.5	0.5	0.63	Surfynol 104H surfact	Add	0.00
Totals	100.0	100.0	100.00			60.16

Category	Quantity (lbs)	Weight (%)	Quantity (gal)	Volume (%)
Pigment	30	30	1.53296	15.20181
Solvent	58	58	7.08614	70.27070
Additive	5	5	0.62462	6.19413
Resin	7	7	0.84034	7.33335
Totals	100		10.08406	

(b) **Blue base, formula analysis**

CAS number	Description	Wt (%)	Haz (%)	Haz (tot)	Hazard type	Org.
7732-18-5	Water	53.00	0	0.00	Not a hazard	Non
57455-37-5	Pigment blue 29	30.00	0	0.00	Not a hazard	Non
99999-99-9	Not known	9.70	0	0.00	Not a hazard	Non
64-17-5	Ethyl alcohol	4.49	100	4.50	Fire hazard	CG1
1336-21-6	Ammonium hydroxide	2.00	100	2.00	Hazard	PS1
109-60-4	*n*-Propyl acetate	0.27	100	0.28	Fire hazard	Skn
67-63-0	*iso*-Propyl alcohol	0.22	100	0.22	Fire hazard	Non
107-21-1	Ethylene glycol	0.16	100	0.17	Acute health	PS1
111-90-0	Glycoletherde	0.14	100	0.14	Fire hazard	Eye
	Totals	99.98		7.31		

Table 9.A.1 *Continued*

Material	Quantity	Wt (%)	Vol (%)	Description	Cat.	Volatile
7732-18-5	53.000	53.00	63.08	Water	Sol	53.000 n
57455-37-5	30.000	30.00	15.20	Pigment blue 29	Pig	0.000 n
9999999-99-9	9.705	9.71	11.66	Not known	Sol	0.157 n
64-17-5	4.495	4.50	6.45	Ethyl alcohol	Sol	4.495 f
1336-21-6	2.000	2.00	2.53	Ammonium hydroxide	Add	2.000 y
109-60-4	0.275	0.28	0.39	*n*-Propyl acetate	Sol	0.275 f
67-63-0	0.220	0.22	0.33	*iso*-Propyl alcohol	Cas	0.220 f
107-21-1	0.165	0.17	0.19	Ethylene glycol	Add	0.018 a
111-90-0	0.140	0.14	0.17	Glycoletherde	Res	0.000 f
Totals	100.000	100.00	99.86			60.165

Category	Qty (lbs)	Wt (%)	Qty (gal)	Vol (%)
Pigment	30.00	30.00	1.53295	15.19933
Solvent	57.78	57.78	7.05429	69.94356
CAS number	0.22	0.22	0.03348	0.33201
Additive	5.00	5.00	0.62462	6.19313
Resin	7.00	7.00	0.84033	8.33195
Totals	100.00		10.08567	

VOC: 1.48 lb/gal, Volatile: 60.17 lbs, Exempt volatile: 55.00 lbs
Total VOC: 0.51 lb/gal, Volume: 10.09 gal, Exempt volatile: 6.59 gal

(c) **Blue base, formula attributes**

Attribute	Blue base
Quantity (lbs)	100.000
Quantity (gal)	10.084
Unit cost (lbs)	0.991
Unit cost (gal)	9.823
Volatile (lbs)	60.165
Volatile (gal)	72.986
Solids (lbs)	39.835
Solids (gal)	27.014
Pigment (lbs)	30.000
Pigment (gal)	15.202
Exempt (lbs)	55.000
Exempt (gal)	65.625
Denisty (lb/gal)	9.917
VOC (lbs/gal)	1.490
PB ratio	3.050
PVC (%)	56.274
Mileage-mil dry	433.305
RWM cost/sq ft	$0.0227

Table 9.A.1 *Continued*

(d) **Blue base, structure**

				Percent
⟶ PBD-65		30.00% lb		
·	57455-35-5	100.00% lb	(Pigment blue 29)	
⟶ S-2-7		53.00% lb		
·	7732-18-5	100.00% lb	(Water)	
⟶ S-2-2		5.00% lb		
·	64-17-5	89.90% lb	(Ethyl alcohol)	(4.495)
·	67-63-0	4.40% lb	(*iso*-Propyl alcohol)	(0.220)
·	67-63-0	100.00% lb	(*iso*-Propyl alcohol)	
·	109-60-4	5.50% lb	(*n*-Propyl acetate)	(0.275)
·	9999999-99-9	0.20% lb	(Not known)	(0.010)
⟶ XA-495		0.50% lb		
·	9999999-99-9	100.00% lb	(Not known)	
⟶ XA-518		1.00% lb		
·	9999999-99-9	100.00% lb	(Not known)	
⟶ XA-537		1.50% lb		
·	107-21-1	11.00% lb	(Ethylene glycol)	(0.165)
·	9999999-99-9	89.00% lb	(Not known)	(1.335)
⟶ WRA-1-55		7.00% lb		
·	111-90-0	2.00% lb	(Glycoletherde)	(0.140)
·	9999999-99-9	98.00% lb	(Not known)	(6.860)
⟶ A-6-6		2.00% lb		
·	1336-21-6	100.00% lb	(Ammonium hydroxide)	

(e) **Blue base, physical data**

Item	Description	Unit	lb/gal	TNV (%)	Pig. (%)	Exm (%)	Cat.
S-2-7	Water	lb	8.33	0	0	100	Sol *
PBD-65	Pigment blue 29	lb	19.57	100	100	0	Pig *
WRA-1-55	Joncryl 678	lb	8.33	100	0	0	Res *
S-2-2	Inksolv sol 6264, 190 proof	lb	6.91	0	0	0	Sol *
A-6-6	Aqua ammonia 26%	lb	7.84	0	0	100	Add *
XA-537	Surfynol CT-136	lb	8.82	89	0	0	Add *
XA-518	Dee fo 1020F	lb	7.34	100	0	0	Add *
XA-495	Surfynol 104H surfactant	lb	7.91	100	0	0	Add *

CAS number	Description	Flash point (%)	Boiling range (deg) Lower	Boiling range (deg) Upper	Expl limits Lower	Expl limits Upper
7732-18-5	Water	NA	212.0	212.0 f	NA	NA
57455-37-5	Pigment blue 29	NA	NA	NA	NA	NA
99999-99-9	Not known	0 f	NA	NA	NA	NA
64-17-5	Ethyl alcohol	54 f	165.0	176.0 f	3.3	19.0
1336-21-6	Ammonium hydroxide	NA	NA	NA	NA	NA
109-60-4	*n*-Propyl acetate	55 f	215.0	215.0 f	2.0	8.0
67-63-0	*iso*-Propyl alcohol	54 f	180.0	181.0 f	2.0	12.0
107-21-1	Ethylene glycol	232 f	197.5	197.5 c	3.2	3.2
111-90-0	Glycoletherde	201 f	201.0	201.0 c	NA	NA

Table 9.A.1 *Continued*

(f) **Blue base, health data**

		ACGIH				OSHA
CAS number	Description	TWA (ppm)	Stel	TWA (mg/m^3)	Stel	Pel (ppm)
7732-18-5	Water	NA	NA	NA	NA	NA
57455-37-5	Pigment blue 29	NA	NA	10	0	NA
99999-99-9	Not known	NA	NA	NA	NA	NA
64-17-5	Ethyl alcohol	1000	0	1880	0	200
1336-21-6	Ammonium hydroxide	25	0	NA	NA	50
109-60-4	*n*-Propyl acetate	200	250	840	1050	200
67-63-0	*iso*-Propyl alcohol	400	500	983	1230	400
107-21-1	Ethylene glycol	50	0	127	0	50
111-90-0	Glycoletherde	30	0	NA	NA	NA

Table 9.A.2 A simplistic formula for an extender, with physical and chemical properties

(a) **Extender, master formula**

Ingredient	Description	Qty (lbs)	Qty (gal)	lb/gal	TNV	PCT
S-2-7	Water	50.5	6.06	8.33	0	51
A-6-6	Aqua ammonia 26%	1.5	0.19	7.84	0	2
WRA-1-55	Joncryl 678	5.0	0.60	8.33	100	5
XWRA-259	Acrysol 1-98	5.0	0.51	9.82	32	5
XWRA-401	Joncryl 89, 16360-4-6	10.0	1.15	8.66	45	10
XWRA-638	Zinpol 295	20.0	2.30	8.71	49	20
S-2-2	Inksolv sol 6264, 190 proof	5.0	0.72	6.91	0	5
S-2-56	PM solvent	1.0	0.13	7.69	0	1
XA-323	Johnwax 26	2.0	0.24	8.20	25	2

Attribute	Weight	Volume	Attribute	Value
Quantity	100.000	11.912	Density (lb/gal)	8.395
Unit cost	0.427	3.585	VOC (lbs/gal)	1.890
Volatile (%)	78.600	93.976	PB ratio	0.000
Solids (%)	21.400	6.024	PVC (%)	0.000
Pigment (%)	0.000	0.000	Mileage-mil dry	96.620 sq. ft
Exempt solv. (%)	75.520	72.998	RWM cost/sq. ft	$0.0371 per sq. ft

Material	Quantity	Wt (%)	Vol (%)	Description	Cat.	Volatile
S-2-7	50.5	50.5	50.90	Water	Sol.	50.5
XWRA-638	20.0	20.0	19.28	Zinpol 295	Res.	10.2
XWRA-401	10.0	10.0	9.69	Joncryl 89 16360-4-	Res.	5.5
S-2-2	5.0	5.0	6.07	Inksolv sol 6264, 190 proof	Sol.	5.0
XWRA-259	5.0	5.0	4.27	Acrysol 1-98	Res.	3.4
WRA-1-55	5.0	5.0	5.04	Joncryl 678	Res.	0.0
XA-323	2.0	2.0	2.05	Johnwax 26	Add.	1.5
A-6-6	1.5	1.5	1.61	Aqua ammonia 26%	Add.	1.5
A-2-56	1.0	1.0	1.09	OM solvent	Sol.	1.0
Totals	100.0	100.0	100.00			78.6

Table 9.A.2 *Continued*

Category	Quantity (lbs)	Weight (%)	Quantity (gal)	Volume (%)
Solvent	56.5	56.5	6.91605	58.06137
Additive	3.5	3.5	0.43523	3.65383
Resin	40.0	40.0	4.56034	38.28480
Totals	100		11.91162	

(b) **Extender, formula analysis**

Material	Quantity	Wt (%)	Vol (%)	Description	Cat.	Volatile
7732-18-5	69.820	69.82	69.25	Water	Sol.	61.471 n
9999999-99-9	20.850	20.85	20.16	Not known	Res.	8.591 n
64-17-5	4.495	4.50	5.46	Ethyl alcohol	Sol.	4.495 f
1336-21-6	2.440	2.44	2.51	Ammonium hydroxide	Add.	1.991 y
107-98-2	1.000	1.00	1.09	1-Methoxy-2-propanol	Sol.	1.000 f
9999999-99-6	0.580	0.58	0.59	Non-hazardous	Add.	0.435 n
109-60-4	0.275	0.28	0.33	*n*-Propyl acetate	Sol.	0.275 f
67-63-0	0.220	0.22	0.28	*iso*-Propyl alcohol	Cas.	0.220 f
25322-69-4	0.200	0.20	0.19	Polypropylene glycol	Res.	0.110 a
111-90-0	0.100	0.10	0.10	Glycoletherde	Res.	0.000 f
100-42-5	0.020	0.02	0.02	Styrene monomer	Res.	0.011 f
Totals	100.000	100.00	97.89			78.600

Category	Qty (lbs)	Wt (%)	Qty (gal)	Vol (%)
Solvent	56.28	56.28	6.88421	57.78605
Additive	3.50	3.50	0.43522	3.65330
Resin	40.00	40.00	4.56035	38.27955
CAS number	0.22	0.22	0.03348	0.28107
Totals	100.00		11.91326	

VOC:	1.89 lb/gal,	Volatile:	78.60 lbs,	Exempt volatile:	72.52 lbs
Total VOC:	0.51 lb/gal,	Volume:	11.91 gal,	Exempt volatile:	8.70 gal

(c) **Extender, formula attributes**

Attribute	Extender
Quantity (lbs)	100.000
Quantity (gal)	11.912
Unit cost (lbs)	0.427
Unit cost (gal)	3.588
Volatile (lbs)	78.600
Volatile (gal)	79.273
Solids (lbs)	21.400
Solids (gal)	20.727
Pigment (lbs)	0.000
Pigment (gal)	0.000
Exempt (lbs)	72.520
Exempt (gal)	72.025
Density (lb/gal)	8.395
VOC (lbs/gal)	1.825
PB ratio	0.000
PVC (%)	0.000
Mileage-mil dry	332.461
RWM cost/sq ft	$0.0108

Table 9.A.2 *Continued*

(d) **Extender, formula structure**

			Percent
⟶ S-2-7	50.50% lb		
· 7732-18-5	100.00% lb (Water)		
⟶ A-6-6	1.50% lb		
· 1336-21-6	100.00% lb (Ammonium hydroxide)		
⟶ WRA-1-55	5.00% lb		
· 111-90-0	2.00% lb (Glycoletherde)		(0.100)
· 9999999-99-9	98.00% lb (Not known)		(4.900)
⟶ XWRA-259	5.00% lb		
· 9999999-99-9	32.00% lb (Not known)		(1.600)
· 7732-18-5	68.00% lb (Water)		(3.400)
⟶ XWRA-401	10.00 lb		
· 9999999-99-9	45.00% lb (Not known)		(4.500)
· 1336-21-6	3.00% lb (Ammonium hydroxide)		(0.300)
· 25322-69-4	2.00% lb (Polypropylene glycol)		(0.200)
· 7732-18-5	49.80% lb (Water)		(4.980)
· 100-42-5	0.20% lb (Styrene monomer)		(0.020)
⟶ XWRA-638	20.00% lb		
· 9999999-99-9	49.00% lb (Not known)		(9.800)
· 9999999-99-9	0.20% lb (Not known)		(0.040)
· 1336-21-6	3.20% lb (Ammonium hydroxide)		(0.640)
· 7732-18-5	47.60% lb (Water)		(9.520)
⟶ S-2-2	5.00% lb		
· 64-17-5	89.90% lb (Ethyl alcohol)		(4.495)
· 67–63–0	4.40% lb (*iso*-Propyl alcohol)		(0.220)
· 67-63-0	100.00% lb (*iso*-Propyl alcohol)		
· 109-60-4	5.50% lb (*n*-Propyl acetate)		(0.275)
· 9999999-99-9	0.20% lb (Not known)		(0.010)
⟶ S-2-56	1.00% lb		
· 107-98-2	100.00% lb (1-Methoxy-2-propanol)		
⟶ XA-323	2.00% lb		
· 7732-18-5	71.00% lb (Water)		(1.420)
· 9999999-99-6	29.00% lb (Non-hazardous)		(0.580)

(e) **Extender, physical data**

Item	Description	Unit	lb/gal	TNV (%)	Pig (%)	Exm (%)	Cat.
S-2-7	Water	lb	8.33	0	0	100	Sol. *
XWRA-638	Zinpol 295	lb	8.71	49	0	51	Res. *
XWRA-401	Joncryl 89 16360-4-6	lb	8.66	45	NA	55	Res. *
S-2-2	Inksolv sol 6264, 190 proof	lb	6.91	0	0	0	Sol. *
XWRA-259	Acrysol 1-98	lb	9.82	32	0	68	Res. *
WRA-1-55	Joncryl 678	lb	8.33	100	0	0	Res. *
XA-323	Johnwax 26	lb	8.20	25	0	71	Add. *
A-6-6	Aqua ammonia 26%	lb	7.84	0	0	100	Add. *
S-2-56	PM solvent	lb	7.69	0	0	0	Sol. *

Table 9.A.2 *Continued*

(f) **Extender, health data**

CAS number	Description	ACGIH TWA (ppm)	ACGIH Stel	ACGIH TWA (mg/m^3)	ACGIH Stel	OSHA Pel (ppm)
7732-18-5	Water	NA	NA	NA	NA	NA
99999-99-9	Not known	NA	NA	NA	NA	NA
64-17-5	Ethyl alcohol	1000	0	1880	0	200
1336-21-6	Ammonium hydroxide	25	0	NA	NA	50
107-98-2	1-Methoxy-2-propanol	100	150	NA	NA	100
99999-99-6	Non-hazardous	NA	NA	NA	NA	NA
109-60-4	*n*-Propyl acetate	200	250	840	1050	200
67-63-0	*iso*-Propyl alcohol	400	500	983	1230	400
25322-69-4	Polypropylene glycol	NA	NA	NA	NA	NA
111-90-0	Glycoletherde	30	0	NA	NA	NA
100-42-5	Styrene monomer	50	100	213	426	50

CAS numer	Description	Wt (%)	Haz (%)	Haz (tot)	Hazard type	Org
7732-18-5	Water	69.82	0	0.00	Not a hazard	Non
99999-99-9	Not known	20.85	0	0.00	Not a hazard	Non
64-17-5	Ethyl alcohol	4.49	100	4.50	Fire hazard	CG1
1336-21-6	Ammonium hydroxide	2.44	100	2.44	Hazard	PS1
107-98-2	1-Methoxy-2-propanol	1.00	100	1.00	Fire hazard	Non
99999-99-6	Non-hazardous	0.58	0	0.00	Not a hazard	Non
109-60-4	*n*-Propyl acetate	0.27	100	0.28	Fire hazard	Skn
67-63-0	*iso*-Propyl alcohol	0.22	100	0.22	Fire hazard	Non
25322-69-4	Polypropylene glycol	0.20	100	0.20	Acute health	PS1
111-90-0	Glycoletherde	0.10	100	0.10	Fire hazard	Eye
100-42-5	Styrene monomer	0.02	100	0.02	Fire hazard	CG2
	Totals	99.99		8.76		

Table 9.A.3 A formula for a heavy ink system, with physical and chemical variables

(a) **Heavy ink, master formula**

Ingredient	Description	Qty (lbs)	Qty (gal)	lb/gal	TNV	PCT
Extender	Water based	15	1.79	8.395	21.40	15
Blue base	Water based	85	8.57	9.917	39.83	85

Attribute	Weight	Volume	Attribute	Value
Quantity	100.000	10.358	Density (lbs/gal)	9.654
Unit cost	0.906	8.747	VOC (lbs/gal)	1.539
Volatile (%)	62.935	74.071	PB ratio	2.205
Solids (%)	37.065	25.929	PVC (%)	48.507
Pigment (%)	25.500	12.577	Mileage-mil dry	415.900 sq. ft
Exempt solv. (%)	57.628	66.707	RWM cost/sq. ft	$0.210 per sq. ft

Table 9.A.3 *Continued*

Material	Quantity	Wt (%)	Vol (%)	Description	Cat.	Volatile
S-2-7	52.625	52.63	60.99	Water	Sol.	52.62
PBD-65	25.500	25.50	12.58	Pigment blue 29	Pig.	0.00
WRA-1-55	6.700	6.70	7.77	Joncryl 678	Res.	0.00
S-2-2	5.000	5.00	6.99	Inksolv sol 6264, 190 proof	Sol.	5.00
XWRA-638	3.000	3.00	3.33	Zinpol 295	Res.	1.53
A-6-6	1.925	1.93	2.37	Aqua ammonia 26%	Add.	1.92
XWRA-401	1.500	1.50	1.67	Joncryl 89 16360-4-	Res.	0.82
XA-537	1.275	1.28	1.40	Surfynol CT-136	Add.	0.14
XA-518	0.850	0.85	1.12	Dee fo 1020f	Add.	0.00
XWRA-259	0.750	0.75	0.74	Acrysol 1-98	Res.	0.51
XA-495	0.425	0.43	0.52	Surfynol 104h surfactant	Add.	0.00
XA-323	0.300	0.30	0.35	Johnwax 26	Add.	0.22
S-2-56	0.150	0.15	0.19	PM solvent	Sol.	0.15
Totals	100.000	100.00	100.00			62.93

Category	Quantity (lbs)	Weight (%)	Quantity (gal)	Volume (%)
Solvent	57.775	57.775	7.06062	68.16461
Additive	4.775	4.775	0.59622	5.75602
Resin	11.950	11.950	1.39834	13.49985
Pigment	25.500	25.500	1.30301	12.57951
Totals	100.000		10.35819	

(b) **Heavy ink, formula analysis**

Material	Quantity	Wt (%)	Vol (%)	Description	Cat.	Volatile
7732-18-5	55.52300	55.52	64.15	Water	Sol.	54.271 n
57455-37-5	25.50000	25.50	12.58	Pigment blue 29	Pig.	0.000 n
9999999-99-9	11.37675	11.38	13.12	Not known	Res.	1.422 n
64-17-5	4.49500	4.50	6.28	Ethyl alcohol	Sol.	4.495 f
1336-21-6	2.06600	2.07	2.53	Ammonium hydroxide	Add.	1.999 y
109-60-4	0.27500	0.28	0.38	*n*-Propyl acetate	Sol.	0.275 f
67-63-0	0.22000	0.22	0.32	*iso*-Propyl alcohol	Cas.	0.220 f
107-98-2	1.15000	0.15	0.19	1-Methoxy-2-propanol	Sol.	0.150 f
107-21-1	0.14025	0.14	0.15	Ethylene glycol	Add.	0.015 a
111-90-0	0.13400	0.13	0.16	Glycoletherde	Res.	0.000 f
9999999-99-6	0.08700	0.09	0.10	Non-hazardous	Add.	0.065 n
25322-60-4	0.03000	0.03	0.03	Polypropylene glycol	Res.	0.017 a
100-42-5	0.00300	0.00	0.00	Styrene monomer	Res.	0.002 f
Totals	100.00000	100.00	99.52			62.930

Category	Qty (lbs)	Wt (%)	Qty (gal)	Vol (%)
Solvent	57.555	57.555	7.02878	67.84648
Additive	4.775	4.775	0.59621	5.75503
Resin	11.950	11.950	1.39833	13.49768
CAS number	0.220	0.220	0.03348	0.32322
Pigment	25.500	25.500	1.30301	12.57756
Totals	100.000		10.35981	

VOC:	1.54 lb/gal,	Volatile:	62.93 lbs,	Exempt volatile:	57.63 lbs
Total VOC:	0.51 lb/gal,	Volume:	10.36 gal,	Exempt volatile:	6.91 gal

Table 9.A.3 *Continued*

(c) **Heavy ink, formula structure**

			Percent
⟶ Extender		15.00% lb	
· S-2-7		50.50% lb	(7.575)
··	7732-18-5	100.00% lb (Water)	
· A-6-6		1.50% lb	(0.225)
··	1336-21-6	100.00% lb (Ammonium hydroxide)	
· WRA-1-55		5.00% lb	(0.750)
··	111-90-0	2.00% lb (Glycoletherde)	(0.015)
··	9999999-99-9	98.00% lb (Not known)	(0.735)
· XWRA-259		5.00% lb	(0.750)
··	9999999-99-9	32.00% lb (Not known)	(0.240)
··	7732-18-5	68.00% lb (Water)	(0.510)
· XWRA-401		10.00% lb	(1.500)
··	9999999-99-9	45.00% lb (Not known)	(0.675)
··	1336-21-6	3.00% lb (Ammonium hydroxide)	(0.045)
··	25322-69-4	2.00 lb (Polypropylene glycol)	(0.030)
··	7732-18-5	49.80% lb (Water)	(0.747)
··	100-42-5	0.20% lb (Styrene monomer)	(0.003)
· XWRA-638		20.00% lb	(3.000)
··	9999999-99-9	49.00% lb (Not known)	(1.470)
··	9999999-99-9	0.20% lb (Not known)	(0.006)
··	1336-21-6	3.20% lb (Ammonium hydroxide)	(0.096)
··	7732-18-5	47.60% lb (Water)	(1.428)
· S-2-2		5.00% lb	(0.750)
··	64-17-5	89.90% lb (Ethyl alcohol)	(0.674)
··	67-63-0	4.40% lb (*iso*-Propyl alcohol)	(0.033)
···	67-63-0	100.00% lb (*iso*-Propyl alcohol)	
··	109-60-4	5.50% lb (*n*-Propyl acetate)	(0.041)
··	9999999-99-9	0.20% lb (Not known)	(0.002)
· S-2-56		1.00% lb	(0.150)
··	107-98-2	100.00% lb (1-Methoxy-2-propanol)	
· XA-323		2.00% lb	(0.300)
··	7732-18-5	71.00% lb (Water)	(0.213)
··	9999999-99-6	29.00% lb (Non-hazardous)	(0.087)

Table 9.A.3 *Continued*

(c) **Heavy ink, formula structure** *continued*

			Percent
⟶ Blue base		85.00% lb	
· PBD-65		30.00% lb	(25.500)
··	57455-37-5	100.00% lb (Pigment blue 29)	
· S-2-7		53.00% lb	(45.050)
··	7732-18-5	100.00% lb (Water)	
· S-2-2		5.00% lb	(4.250)
··	64-17-5	89.90% lb (Ethyl alcohol)	(3.821)
··	67-63-0	4.40 lb (*iso*-Propyl alcohol)	(0.187)
···	67-63-0	100.00% lb (*iso*-Propyl alcohol)	
··	109-60-4	5.50 lb (*n*-Propyl acetate)	(0.234)
··	9999999-99-9	0.20% lb (Not known)	(0.009)
· XA-495		0.50% lb	(0.425)
··	9999999-99-9	100.00% lb (Not known)	
· XA-518		1.00% lb	(0.850)
··	9999999-99-9	100.00% lb (Not known)	
· XA-537		1.50% lb	(1.275)
··	107-21-1	11.00% lb (Ethylene glycol)	(0.140)
··	9999999-99-9	89.00% lb (Not known)	(1.135)
· WRA-1-55		7.00% lb	(5.950)
··	111-90-0	2.00% lb (Glycoletherde)	(0.119)
··	9999999-99-9	98.00% lb (Not known)	(5.831)
· A-6-6		2.00% lb	(1.700)
··	1336-21-6	100.00% lb (Ammonium hydroxide)	

(d) **Heavy ink, physical data**

Item	Description	Unit	lb/gal	TNV (%)	Pig (%)	Exm (%)	Cat.
S-2-7	Water	lb	8.33	0	0	100	Sol. *
PBD-65	Pigment blue 29	lb	19.57	100	100	0	Pig. *
WRA-1-55	Joncryl 678	lb	8.33	100		0	Res. *
S-2-2	Inksolv sol 6264, 190 proof	lb	6.91	0		0	Sol. *
XWRA-638	Zinpol 295	lb	8.71	49		51	Res. *
A-6-6	Aqua ammonia 26%	lb	7.84	0		100	Add. *
XWRA-401	Joncryl 89 16360-4-6	lb	8.66	45	NA	55	Res. *
XA-537	Surfynol CT-136	lb	8.82	89		0	Add. *
XA-518	Dee fo 1020F	lb	7.34	100		0	Add. *
XWRA-259	Acrysol 1-98	lb	9.82	32		68	Res. *
XA-495	Surfynol 104H surfactant	lb	7.91	100		0	Add. *
XA-323	Johnwax 26	lb	8.20	25		71	Add. *
S-2-56	PM solvent	lb	7.69	0		0	Sol. *

Table 9.A.3 *Continued*

CAS number	Description	Flash point (%)	Boiling range (deg)		Expl limits	
			Lower	Upper	Lower	Upper
7732-18-5	Water	NA	212.0	212.0 f	NA	NA
57455-37-5	Pigment blue 29	NA	NA	NA	NA	NA
99999-99-9	Not known	0 f	NA	NA	NA	NA
64-17-5	Ethyl alcohol	54 f	165.0	176.0 f	3.30	19.0%
1336-21-6	Ammonium hydroxide	NA	NA	NA	NA	NA
109-60-4	*n*-Propyl acetate	55 f	215.0	215.0 f	2.00	8.0%
67-63-0	*iso*-Propyl alcohol	54 f	180.0	181.0 f	2.00	12.0%
107-98-2	1-Methoxy-2-propanol	100 f	250.0	250.0 f	1.10	13.1%
107-21-1	Ethylene glycol	232 f	197.5	197.5 c	3.20	3.2%
111-90-0	Glycoletherde	201 f	201.0	201.0 c	NA	NA
99999-99-6	Non-hazardous	NA	NA	NA	NA	NA
25322-69-4	Polypropylene glycol	NA	NA	NA	NA	NA
100-42-5	Styrene monomer	88 f	146.0	146.0 c	1.10	6.1%

(e) **Heavy ink, health data**

CAS number	Description	ACGIH				OSHA
		TWA (ppm)	Stel	TWA (mg/m^3)	Stel	Pel (ppm)
7732-18-5	Water	NA	NA	NA	NA	NA
57455-37-5	Pigment blue 29	NA	NA	10	0	NA
99999-99-9	Not known	NA	NA	NA	NA	NA
64-17-5	Ethyl alcohol	1000	0	1880	0	200
1336-21-6	Ammonium hydroxide	25	0	NA	NA	50
109-60-4	*n*-Propyl acetate	200	250	840	1050	200
67-63-0	*iso*-Propyl alcohol	400	500	983	1230	400
107-98-2	1-Methoxy-2-propanol	100	150	NA	NA	100
107-21-1	Ethylene glycol	50	0	127	0	50
111-90-0	Glycoletherde	30	0	NA	NA	NA
99999-99-6	Non-hazardous	NA	NA	NA	NA	NA
25322-69-4	Polypropylene glycol	NA	NA	NA	NA	NA
100-42-5	Styrene monomer	50	100	213	426	50

Table 9.A.4 Data for a flexographic standard ink

(a) **Flexographic standard ink, master formula**

Ingredient	Description	Qty (lbs)	Qty (gal)	lb/gal	TNV	PCT
Extender	Water based	50	5.96	8.395	21.40	50
Blue base	Water based	50	5.04	9.917	39.83	50

Attribute	Weight	Volume	Attribute	Value
Quantity	100.000	10.999	Density (lbs/gal)	9.092
Unit cost	0.709	6.446	VOC (lbs/gal)	0.000
Volatile (%)	69.385	76.386	PB ratio	0.961
Solids (%)	30.615	23.614	PVC (%)	29.506
Pigment (%)	15.000	6.968	Mileage-mil dry	378.776 sq. ft
Exempt solv. (%)	121.388	132.333	RWM cost/sq. ft	$0.017 per sq. ft

Table 9.A.4 *Continued*

Material	Quantity	Wt (%)	Vol (%)	Description	Cat.	Volatile
S-2-7	51.75	51.75	56.49	Water	Sol.	51.75
PBD-65	15.00	15.00	6.97	Pigment blue 29	Pig.	0.00
XWRA-638	10.00	10.00	10.44	Zinpol 295	Res.	5.10
WRA-1-55	6.00	6.00	6.55	Joncryl 678	Res.	0.00
S-2-2	5.00	5.00	6.58	Inksolv sol 6264, 190 proof	Sol.	5.00
XWRA-401	5.00	5.00	5.25	Joncryl 89 16360-4-	Res.	2.75
XWRA-259	2.50	2.50	2.31	Acrysol 1-98	Res.	1.70
A-6-6	1.75	1.75	2.03	Aqua ammonia 26%	Add.	1.75
XA-323	1.00	1.00	1.11	Johnwax 26	Add.	0.75
XA-537	0.75	0.75	0.77	Surfynol CT-136	Add.	0.08
XA-518	0.50	0.50	0.62	Dee fo 1020F	Add.	0.00
S-2-56	0.50	0.50	0.59	PM solvent	Sol.	0.50
XA-495	0.25	0.25	0.29	Surfynol 104H surfactant	Add.	0.00
Totals	100.00	100.00	100.00			69.38

Category	Quantity (lbs)	Weight (%)	Quantity (gal)	Volume (%)
Solvent	57.25	57.25	7.00108	63.65874
Additive	4.25	4.25	0.52992	4.81841
Resin	23.50	23.50	2.70035	24.55348
Pigment	15.00	15.00	0.76648	6.96937
Totals	100.00		10.99783	

(b) **Flexographic standard ink, formula analysis**

Material	Quantity	Wt (%)	Vol (%)	Description	Cat.	Volatile
7732-18-5	61.4100	61.41	66.42	Water	Sol.	57.236 n
9999999-99-9	15.2775	15.28	16.26	Not known	Res.	4.374 n
57455-37-5	15.0000	15.00	6.97	Pigment blue 29	Pig.	0.000 n
64-17-5	4.4950	4.50	5.91	Ethyl alcohol	Sol.	4.495 f
1336-21-6	2.2200	2.22	2.52	Ammonium hydroxide	Add.	1.996 y
107-98-2	0.5000	0.50	0.59	1-Methoxy-2-propanol	Sol.	0.500 f
9999999-99-6	0.2900	0.29	0.32	Non-hazardous	Add.	0.218 n
109-60-4	0.2750	0.28	0.36	*n*-Propyl acetate	Sol.	0.275 f
67-63-0	0.2200	0.22	0.30	*iso*-Propyl alcohol	Cas.	0.220 f
111-90-0	0.1200	0.12	0.13	Glycoletherde	Res.	0.000 f
25322-69-4	0.1000	0.10	0.11	Polypropylene glycol	Res.	0.055 a
107-21-1	0.0825	0.08	0.09	Ethylene glycol	Add.	0.009 a
100-42-5	0.0100	0.01	0.01	Styrene monomer	Res.	0.006 f
Totals	100.0000	100.00	98.79			69.382

Category	Qty (lbs)	Wt (%)	Qty (gal)	Vol (%)
Solvent	57.03	57.03	6.96925	63.35981
Additive	4.25	4.25	0.52992	4.81772
Resin	23.50	23.50	2.70034	24.54971
CAS number	0.22	0.22	0.03348	0.30442
Pigment	15.00	15.00	0.76647	6.96831
Totals	100.00		10.99946	

VOC:	1.68 lb/gal,	Volatile:	69.38 lbs,	Exempt volatile:	63.76 lbs
Total VOC:	0.51 lb/gal,	Volume:	11.00 gal,	Except volatile:	7..65 gal

Table 9.A.4 *Continued*

(c) **Flexographic standard ink, formula structure**

			Percent
→ Extender		50.00% lb	
· S-2-7		50.50% lb	(25.250)
··	7732-18-5	100.00% lb (Water)	
· A-6-6		1.50% lb	(0.750)
··	1336-21-6	100.00% lb (Ammonium hydroxide)	
· WRA-1-55		5.00% lb	(2.500)
··	111-90-0	2.00% lb (Glycoletherde)	(0.050)
··	9999999-99-9	98.00% lb (Not known)	(2.450)
· XWRA-259		5.00% lb	(2.500)
··	9999999-99-9	32.00% lb (Not known)	(0.800)
··	7732-18-5	68.00% lb (Water)	(1.700)
· XWRA-401		10.00% lb	(5.000)
··	9999999-99-9	45.00% lb (Not known)	(2.250)
··	1336-21-6	3.00% lb (Ammonium hydroxide)	(0.150)
··	25322-69-4	2.00 lb (Polypropylene glycol)	(0.100)
··	7732-18-5	49.80% lb (Water)	(2.490)
··	100-42-5	0.20% lb (Styrene monomer)	(0.010)
· XWRA-638		20.00% lb	(10.000)
··	9999999-99-9	49.00% lb (Not known)	(4.900)
··	9999999-99-9	0.20% lb (Not known)	(0.020)
··	1336-21-6	3.20% lb (Ammonium hydroxide)	(0.320)
··	7732-18-5	47.60% lb (Water)	(4.760)
· S-2-2		5.00% lb	(2.500)
··	64-17-5	89.90% lb (Ethyl alcohol)	(2.248)
··	67-63-0	4.40% lb (*iso*-Propyl alcohol)	(0.110)
···	67-63-0	100.00% lb (*iso*-Propyl alcohol)	
··	109-60-4	5.50% lb (*n*-Propyl acetate)	(0.138)
··	9999999-99-9	0.20% lb (Not known)	(0.005)
· S-2-56		1.00% lb	(0.500)
··	107-98-2	100.00% lb (1-Methoxy-2-propanol)	
· XA-323		2.00% lb	(1.000)
··	7732-18-5	71.00% lb (Water)	(0.710)
··	9999999-99-6	29.00% lb (Non-hazardous)	(0.290)

Table 9.A.4 *Continued*

(c) **Flexographic standard ink, formula structure** *continued*

		Percent
⟶ Blue base	50.00% lb	
· PBD-65	30.00% lb	(15.000)
·· 57455-37-5	100.00% lb (Pigment blue 29)	
· S-2-7	53.00% lb	(26.500)
·· 7732-18-5	100.00% lb (Water)	
· S-2-2	5.00% lb	(2.500)
·· 64-17-5	89.90% lb (Ethyl alcohol)	(2.248)
·· 67-63-0	4.40 lb (*iso*-Propyl alcohol)	(0.110)
··· 67-63-0	100.00% lb (*iso*-Propyl alcohol)	
·· 109-60-4	5.50 lb (*n*-Propyl acetate)	(0.138)
·· 9999999-99-9	0.20% lb (Not known)	(0.005)
· XA-495	0.50% lb	(0.250)
·· 9999999-99-9	100.00% lb (Not known)	
· XA-518	1.00% lb	(0.500)
·· 9999999-99-9	100.00% lb (Not known)	
· XA-537	1.50% lb	(0.750)
·· 107-21-1	11.00% lb (Ethylene glycol)	(0.083)
·· 9999999-99-9	89.00% lb (Not known)	(0.668)
· WRA-1-55	7.00% lb	(3.500)
·· 111-90-0	2.00% lb (Glycoletherde)	(0.070)
·· 9999999-99-9	98.00% lb (Not known)	(3.430)
· A-6-6	2.00% lb	(1.000)
·· 1336-21-6	100.00% lb (Ammonium hydroxide)	

(d) **Flexographic standard ink, physical data**

Item	Description	Unit	lb/gal	TNV (%)	Pig (%)	Exm (%)	Cat.
S-2-7	Water	lb	8.33	0	0	100	Sol. *
PBD-65	Pigment blue 29	lb	19.57	100	100	0	Pig. *
XWRA-638	Zinpol 295	lb	8.71	49	0	51	Res. *
WRA-1-55	Joncryl 678	lb	8.33	100	0	0	Res. *
S-2-2	Inksolv sol 6264, 190 proof	lb	6.91	0	0	0	Sol. *
XWRA-401	Joncryl 89 16360-4-6	lb	8.66	45	NA	55	Res. *
XWRA-259	Acrysol 1-98	lb	9.82	32	0	68	Res. *
A-6-6	Aqua ammonia 26%	lb	7.84	0	0	100	Add. *
XA-323	Johnwax 26	lb	8.20	25	0	71	Add. *
XA-537	Surfynol CT-136	lb	8.82	89	0	0	Add. *
XA-518	Dee fo 1020F	lb	7.34	100	0	0	Add. *
S-2-56	PM solvent	lb	7.69	0	0	0	Sol. *
XA-495	Surfynol 104H surfactant	lb	7.91	100	0	0	Add. *

Table 9.A.4 *Continued*

CAS number	Description	Flash point (%)	Boiling range (deg)		Expl limits	
			Lower	Upper	Lower	Upper
7732-18-5	Water	NA	212.0	212.0 f	NA	NA
99999-99-9	Not known	0 f	NA	NA	NA	NA
57455-37-5	Pigment blue 29	NA	NA	NA	NA	NA
64-17-5	Ethyl alcohol	54 f	165.0	176.0 f	3.3	19.0%
1336-21-6	Ammonium hydroxide	NA	NA	NA	NA	NA
107-98-2	1-Methoxy-2-propanol	100 f	250.0	250.0 f	1.10	13.1%
99999-99-6	Non-hazardous	NA	NA	NA	NA	NA
109-60-4	*n*-Propyl acetate	55 f	215.0	215.0 f	2.00	8.0%
67-63-0	*iso*-Propyl alcohol	54 f	180.0	181.0 f	2.00	12.0%
111-90-0	Glycoletherde	201 f	201.0	201.0 c	NA	NA
25322-69-4	Polypropylene glycol	NA	NA	NA	NA	NA
107-21-1	Ethylene glycol	232 f	197.5	197.5 c	3.20	3.2%
100-42-5	Styrene monomer	88 f	146.0	146.0 c	1.10	6.1%

(e) **Flexographic standard ink, health data**

CAS number	Description	ACGIH				OSHA
		TWA (ppm)	Stel	TWA (mg/m^3)	Stel	Pel (ppm)
7732-18-5	Water	NA	NA	NA	NA	NA
99999-99-9	Not known	NA	NA	NA	NA	NA
57455-37-5	Pigment blue 29	NA	NA	10	0	NA
64-17-5	Ethyl alcohol	1000	0	1880	0	200
1336-21-6	Ammonium hydroxide	25	0	NA	NA	50
107-98-2	1-Methoxy-2-propanol	100	150	NA	NA	100
99999-99-6	Non-hazardous	NA	NA	NA	NA	NA
109-60-4	*n*-Propyl acetate	200	250	840	1050	200
67-63-0	*iso*-Propyl alcohol	400	500	983	1230	400
111-90-0	Glycoletherde	30	0	NA	NA	NA
25322-69-4	Polypropylene glycol	NA	NA	NA	NA	NA
107-21-1	Ethylene glycol	50	0	127	0	50
100-42-5	Styrene monomer	50	100	213	426	50

Table 9.A.5 Comparison of the physical constants for the five ink systems

Attribute	Blue base	Heavy ink	Standard ink	Standard ink G	Extender
Quantity (lbs)	100.000	100.000	100.000	100.000	100.000
Quantity (gal)	10.084	10.358	10.998	11.181	11.912
Unit cost (lbs)	0.991	0.906	0.709	0.653	0.427
Unit cost (gal)	9.823	8.748	6.447	5.837	3.588
Volatile (lbs)	60.165	62.930	69.383	71.226	78.600
Volatile (gal)	72.986	74.070	76.391	77.005	79.273
Solids (lbs)	39.835	37.070	30.617	28.774	21.400
Solids (gal)	27.014	25.930	23.609	22.995	20.727
Pigment (lbs)	30.000	25.500	15.000	12.000	0.000
Pigment (gal)	15.202	12.580	6.969	5.484	0.000
Exempt (lbs)	55.000	57.628	63.760	65.512	72.520
Exempt (gal)	65.625	66.729	69.091	69.716	72.025
Density (lbs/gal)	9.917	9.654	9.093	8.944	8.395
VOC (lbs/gal)	1.490	1.539	1.654	1.688	1.825
PB ratio	3.050	2.204	0.960	0.715	0.000
PVC (%)	56.274	48.514	29.520	23.850	0.000
Mileage-mil dry	433.305	415.910	378.695	368.841	332.461
RWM cost/sq ft	$0.0227	$0.0210	£0.0170	$0.0158	$0.0108

Material	Quantity	Wt (%)	Vol (%)	Description	Cat.	Volatile
7732-18-5	63.092	63.09	67.03	Water	Sol.	58.083 n
9999999-99-9	16.392	16.39	17.09	Not known	Res.	5.218 n
57455-37-5	12.000	12.00	5.48	Pigment blue 29	Pig.	0.000 n
64-17-5	4.495	4.50	5.82	Ethyl alcohol	Sol.	4.495 f
1336-21-6	2.264	2.26	2.52	Ammonium hydroxide	Add.	1.995 y
107-98-2	0.600	0.60	0.70	1-Methoxy-2-propanol	Sol.	0.600 f
9999999-99-6	0.348	0.35	0.38	Non-hazardous	Add.	0.261 n
109-60-4	0.275	0.28	0.36	*n*-Propyl acetate	Sol.	0.275 f
67-63-0	0.220	0.22	0.30	*iso*-Propyl alcohol	Cas.	0.220 f
25322-69-4	0.120	0.12	0.12	Polypropylene glycol	Res.	0.066 a
111-90-0	0.116	0.12	0.12	Glycoletherde	Res.	0.000 f
107-21-1	0.066	0.07	0.07	Ethylene glycol	Add.	0.007 a
100-42-5	0.012	0.01	0.01	Styrene monomer	Res.	0.007 f
Totals	100.000	100.00	98.60			71.226

Category	Qty (lbs)	Wt (%)	Qty (gal)	Vol (%)
Solvent	56.88	56.88	6.95224	62.17218
Additive	4.10	4.10	0.51098	4.56961
Resin	26.80	26.80	3.07234	27.47520
CAS number	0.22	0.22	0.03348	0.29945
Pigment	12.00	12.00	0.61318	5.48354
Totals	100.00		11.18222	

VOC:	1.72 lb/gal,	Volatile:	71.23 lbs,	Exempt volatile:	65.51 lbs
Total VOC:	0.51 lb/gal,	Volume:	11.18 gal,	Exempt volatile:	7.86 gal

Table 9.A.6 Data for a gravure standard ink

(a) **Gravure standard ink, master formula**

Ingredient	Description	Qty (lbs)	Qty (gal)	lb/gal	TNV	PCT
Extender	Water based	60	7.15	8.395	21.40	60
Blue base	Water base	40	4.03	9.917	39.83	40

Attribute	Weight	Volume	Attribute	Value
Quantity	100.000	11.181	Density (lbs/gal)	8.944
Unit cost	0.653	5.840	VOC (lbs/gal)	1.719
Volatile (%)	71.228	77.005	PB ratio	0.715
Solids (%)	28.772	22.995	PVC (%)	23.846
Pigment (%)	12.000	5.483	Mileage-mil dry	368.847 sq. ft
Exempt solv. (%)	65.512	70.256	RWM cost/sq. ft	$0.0158 per sq. ft

Material	Quantity	Wt (%)	Vol (%)	Description	Cat.	Volatile
S-2-7	51.5	51.5	55.30	Water	Sol.	51.50
PBD-65	12.0	12.0	5.48	Pigment blue 29	Pig.	0.00
XWRA-638	12.0	12.0	12.32	Zinpol 295	Res.	6.12
XWRA-401	6.0	6.0	6.20	Joncryl 89 16360-4-	Res.	3.30
WRA-1-55	5.8	5.8	6.23	Joncryl 678	Res.	0.00
S-2-2	5.0	5.0	6.47	Inksolv sol 6264, 190 proof	Sol.	5.00
XWRA-259	3.0	3.0	2.73	Acrysol 1-98	Res.	2.04
A-6-6	1.7	1.7	1.94	Aqua ammonia 26%	Add.	1.70
XA-323	1.2	1.2	1.31	Johnwax 26	Add.	0.90
XA-537	0.6	0.6	0.61	Surfynol CT-136	Add.	0.06
S-2-56	0.6	0.6	0.70	PM solvent	Sol.	0.60
XA-518	0.4	0.4	0.49	Dee fo 1020f	Add.	0.00
XA-495	0.2	0.2	0.23	Surfynol 104h surfactant	Add.	0.00
Totals	100.0	100.0	100.00			71.22

Category	Quantity (lbs)	Weight (%)	Quantity (gal)	Volume (%)
Solvent	57.1	57.1	6.98408	62.46611
Additive	4.1	4.1	0.51099	4.57033
Resin	26.8	26.8	3.07234	27.47923
Pigment	12.0	12.0	0.61318	5.48433
Totals	100.0		11.18059	

Table 9.A.6 *Continued*

(b) **Gravure standard ink, formula structure**

			Percent
⟶ Extender		60.00% lb	
· S-2-7		50.50% lb	(30.300)
··	7732-18-5	100.00% lb (Water)	
· A-6-6		1.50% lb	(0.900)
··	1336-21-6	100.00% lb (Ammonium hydroxide)	
· WRA-1-55		5.00% lb	(3.000)
··	111-90-0	2.00% lb (Glycoletherde)	(0.060)
··	9999999-99-9	98.00% lb (Not known)	(2.940)
· XWRA-259		5.00% lb	(3.000)
··	9999999-99-9	32.00% lb (Not known)	(0.960)
··	7732-18-5	68.00% lb (Water)	(2.040)
· XWRA-401		10.00% lb	(6.000)
··	9999999-99-9	45.00% lb (Not known)	(2.700)
··	1336-21-6	3.00% lb (Ammonium hydroxide)	(0.180)
··	25322-69-4	2.00 lb (Polypropylene glycol)	(0.120)
··	7732-18-5	49.80% lb (Water)	(2.988)
··	100-42-5	0.20% lb (Styrene monomer)	(0.012)
· XWRA-638		20.00% lb	(12.000)
··	9999999-99-9	49.00% lb (Not known)	(5.880)
··	9999999-99-9	0.20% lb (Not known)	(0.024)
··	1336-21-6	3.20% lb (Ammonium hydroxide)	(0.384)
··	7732-18-5	47.60% lb (Water)	(5.712)
· S-2-2		5.00% lb	(3.000)
··	64-17-5	89.90% lb (Ethyl alcohol)	(2.697)
··	67-63-0	4.40% lb (*iso*-Propyl alcohol)	(0.132)
···	67-63-0	100.00% lb (*iso*-Propyl alcohol)	
··	109-60-4	5.50% lb (*n*-Propyl acetate)	(0.165)
··	9999999-99-9	0.20% lb (Not known)	(0.006)
· S-2-56		1.00% lb	(0.600)
··	107-98-2	100.00% lb (1-Methoxy-2-propanol)	
· XA-323		2.00% lb	(1.200)
··	7732-18-5	71.00% lb (Water)	(0.852)
··	9999999-99-6	29.00% lb (Non-hazardous)	(0.348)

Table 9.A.6 *Continued*

(b) **Gravure standard ink, formula structure** *continued*

			Percent
⟶ Blue base	40.00% lb		
· PBD-65	30.00% lb		(12.000)
··	57455-37-5	100.00% lb (Pigment blue 29)	
· S-2-7	53.00% lb		(21.200)
··	7732-18-5	100.00% lb (Water)	
· S-2-2	5.00% lb		(2.000)
··	64-17-5	89.90% lb (Ethyl alcohol)	(1.798)
··	67-63-0	4.40 lb (*iso*-Propyl alcohol)	(0.088)
···	67-63-0	100.00% lb (*iso*-Propyl alcohol)	
··	109-60-4	5.50 lb (*n*-Propyl acetate)	(0.110)
··	9999999-99-9	0.20% lb (Not known)	(0.004)
· XA-495	0.50% lb		(0.200)
··	9999999-99-9	100.00% lb (Not known)	
· XA-518	1.00% lb		(0.400)
··	9999999-99-9	100.00% lb (Not known)	
· XA-537	1.50% lb		(0.600)
··	107-21-1	11.00% lb (Ethylene glycol)	(0.066)
··	9999999-99-9	89.00% lb (Not known)	(0.534)
· WRA-1-55	7.00% lb		(2.800)
··	111-90-0	2.00% lb (Glycoletherde)	(0.056)
··	9999999-99-9	98.00% lb (Not known)	(2.744)
· A-6-6	2.00% lb		(0.800)
··	1336-21-6	100.00% lb (Ammonium hydroxide)	

(c) **Gravure standard ink, physical data**

Item	Description	Unit	lb/gal	TNV (%)	Pig (%)	Exm (%)	Cat.
S-2-7	Water	lb	8.33	0	0	100	Sol. *
PBD-65	Pigment blue 29	lb	19.57	100	100	0	Pig. *
XWRA-638	Zinpol 295	lb	8.71	49	0	51	Res. *
XWRA-401	Joncryl 89 16360-4-6	lb	8.66	45	NA	55	Res. *
WRA-1-55	Joncryl 678	lb	8.33	100	0	0	Res. *
S-2-2	Inksolv sol 6264, 190 proof	lb	6.91	0	0	0	Sol. *
XWRA-259	Acrysol 1-98	lb	9.82	32	0	68	Res. *
A-6-6	Aqua ammonia 26%	lb	7.84	0	0	100	Add. *
XA-323	Johnwax 26	lb	8.20	25	0	71	Add. *
XA-537	Surfynol CT-136	lb	8.82	89	0	0	Add. *
S-2-56	PM solvent	lb	7.69	0	0	0	Sol. *
XA-518	Dee fo 1020F	lb	7.34	100	0	0	Add. *
XA-495	Surfynol 104H surfactant	lb	7.91	100	0	0	Add. *

Table 9.A.6 *Continued*

CAS number	Description	Flash point (%)	Boiling range (deg)		Expl limits	
			Lower	Upper	Lower	Upper
7732-18-5	Water	NA	212.0	212.0 f	NA	NA
99999-99-9	Not known	0 f	NA	NA	NA	NA
57455-37-5	Pigment blue 29	NA	NA	NA	NA	NA
64-17-5	Ethyl alcohol	54 f	165.0	176.0 f	3.3	19.0%
1336-21-6	Ammonium hydroxide	NA	NA	NA	NA	NA
107-98-2	1-Methoxy-2-propanol	100 f	250.0	250.0 f	1.1	13.1%
99999-99-6	Non-hazardous	NA	NA	NA	NA	NA
109-60-4	*n*-Propyl acetate	55 f	215.0	215.0 f	2.0	8.0%
67-63-0	*iso*-Propyl alcohol	54 f	180.0	181.0 f	2.0	12.0%
25322-69-4	Polypropylene glycol	NA	NA	NA	NA	NA
111-90-0	Glycoletherde	201 f	201.0	201.0 c	NA	NA
107-21-1	Ethylene glycol	232 f	197.5	197.5 c	3.2	3.2%
100-42-5	Styrene monomer	88 f	146.0	146.0 c	1.1	6.1%

(d) **Gravure standard ink, health data**

CAS number	Description	ACGIH				OSHA
		TWA (ppm)	Stel	TWA (mg/m^3)	Stel	Pel (ppm)
7732-18-5	Water	NA	NA	NA	NA	NA
99999-99-9	Not known	NA	NA	NA	NA	NA
57455-37-5	Pigment blue 29	NA	NA	10	0	NA
64-17-5	Ethyl alcohol	1000	0	1880	0	200
1336-21-6	Ammonium hydroxide	25	0	NA	NA	50
107-98-2	1-Methoxy-2-propanol	100	150	NA	NA	100
99999-99-6	Non-hazardous	NA	NA	NA	NA	NA
109-60-4	*n*-Propyl acetate	200	250	840	1050	200
67-63-0	*iso*-Propyl alcohol	400	500	983	1230	400
25322-69-4	Polypropylene glycol	NA	NA	NA	NA	NA
111-90-0	Glycoletherde	30	0	NA	NA	NA
107-21-1	Ethylene glycol	50	0	127	0	50
100-42-5	Styrene monomer	50	100	213	426	50

10 Particle size reduction of pigments using a small media mill

H.W. WAY

10.1 Introduction

A small media mill (Figure 10.1) is used for particle size reduction and dispersion of particles in liquid systems. The term 'small media mill' refers to the size of the grinding balls (the medium) used in the machine, generally in the 1 to 2 mm nominal range. Agitation of the medium breaks apart the agglomerates and aggregates of ceramic powders predispersed in a liquid carrier system. The small media mill has a horizontally arranged vessel, or chamber filled with the grinding medium. A variable rotational speed rotor or agitator transfers energy to the media. The agitator design controls the shearing and impacting forces of the media. The machine uses a pump to feed the slurry to be dispersed to the enclosed grinding vessel. This method of operation results in a continuous grinding system. The machine elements can be manufactured using any commercially available ceramic or polymer material. Therefore high purity or low contamination dispersions are produced.

The machines discussed in this article are primarily used for dispersing and 'grinding' (particle size reduction) pigments. This equipment is used in many industries, primarily for the manufacture of industrial and automotive paints and coatings, magnetic coatings for audio tape, video tape and data storage, the production of pigments and pigment dispersions, and printing inks.

Historically the primary method for obtaining dispersion of solids in liquids has been to use ball mills, vertical sand mills or attrition mills. The small media mill is the latest progression in milling technology. The small media mill offers advantages in higher grinding efficiency through the use of finer grinding media, greater energy input, and more precise control of processing parameters. Higher productivity and tighter particle size distributions in the micron and submicron range result.

The machine, essentially, is a tank with an agitator. The tank is filled with the grinding medium, generally small balls or spheres. The agitator imparts kinetic energy to the medium through direct impact, adhesion to the discs and cohesion within the medium [1]. The interaction of the medium with the particles results in bead to bead contact, with the particles being caught between the beads. This crushing action separates the particles.

Shearing action within the medium forces the particles into the liquid grinding medium resulting in stabilized dispersions of fine particles. Feed particle sizes

Figure 10.1 The LMC Series mills are available in 14 to 60 l capacity models.

typically range from 44 to 200 microns. Particle sizes obtained from milling range from 0.25 micron to 10 microns depending on the requirement. Productivity from the machines can vary based on the fineness required and machine size used.

This chapter is an overview of the mechanics of a small media mill, the operating parameters and the applications for the machine.

10.2 Machine design

The horizontal mill (Figure 10.2) is, in essence, a very simple machine doing a complex job. The machine consists of several subsystems relating to the final operation of the unit. Looking at Figure 10.2 we can see the progression of how the agitator is driven and the subsystems used to make the horizontal media mill operation trouble free. The process begins with the electric motor, (A). The motor size selected is based on the mill chamber volume and ranges from 3 horsepower (HP) to 250 HP. The electric motor drives the agitator shaft through the use of a variable speed drive, either a split pulley or a frequency inverter, or if the machine is larger than 60 l chamber volume a fixed speed v-belt drive is used, (B). The agitator shaft is supported by a set of bearings, (C). One roller bearing near the drive shaft and a double row ball bearing near the chamber rigidly fix the shaft to prevent any vibration.

To prevent product that is being processed from entering the bearings a double acting mechanical seal is installed between the grinding vessel and the bearings, (C). This device consists of a rotating tungsten carbide or silicon carbide 'face' sealing against a stationary silicon carbide face on the chamber side and a high chrome steel rotating face running on a carbon impregnated polymer stationary face on the atmospheric or bearing side. A liquid is filled into the seal reservoir, (D) that is compatible with the product to be ground. This is necessary for lubrication of the seal faces, a compatible material (water, organic solvents, etc.) is used because there is a small amount of this seal flush entering the grinding chamber. A pumping device is incorporated within the seal assembly to circulate the fluid. Heat is removed from the seal faces by the fluid. A

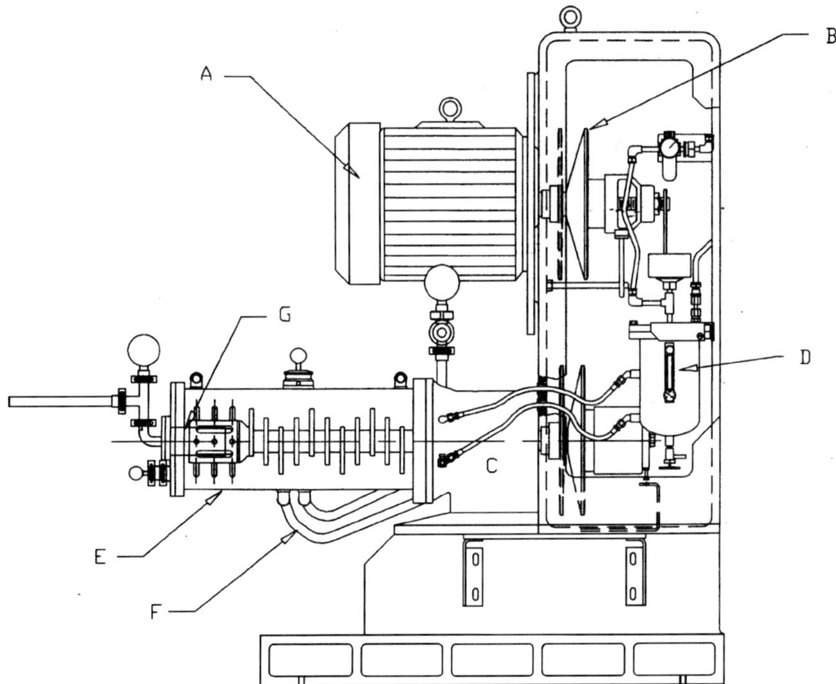

Figure 10.2 AutoCad rendering of an LMC60 Netzsch Small Media Mill.

cooling coil in the reservoir removes the heat from the seal fluid. The seal is pressurized with a static head of compressed gas to prevent product from entering the seal. Safety switches are supplied as a standard control on the mill to protect the mechanical seal from high chamber pressures, fluid loss, and low static air pressure.

The grinding chamber, (E), clearly serves as the vessel for containing the media and agitator shaft. The chamber has a specific volume; this is the basis for the machine designation. This mill volume is calculated to include the volume taken by the agitator shaft and separator.

The standard materials of construction for the chamber are high chrome wear resistant steel. Stainless steel components and hard chrome plating are common options. The lining of the chamber has also been supplied with neoprene, UHMW polyethylene, nylon, urethane, alumina oxide, silicon nitride, or zirconium oxide. The mechanical seal housing and end plate or cover for the chamber are also faced with these materials. This results in either contamination-free dispersions or small quantities of acceptable contaminants in the parts per million range.

The grinding chamber is constructed to include a cooling jacket. This is necessary due to the heat generated during the grinding process. Many designs are available, the most efficient use a two zone spiralled jacket for increased cooled surface area and increased flow through with reduced pressure drop. This allows maximum turbulence required for efficient cooling.

The most important part of the machine is the agitator design. There are many variations; however, the most effective are either eccentric disc designs, pegged agitator designs and in some cases a concentric disc design.

10.3 Machine operation

The basic operation of any horizontal mill is simply pumping a predispersed slurry to the grinding mill (Figure 10.3). A pump is supplied with the machine. The standard pump is a double gear; however, for applications that are very large in initial particle size or cannot tolerate the potential steel contamination from a gear pump either a progressive cavity pump or peristaltic pump is supplied.

As we can see in Figure 10.3 the premixed slurry is prepared using the high speed dissolver shown. The preparation of the premix is a separate field; many papers have been written on this process. Basically a liquid is placed in the tank, additives such as dispersants and surfactants or resin solutions are added and the solids to be ground are then added. There are systems for continuously feeding the machine using metering pumps and dry solids feeders. Although the capital equipment cost is expensive, the long-term benefits as far as labor costs and reduction of operator exposure to hazardous materials will justify this expenditure.

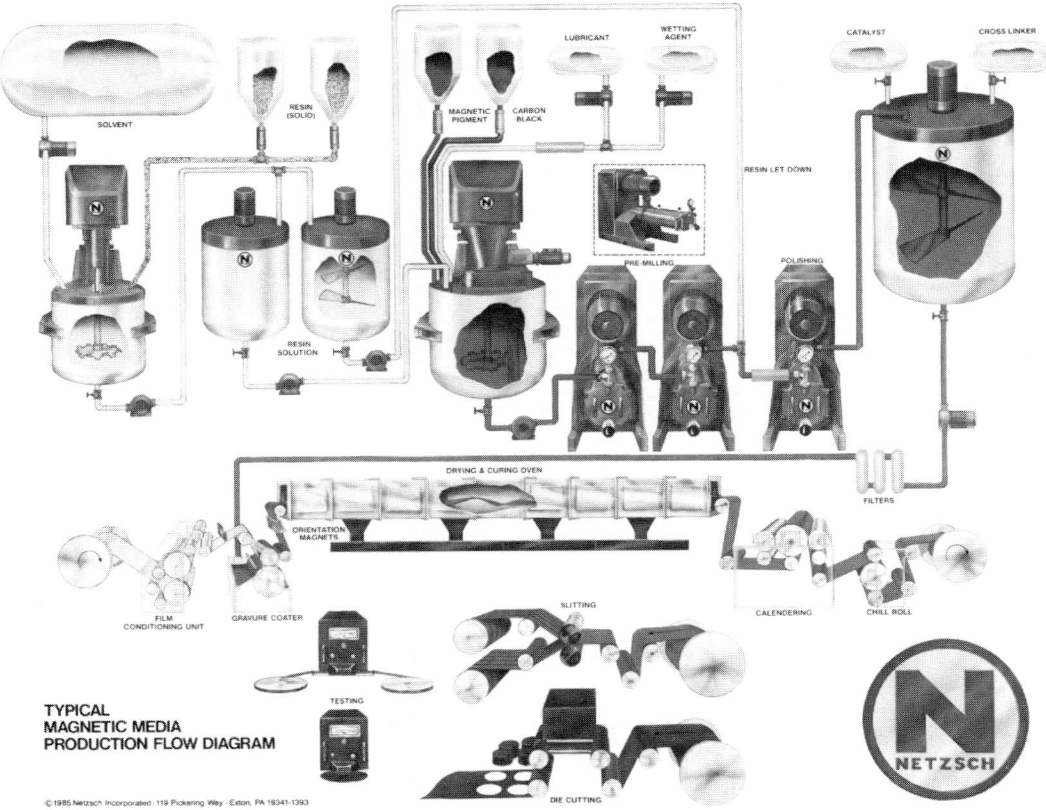

Figure 10.3 Typical magnetic media production flow diagram.

10.4 Controlling factors in milling

When the premix is supplied to the mill there are many factors controlling the dispersing action inside the chamber. The three most important parameters, apart from machine design, are the grinding media, pumping rate (flow rate or throughput rate) and the agitator speed.

Selecting the grinding media is the most important of these three parameters. The first consideration must be the material of the media. This directly affects acceptable product contamination or lack of contamination and the density. The three main categories are steel, glass, and ceramic. There are many subdivisions to each of these categories.

Steel can be purchased as high chrome, ball bearing grade as the most expensive, highest quality and longest wearing to cast steel shot which is basically a tighter size distribution of blasting shot. This shot is usually hardened and preconditioned for use as a grinding medium. Appropriate applications of this medium are for grinding dark colors and the manufacture of printing inks. This is due to the steel contamination that will be generated from this medium. Steel is usually the best grinding medium from a density standpoint. The use of steel media seems to be declining due to the contamination problems and the steps required to remove the steel contamination.

Glass can be either barium titanate, leaded, soda lime unleaded, or sand. For dispersing agglomerates and light colors in low viscosity systems glass is best. This is mainly due to the low cost, low density and tight size ranges available. These factors will result in low capital cost, low power cost and efficient dispersing through high peripheral speeds imparted from the agitator. The low density of glass usually cannot grind aggregates to their primary particle size. Products that have high solids and high viscosities seldom disperse well with glass media. The high solids require high kinetic energy in the media for uniform dispersion. The low kinetic energy in a glass bead is not uniformly transferred to all the media in the chamber, only the product closest to the shaft receives the high energy required.

Ceramic media have the greatest selection. Zirconium silicate is widely used for coatings and ceramic application. ZrO_2SiO_2 is available in very close size distributions, it is non-porous, has very low wear, cleans easily and is relatively cost effective.

Zirconium oxide media are very commonly used; they are, however, somewhat more expensive than zirconium silicates. The trade-off is that recently the quality of this medium has been greatly improved. This is a result of using higher purity zirconium oxide. A high purity magnesium stabilized zirconium oxide is being produced in this country that initial tests have shown to be much lower in wear than previous grades, and it is very cost effective. There are high purity yttrium stabilized grades available that show very low wear. However these media are approximately ten times the cost of zirconium silicates and the mass produced zirconium oxides.

Alumina oxides are available in various degrees of purity, roundness and size. Silicon nitride medium in the 2 mm nominal range has been produced but is expensive. Titanium dioxide medium shows promise in some areas, although its cost is high. Mullite media are available; however, there is a very high wear factor associated with these media.

Media should be selected based on the surface of the bead, whether it is porous or has a smooth surface, the size distribution available, and the size required. The porosity is very important as far as wear and cleaning are concerned. Cheap grades will generally be very porous resulting in high media wear, high mill wear and product contamination. This is a result of the media surface continually abrading.

The second factor in selecting a medium is the size distribution. The size distributions are generally given in ranges, for example 0.3–0.42, 0.6–0.8 and 0.8–1.0 mm for zirconium silicate. Steel is generally specified by its nominal size, 0.25 mm, 0.45 mm, 1 mm, 2 mm, etc. Glass is specified by the mesh range by which it is classified. Size distribution is important, based on the fact that the tighter the size range, the more efficient energy transfer within the media range [2].

A small bead diameter will result in a lower particle size distribution (PSD). This relates to the exponential increase in the number of beads per unit volume and the associated increase in media contact points. In literature available from media manufacturers we find that if a 2.0–2.8 mm bead is used, the average particle count per pound of medium is 29 000 beads. If a 0.85–1.18 mm bead is used, there are 380 000 beads per pound [3]. There are approximately 12 contact points per bead with the surrounding medium.

Clearly, the higher the number of contact points per unit volume, the higher the probability the particles will be caught between two beads, crushed, sheared and dispersed (comminution). This is illustrated in Figure 10.4 where a product is run through media of the same material at identical milling conditions.

A decrease in the packing dimension between the beads is also associated with finer media. If a very fine medium is used and a finer particle (a ratio of 10 : 1 has been used as a rule of thumb) is fed into the medium the finer particles are caught between the medium. If the particles fed into the medium are as large or larger than those of the medium we can see that the opportunity for the medium to grind them is diminished. This is generally an inefficient method of operation if we are to rely on the medium for grinding and dispersion. Therefore, we can see that if we are to grind a material that is relatively large (greater than 100 microns) to a submicron particle size a two stage operation is generally required. This means processing the material through a 2 mm or larger medium first. Processing the material in smaller beads then gives a micron or submicron particle size. We feel this is illustrated in Figure 10.5 where a ceramic tile glaze required a minimum quantity of grit on a 325 mesh screen. Progressively larger beads produced a tighter distribution.

The second two factors of mill operation, throughput rate and agitator speed

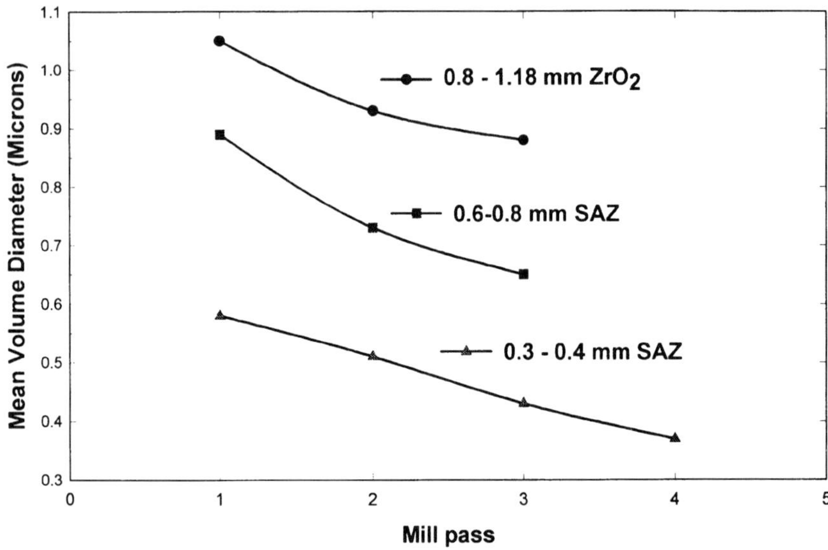

Figure 10.4 Particle size reduction as a function of media size for 25% cerium oxide in water.

can be clearly understood. Throughput rate is inversely proportional to residence time in the mill. The residence time (dwell time) is simply calculated by determining the grinding volume available in the mill and dividing this figure by the throughput rate. Residence time is the amount of time the material is in the grinding chamber and is exposed to the process. Therefore the higher the

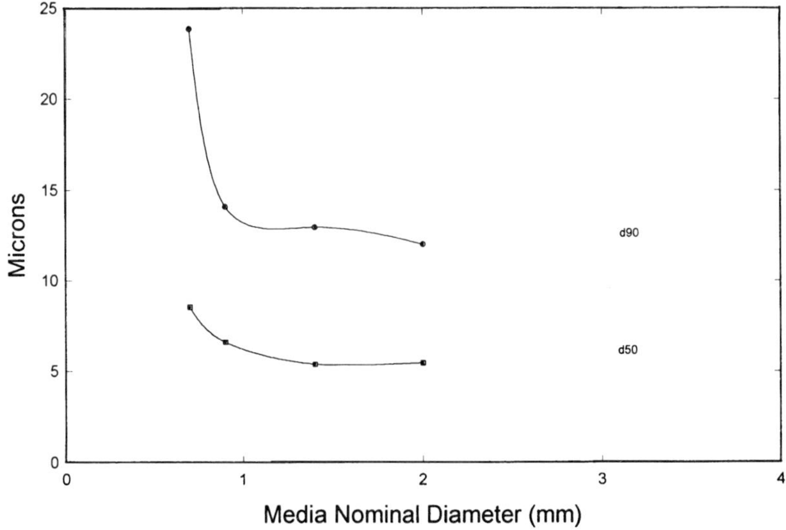

Figure 10.5 Top size reduction as a function of media size for 40% talc in water.

throughput rate the shorter the residence time. The shorter the residence time, the less grinding takes place.

An important consideration in operation of the horizontal mill is the flow rate through the machine. If a material is relatively easy to grind, obviously a high flow rate is desired. However, if the flow rate is too high, hydraulic packing of the medium can occur. This packing or pressure grinding of the medium results from the movement of the medium's mass from being evenly distributed in the grinding chamber to the discharge end. A concentrated mass of medium develops. This compacted mass of medium results in high product temperatures and high current draw on the motor through frictional contact with the medium. Fine particle sizes can be obtained at the expense of the medium and wear on the mill.

The agitator speed is the controlling factor of power input to the media. The higher the agitator speed, the more energy is applied to the volume of the medium. Higher kinetic energy in the medium through high agitator speed results in higher crushing and dispersion of the product. The agitator speed is limited by the motor power available and the solids (viscosity) of the material being ground. There is a limited amount of power available on a machine. If a high solids, high viscosity, slurry is processed on a machine with fixed rotor speed then motor overload may occur. This is why laboratory, pilot plant and small production mills are always supplied with variable speed drives.

10.5 Mill design

The media separator is a very important design feature of the small media mill. An effective media separation method can be the controlling factor for successful processing. The separator is relied upon to keep the beads in the chamber. If the separator is not designed properly, hydraulic packing of the medium occurs, resulting in reduced productivity, breakage of the medium or damage to the separator.

The controlling factor for selecting the type of separator and the size of separator is media size. A standard rule of thumb is to specify a separator slot width that is $\frac{1}{3}$ the size of the medium's smallest diameter.

The separator employed on a horizontal mill historically has been a rotating gap (Figure 10.6) or a static screen. Static screens are seldom used today because of the wear and maintenance problems associated with them. The gaps are limited by the size of the machine due to tolerances such as bearing run out and flow rate through the mill. A typical, practical gap size is 0.5 mm. Many manufacturers supply the machine initially with a 0.3 mm gap; however, after running the machine for a time it is more appropriate to say that the gap is 0.5 mm.

The rotating gap design works well for media larger than 1.5 mm, depending on machine size. As machines become larger, and higher flow rates through the

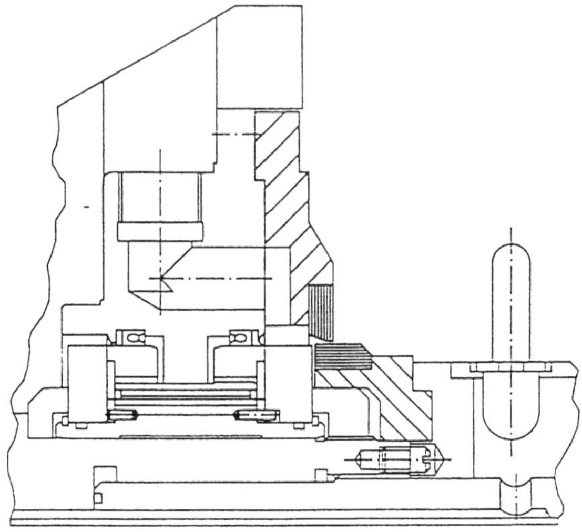

Figure 10.6 Rotating gap.

mill are expected, the limited open surface area of a gap separator results in hydraulic packing of the media. This is due to the increase in flow rate and the non-proportional increase in open surface area of the gap. The advantage to the rotating gap is that it can be made of ceramic materials for reduced product contamination and does not clog with product because it is dynamic.

When media in the 0.2 to 1 mm range are used it is necessary to use a horizontal mill with a dynamic cartridge screen separator (Figure 10.7). In this case, we can use a screen (2A) with a slot width of 0.05 mm or 50 microns, although typically a 0.1 mm or 0.2 mm screen is selected.

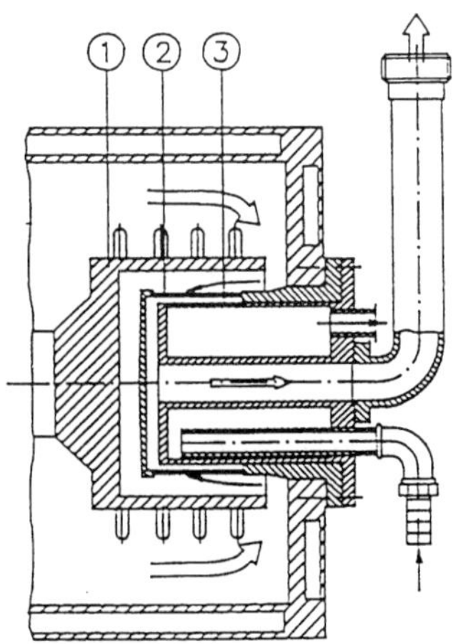

Figure 10.7 Netzsch-patented dynamic cartridge screen separator.

The patented Dynamic Cartridge screen's main advantage is an increase of 40 to 100 times the open surface area of a rotating gap separator. This large surface area allows high flow through rates of high solids, high viscosity dispersions while minimizing the hydraulic packing of the media.

The dynamic cartridge separator operates by creating a media free zone around the discharge end of the mill. When the product is flowing through the mill, the medium sweeps past the screen. The rotating or dynamic slotted pipe attached to the end of the agitator shaft surrounds the screen. With the agitator running, the slotted pipe centrifuges the higher mass media away from the screen. The medium passes through the slots in the pipe and is put back into the grinding zone of the mill. The sweeping action of the medium acts as a screen cleaning device, the centrifugal expulsion of the medium reduces hydraulic packing. This design prevents the screen from being in the direct action of the grinding media, reducing screen wear and increasing screen life five to ten times that of other screen separators. A significant reduction in machine down time for screen changes is a major advantage to this design.

When changing the batch on the mill or doing a color change, it is usually necessary to clean the separator. In a rotating gap design, this would entail removing all the grinding media, removing the chamber and rotating gap assembly. In a flat screen design the medium must be dropped and the screen removed to clean behind it. The dynamic cartridge separator simply requires the removal of its cover plate and accompanying discharge pipe. The screen can be rinsed with a tube brush and solvent. The cover place can then be reinstalled and the machine is ready for the next color.

10.6 Agitation systems

There are several agitation systems available. Each machine is designed for specific types of product or formulation. Machines range in size from 0.5 to 560 l. This broad range gives production rates of 0.05 to 560 l of slurry per minute.

The peg mill or John mill (Figure 10.8) was designed specifically for grinding and dispersing carbon black heatset inks. The main objective was to design a machine that would apply a uniform acceleration to the media. The John mill incorporates a large diameter shaft with short pegs for acceleration of the media. A chamber length to chamber diameter ratio of 3 : 1 is kept constant from laboratory horizontal mills to production size equipment. The result is a long annular grinding zone. The grinding tank has counter pegs attached to it. The effect is then a uniform acceleration of grinding media in a long narrow cylinder. As the medium is accelerated by the pegs on the shaft it streams past the counter peg on the chamber wall creating extremely high shearing and impacting forces to tear agglomerates apart.

As the medium stream collides immediately after the peg, impact grinding occurs for particle size reduction. The large number of pegs on the shaft and

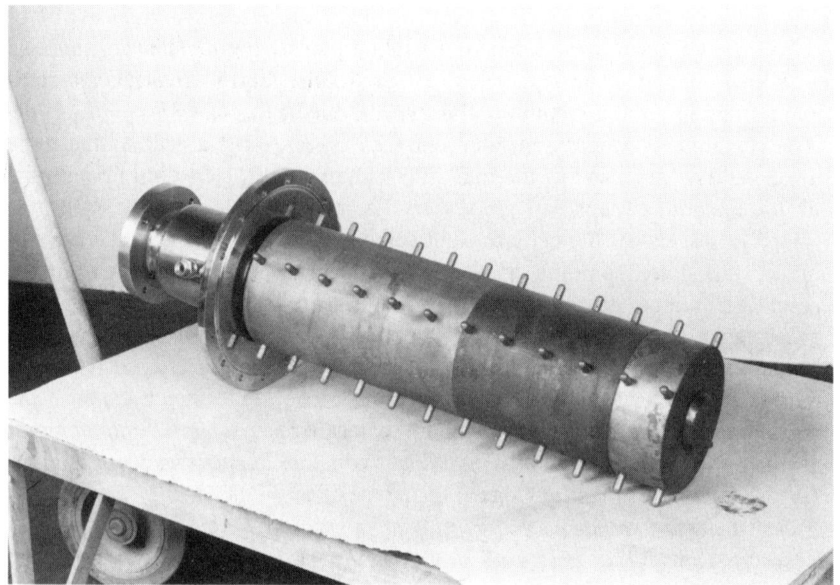

Figure 10.8 Netzsch John mill agitator shaft.

chamber can be considered ideal mixers and this improves the plug flow characteristics of the machine.

The pegs are made of tungsten carbide for extremely long life and low long-term maintenance costs. This system is available in vertical and horizontal configurations. It is most appropriate for very high viscosity, high solids dispersions. Urethane sealants in the range of 1 million centipoise have been processed successfully. A typical application would be a high solids tint base, 90% total solids, for industrial paint. This would then be reduced with a resin solution to a coating viscosity.

The John mill is also appropriate for using very fine media (0.25 mm to 0.45 mm). This relates back to the uniform acceleration required in a high viscosity product. The limitation of the John mill for processing ceramic materials is the steel construction of the machine. To construct the machine with ceramic material could be done, but is extremely expensive.

The eccentric disc or Molinex system is the most common system supplied. The Molinex agitator consists of discs eccentrically mounted to the driven shaft (Figure 10.9). The discs are cast with a half moon shaped hole near the eccentric edge. The action of the disc is to accelerate the grinding media centripetally from the inner edge of the disc and centrifugally from the outer diameter of the disc. This creates shearing and impacting forces between the disc and the medium next to the disc. Medium streams away from the disc in translational motion. The kinetic energy in the medium is transferred to all surrounding beads.

Figure 10.9 Netzsch Molinex mill agitator shaft.

When the shaft has been assembled and is rotated the eccentric mount of the discs forms an auger effect which pulses the grinding media against the product flow. The eccentricity of the discs and the orientation of the discs on the shaft create a pulse in the fluidized media bed. This pulse or wave of medium is either acted upon by the next disc coming around or is allowed to spread. The next disc coming around has a wave of medium related to it and the two waves coming together create high shear and impact forces. The absence of a disc coming around allows the wave to spread. This creates the pulsing effect that counteracts the product flow rate. This ensures the even distribution of medium in the chamber and reduces the hydraulic packing of the medium.

The Molinex system will process a broad range of formulations and the resulting viscosities. One hundred to 75 000 centipoise are the ranges typically measured. Solids levels range from 5% to 85%. This system is available in horizontal and vertical mills. The Molinex system design is very easily adapted for ceramic applications or processes requiring minimum contamination from the machine by manufacturing the discs and other components of the shaft (spacers, endcap and wear collar) from ceramic or polymer materials. This is the most important part of the machine to consider when designing or selecting materials of construction because it is the highest point of wear apart from the media grind.

A subject that has recently seen more study is the residence time distribution in the mill. This is the measurement of where particles are when pumped through a mill. The residence time calculation is related to the flow rate and chamber volume. This is useful information when scaling up work from a lab mill to a production mill. In reality the particle's travel through the chamber is a tortuous journey. Studies on residence time distribution show that though we would like to think that particles are moving through the mill at uniform flow and velocity, plug flow, the particles come through at many different velocities resulting in what is called the residence time distribution. The result is that in a horizontal mill a low percentage of material escapes the grinding process through intermixing created by the agitator shaft [4]. Experiments show that a high agitator speed in a disc mill gives a narrower residence time distribution. This is most likely due to the increased axial velocity of the media.

Typically, the horizontal mill is run in a single pass, continuous operation. For most applications this is sufficient. In some cases there will be a low percentage of large particles and this will be unacceptable. Therefore a second pass through the mill will be run to 'clean up' these oversize particles. On very difficult-to-grind materials, the general method of decreasing the particle size distribution is to increase the residence time. However, an increase in residence time usually results in a broader residence time distribution leading to a broad PSD. Again a second or third pass is required to eliminate the large particles. Therefore it is better to run multiple passes through the mill than to run a single slow pass. There have been many studies done by mill manufacturers and companies using mills to determine the residence time distribution in a mill [4, 5]. We believe Figure 10.10 illustrates this by showing the difference in particle size distributions between slow passes of calcium carbonate and high recirculating passes.

The findings of these studies confirm that there is a percentage of material that bypasses the grinding operation and a percentage that remains in the mill for longer than the calculated residence time. From this work we learn that the methods for obtaining a narrower residence time distribution or uniform grinding and dispersion are controlled by agitator design, agitator speed and, most importantly, flow rate. Therefore if a very fine particle size, below 1 micron, and a tight size distribution are required, a high flow rate with multiple passes would be more efficient than three or more slow flow passes. The limitation of a standard horizontal mill is the possibility of hydraulic packing. The Molinex system has the best chance of eliminating hydraulic packing by increasing agitator speed and increasing the pulsing against product flow. The trade off is that discs mill designs are not as efficient when using media smaller than 0.5 mm as peg mill designs.

The LMZ or Zeta system was developed to grind particles that are very difficult to reduce to submicron particle sizes. A design is required that will increase the open surface area of the discharge and increase the activation of the media. This will allow high flow rates without the resulting hydraulic packing. This mill has been achieved by combining the dynamic media separator and the John mill. The machine is designated the Zeta system (Zeta selected as the first letter from the German word for circulation). The dynamic media separator on the Zeta systems is enlarged to nearly the length of the mill and accordingly has high surface area. The agitator design based on the John system creates very intensive media agitation. These two systems in conjunction create a media bed that is almost completely centrifuged from the screen media separator with the agitation pegs thoroughly activating the media charge with impacting and shearing forces.

Flow rates through the mill are from 8 to 30 l/min for a 10 l mill volume. The residence time in the mill is reduced to 20 to 60 s versus 2 to 10 min on conventional mills. The product being ground is then either passed through the mill in a tank to tank system or recirculated. Lower flow rates are used on the tank to tank system; this system is generally preferred for products that are very large in initial particle size, the recirculating mode uses the high flow rate to ensure turnovers on the batch being ground. This means the theoretical number of times the entire batch passes through the mill in a given amount of time. For example, if a 100 l batch is being run on an LMZ 11 at a recirculating rate of 20 l/min, the theory is that every 5 min the batch has been through the mill or turned over. Obviously there is intermixing in the tank; however, we have found that as long as the minimum number of turnovers is 12 [4], all the material will

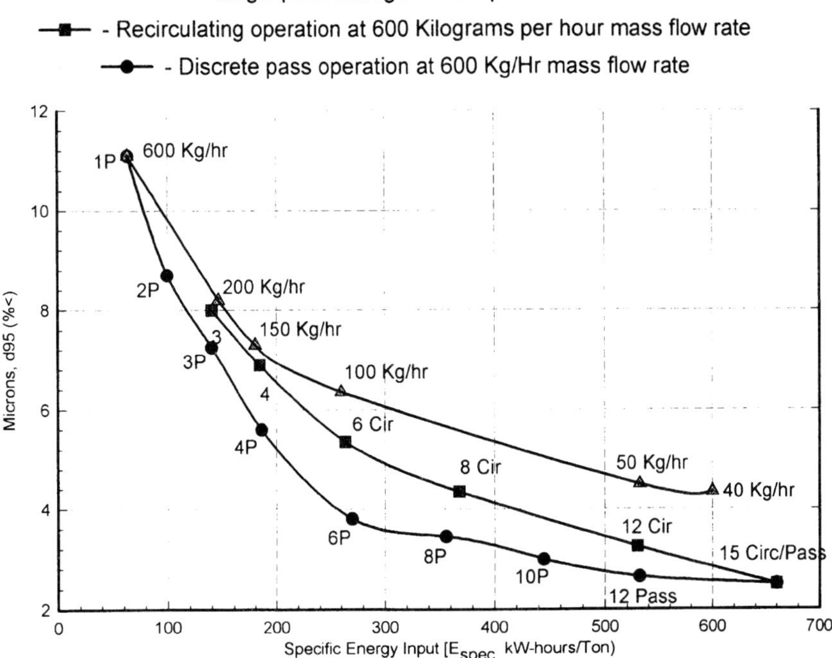

Figure 10.10 ZETA system: single pass, multiple pass and recirculating operation with dependance of d95 on specific energy input and mode of operation.

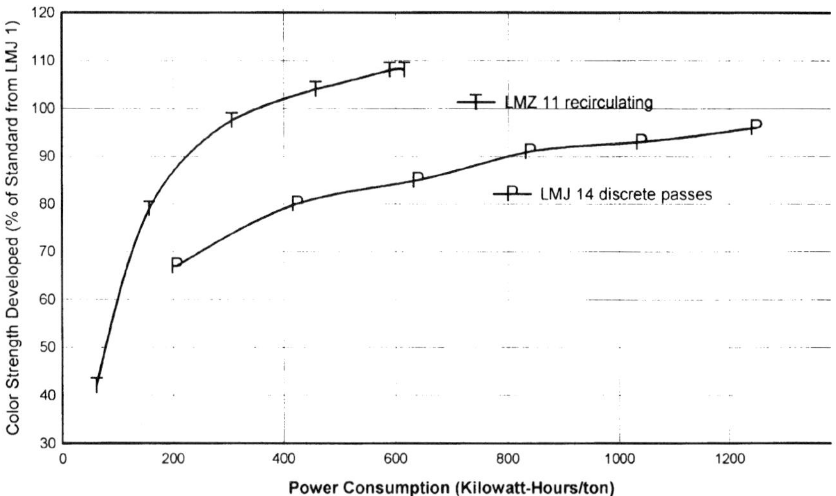

Figure 10.11 Comparison of grinding systems: pin mill vs. Zeta system for crude phthalocyanine blue pigment conversion using 250 micron steel shot as the grinding medium.

be uniformly dispersed. Figure 10.11 is a comparison of the Zeta system versus John system and Zeta system versus Molinex system.

10.7 Applications

The machines handle many types of slurry formulations, ranging in solids levels by weight as low as 5% to 90%. The liquid carrier system ranges from simply water to complex solvent/resin/surfactant formulations. Generally, any formulation that is made in a ball mill or other media milling machine will process through the horizontal mill. Usually the formulation can be optimized to increase both productivity and quality through reduction of resin content and solvent content. The lower viscosities required in the ball mill and other devices are not required in the media mill. Conversely, there is also the fact that if fine media are used, low viscosities are generally required. A high viscosity in a fine medium results in separation of the medium in terms of no point-to-point contact within the medium/mill base mixture. Also when very fine media are used in high viscosity products, separation is inhibited.

During the past few years we have seen interest in using these machines for the production of various ceramic slips. These applications have ranged from ceramic slips for producing ceramic bodies such as wear tiles to slips used in the electronic industry. Materials processed successfully are alumina oxide, alumina monohydrate, alumina trihydrate, barium titanate, ceramic fibers (alumina silica product), cerium oxide, glass powders, graphite, mullite, silicon carbide, silicon dioxide, tile glazes, titanium dioxide, tungsten carbide, zinc oxide and zirconium

Table 10.1 Ceramic milling in Netzsch Small Media Mills: test results

Medium	Feed material (% dispersant/ solids)	Throughput rate (liters per minute)	L&M Microtrac SPA: Feed particle size (microns, mean volume diameter)	L&M Microtrac SPA: Milled particle size (microns, mean volume diameter)	Machine used for test	Grinding medium material size millimeters	Power consumption (kW/ton dry)
Alumina monohydrate	0/30	0.06	20.83	0.91	LMC 4	SA% 0.4–0.6	768
Alumina monohydrate	0/30	0.06	0.91	0.73	LMC 4	SA% 0.3–0.4	126
Alumina oxide (abrasive polishing compound)	10/54	0.71	No data taken	2.0	LMC 4	SA% 1.6–2.5	100
Alumina oxide (magnetic coating additive)	0.5/40	0.16	26.17	1.4	LMC 4	SA% 0.6–0.8	690
Alumina oxide (magnetic coating additive)	0.5/55	0.17	1.86	1.12	LMC 4	SA% 0.6–0.8	410
Alumina oxide (magnetic coating additive)	0.5/55	0.09	1.00	0.68	LMC 4	SA% 0.6–0.8	840
Alumina oxide (alpha, lapping compound)	0.00/50	0.2	27.01	0.79	LMSM 4	SA% 0.6–0.8	440
Alumina oxide (alpha, lapping compound)	1.0/50	0.96	27.01	1.65	LMSM 4	SA% 0.6–0.8	90
Alumina oxide (alpha)	1/60	0.09	12.12	2.26	LMC 4	A120 0.8–1.0	475
Alumina oxide (alpha)	0/40	0.08	12.12	1.44	LMC 4	A120 0.8–1.0	680
Alumina oxide slip (wear tiles)	1.9/67	0.32	30.00	3.86	LME 4	SA% 0.8–1.25	180
Alumina oxide slip (wear tiles)	1.0/76.5	0.42	30.00	4.32	LME 4	A1203 1.5–2.5	130
Alumina trihydrate	0/12.3	0.88	8.40	3.01	LMC 4	SA% 1.2–1.6	420
Alumina trihydrate	0/12.3	4.80	8.40	2.81	LMC 20	SA% 1.2–1.6	310
Automotive slip (Al_2O_3, TiO_2, Fe_2O_3, clay)	5.0/70	1.02	30.00	1.30	LMSM 4	SA% 1.6–2.5	60
Barium titanate	5.00/60	0.14	100	1.17	LMC 4	SA% 1.2–1.6	200
Barium titanate	0.00/88	0.27	no data	1.0	LMSM 4	SA% 0.6–0.8	70
Barium titanate	0.00/60	0.30	no data	1.45	LMSM 4	%r02 0.8–1.18	120
Barium titanate	0.00/60	0.64	no data	2.21	LMSM 4	%r02 0.8–1.18	60
Ceramic tile glaze	3.0/70	3.20	47 grams grit on 325 mesh screen	1 gram on 325 mesh screen	LME 20	Mullite 3 mm	70
Ceramic tile glaze	3.0/70	5.40	47 grams grit on 325 mesh screen	0.48 grams on 325 mesh screen	LME 20	Mullite 3.4–4.8	50
Cerium oxide	0/29	0.10	8.72	0.37	LMC 4	SA% 0.3–0.4	760

oxide. Table 10.1 shows some of the results. Many more products have been processed; however, most of our work is proprietary.

10.8 Conclusions

Our intent is to convey to the reader an understanding of the more essential factors of the horizontal small media mill design. The use of a small media mill for grinding or dispersing ceramic powders is a burgeoning application. We hope to publish more detailed information on the application of the small media mill in this sector of industry. Points to consider are:

1. The small media mill is a viable alternative to the ball mill for particle size reduction and dispersion. The mill has been proven for ceramic applications
2. Finer and narrower particle size distributions can be obtained with a minimization of contaminants through the use of ceramic or polymer machine components. Submicron particle sizes are achieved more efficiently relative to conventional dispersion equipment
3. There are many factors involved in selecting the design conditions for a small media mill. The mill manufacturer has a pilot testing facility and offers the opportunity to test the machines on a no charge basis
4. The machines are available in size range that cover the range from bench scale development to pilot plant studies and production in the tons per hour range
5. Table 10.1 illustrates the point made in this paper concerning particle size reduction and media size. We intend to expand on in a future paper that principles of media milling and formulation in terms of solids content and power consumption

References

1. Chandler, R.H. (1970) Modern grinding and dispersing technology, *Paint Manufacture*, June issue.
2. John, W. *From The Drum Mill to the Annular Chamber Mill.*
3. Potters Industries. (1980) *Technical Quality Glass Spheres.* Conshohocan, PA.
4. Kob, G. and Ott, K. (1992) Efficient agitator bead mills. *European Coatings Journal*, **4**, 212.

11 Dispersion of low viscosity water based inks

J.A. SCHAK

11.1 Introduction

Problems such as lower viscosities, foaming, new heat sensitive surfactants, varying parameters during processing, and the fine tuning of recipes have presented new challenges to formulators, dispersion equipment design engineers and the manufacturing team. The most formidable obstacle is a foaming tendency during the initial wetout of the dry pigments. Entrained air acts as a shock absorber to cushion the energy that would break up the aggregates. The result is impractical downstream multiple media mill passes.

This chapter presents innovative techniques for handling these difficult to deagglomerate concoctions. New designs for a rotor-stator mill have resulted in deaeration while being able to handle varying viscosities at lower operating temperatures. The impact has been so good that in many instances a media mill is not required. The elimination and/or reduction in passes of the media mill suggests shorter processing times, less equipment to maintain, less downtime, less cleanup, less waste and less media contamination.

At the formulation stage a manufacturer can have the best dispersion device ever made and still may not be able to achieve the desired result if he/she does not have the right dispersant package. Too much surfactant may create foaming which will buffer the dispersion and too little will allow the particles to reagglomerate. A successful product will be a marriage of recipe and equipment. Experience assists in determining tendencies but the 'proof is in the pudding'. Actual lab testing is a necessity. Advances in computational fluid dynamics (CFD) have been able to model rotor-stator mills to aid in predicting performance but this should be used to explain empirical results. The test combination of ingredients and equipment arrangements is endless. The formulator will have enough difficulty choosing the right variations of chemistry without having some logical guide for test equipment selection. We will review various equipment designs to minimize guesswork.

11.2 Background

Ink production has been viewed as a three stage approach (Figure 11.1).

1. Premixing. Various wet and dry ingredients are blended together. In a well controlled process, some 'minimum grind' on the NPIRI grind gauge is

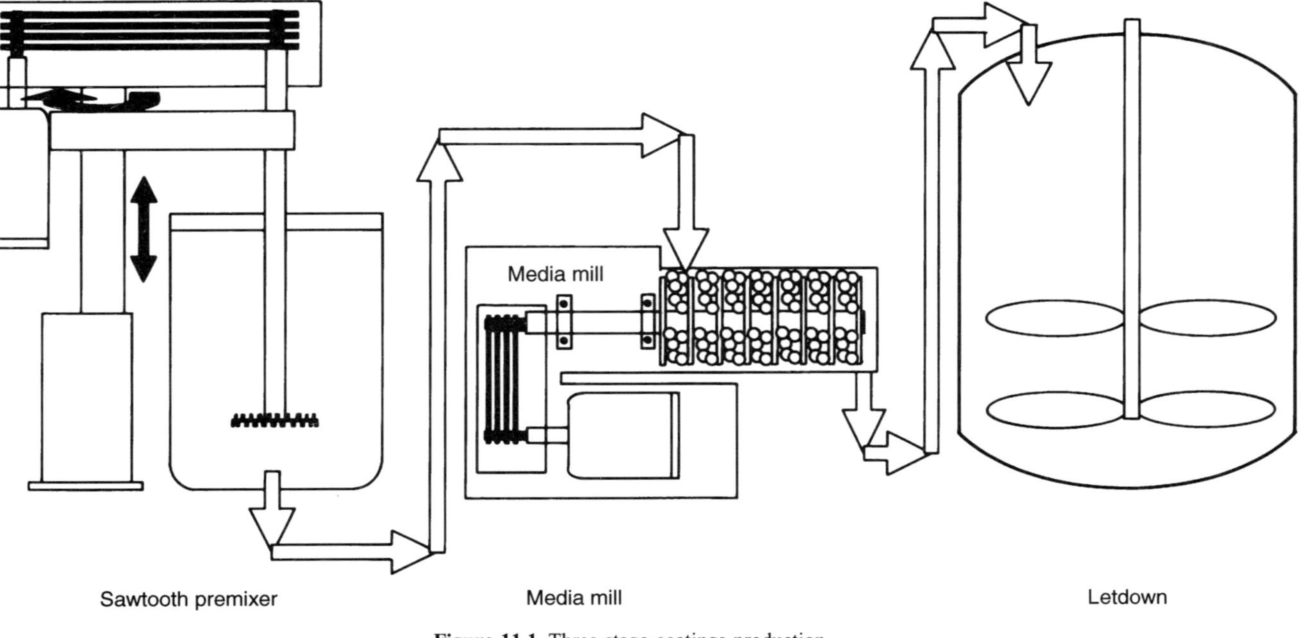

Figure 11.1 Three-stage coatings production.

achieved, but it is more common to run for a predetermined amount of time, such as 30 min.

2. Milling. The pigments are mechanically deagglomerated until a final particle size is reached.
3. Letdown. Various additives are introduced along with any vehicle that was held out during premixing.

11.3 Major dispersion steps

An evaluation of dispersion equipment efficiency will require an understanding of what is needed to obtain the final result. The high surface tension of aqueous based dispersions require higher levels of surfactant to avoid flocculation and achieve a stable product. This is the reason for the high foaming tendencies. The four major steps in achieving a high quality dispersion are:

1. Wetting of pigment. The dispersion process must first evenly distribute the pigment into the vehicle.
2. Deaeration. The replacement of air and moisture in the pigment with vehicle.
3. Deagglomeration of the pigment. Agglomerates are formed either as the pigment was manufactured or at the wetting stage.
4. Stabilization. The particles, after they are wetted and uniformly dispersed, must be stabilized so that they do not reaggregate. This is accomplished by introducing an opposite electrical charge to negate attraction. Surfactants or binders aid in stabilization.

11.4 Dispersion equipment design

The wetting and deagglomeration of the pigment is obtained by exerting mechanical energy into the product and overcoming the forces that bond the particles together. The amount of force required is a function of the final particle size desired and the characteristics of the matrix materials. The particle bonds can be a result of: unbalanced surface charges at the crystal faces, dipole fields (similar to magnetic fields) on the crystal, direct cementation of crystals via reaction by-products or impurities or resins, and mechanical interlocking formed by pigment manufacturing drying process. The transference of bond breaking energy is via impact, attrition, shear or compression. Equipment examples (see Figure 11.2) that use these forces are enclosed impeller, open impeller, sigma blade kneader, roll mills, and media mills. The kneader and the roll mill are designed for high viscosity and will not be addressed in this paper.

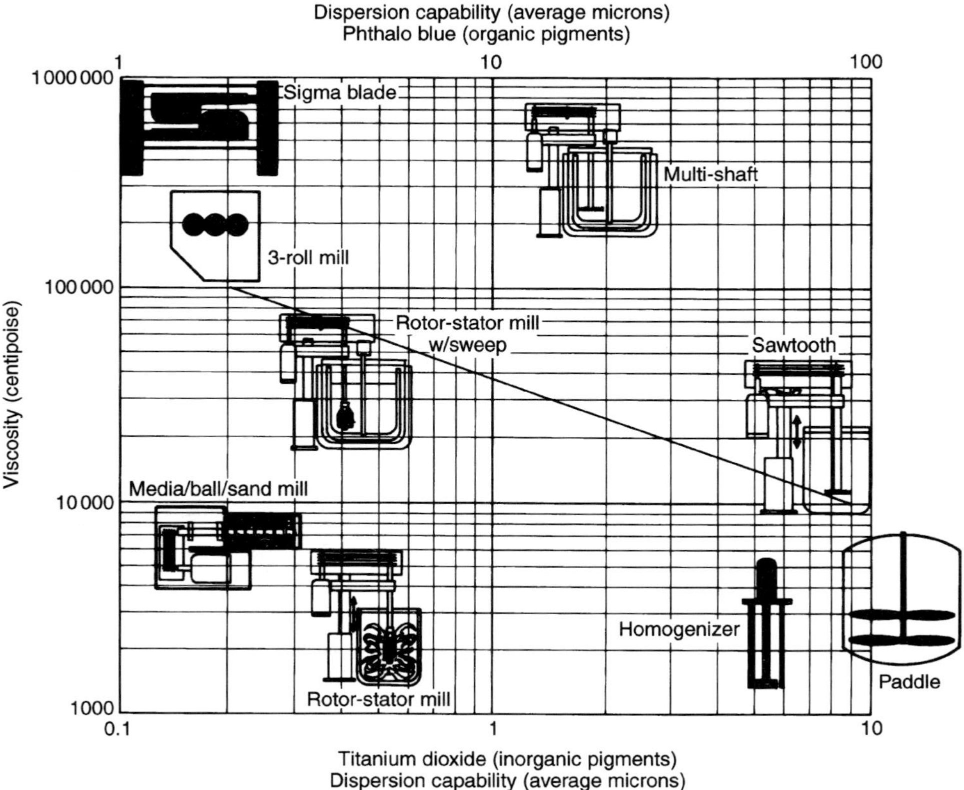

Figure 11.2 Equipment guide (viscosity versus particle size).

11.5 Choosing a dispersion line

There is no singular piece of equipment capable of handling every milling application efficiently. There are too many factors that affect the dispersion process to allow uniform processing such as: viscosity, vehicle, pigment hardness, surface charges, final grind requirements, solids content, color strength, gloss, aeration tendencies, temperature limitations, formulations etc. We will start at the simplest design then graduate to more involved scenarios.

11.6 One-step processes

There is great benefit in limiting the process to the least number of components. The goal of any designer is to do everything in one tank. The advantages of a one step are obvious such as:

1. Vessel can be enclosed or set up for vacuum to contain vapors (a plus for environmental compliance).
2. Cleanup is reduced to one vessel.
3. Minimum operator attention can be required with less chance of operator error.
4. Minimal external pumps and hoses are used.
5. Less chance of cross-contamination.
6. The letdown can be done in the same vessel.
7. Less equipment means less cost and less maintenance.
8. The operation can be easily instrumented and a batch history for ISO 9000 can be recorded with variables.
9. Less chance of foaming.

Typical one step process equipment are as follows:

1. High speed disperser (HSD) or 'sawtooth blade' mixer.
2. Rotor/stator mixer (5000 feet/minute rotor tip speed).
3. Rotor-stator mills (9000+ feet/minute rotor tip speeds).

11.6.1 *'Sawtooth' blade or HSD mixers*

High speed dispersers (or open impeller, Figure 11.3) are distributive mixing device good for initial wetting of pigments at the premix stage. This unit relies on mechanical shear as its primary input with a secondary force of attrition. Key design parameters are as follows:

1. The optimum rotor tip speed is between 4000 and 5200 fpm. Blade tip speed (fmp) = circumference (ft) × shaft speed (rpm).
2. Blade diameter should be 1/3 the tank diameter.
3. 1 horsepower per 10 gallons.
4. Blade distance from the bottom should be 1/2 blade diameter.
5. Relies on vortexing to wetout the pigment.

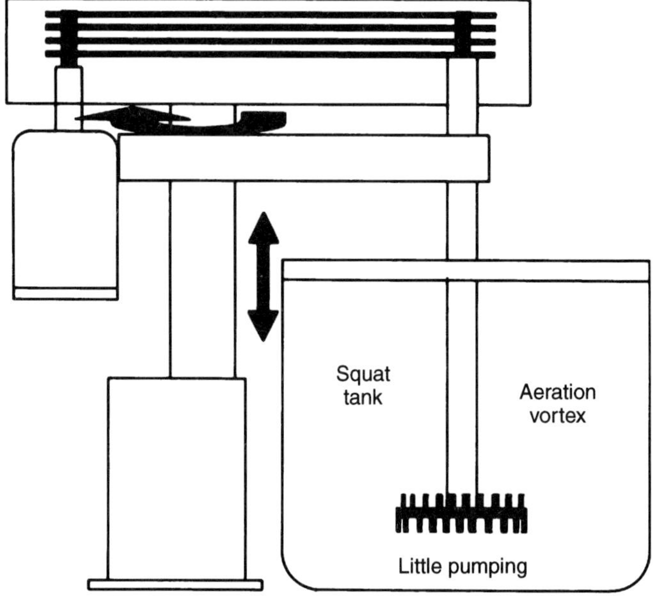

- Needs high viscosity to induce shear for dispersion
- Little shear at low viscosity
- Fixed geometry and fixed batch sizes
- Needs vortexing (aeration)

Figure 11.3 Sawtooth blade mixer (4–5000 ft/min. tip speed).

This sawtooth blade mixer offers cost savings and simplicity. One will probably be able to get a reasonable grind on the softer pigments with higher viscosity (> 10 000 cps). At lower viscosities (< 10 000 cps) the 'sawtooth' mixer tends to push the liquid around without imparting any appreciable shear. The formulator must look at this design to 'hold out liquid' during the mixing stage to achieve higher viscosities and higher shear rates.

Summarizing the advantages of the 'sawtooth' mixer is as follows:

1. Simplicity and cost.
2. Multiple tank servicing. It can be equipped with a lift and a swivel to service one to four tanks.
3. No mechanical seals with minimal cost for parts.
4. Inexpensive device for wetting out prior to media milling.
5. Easy to clean.

The major drawback to the sawtooth blade mixer is its requirement to vortex in the pigment. This allows air to be sucked into the liquid and cause foaming. Other limitations of the 'sawtooth' mixer are:

1. Limited to dispersing of whites to get final grind.
2. Ineffective in handling low viscosity premixes.
3. Difficult to deagglomerate organic pigments requiring multiple media mill passes. Sometimes not even able to get on the NPIRI scale. Inefficient larger media in the media mill are required to break down the +100 micron particles.
4. Additional pre-letdown stage is sometimes required prior to a media mill to lower viscosity and enhance media mill performance.
5. Aeration.
6. Non-repeatable results.

11.6.2 *Rotor-stator homogenizer (5000 feet/min tip speeds)*

The rotor-stator homogenizer is a rotating prop impeller surrounded by a fixed screen. The mixer design was developed for emulsification and homogenization but has been used effectively for wetting out low viscosity dispersions where the 'sawtooth' mixer is ineffective. This unit is designed more for distributive mixing rather than dispersive mixing. The advantages of this design are:

1. Good for low viscosity formulations.
2. Same advantages for one-step processing.
3. Relatively easy to clean.
4. Low cost.
5. No mechanical seal.

The major drawback to homogenizers is that they use a high speed exposed shaft that pumps air into the liquid and creates form (no mechanical seal is used). The limitations are:

1. A bushing located at the head is a maintenance problem.
2. Limited to final grinds on some whites and yellows.
3. Relatively low shear rates (20 000 inverse/s).
4. Ineffective on organic pigments.

11.6.3 *Rotor-stator mills (9000+ feet/min tip speeds)*

These mills differ greatly from the rotor-stator homogenizer/mixers. They not only run at faster tip speeds but Figure 11.4 shows other major differences. A true centrifugal pump impeller is used to accelerate the liquid against the stator causing the pigment to impinge against the fixed stator. This principle is similar to a snowball impacting against a wall. This basic design has been used for many years in the coatings industry but fell out of favor due to the following:

1. Requirements for finer grinds (achieved in the media mill).
2. Temperature limitations of the surfactants used.

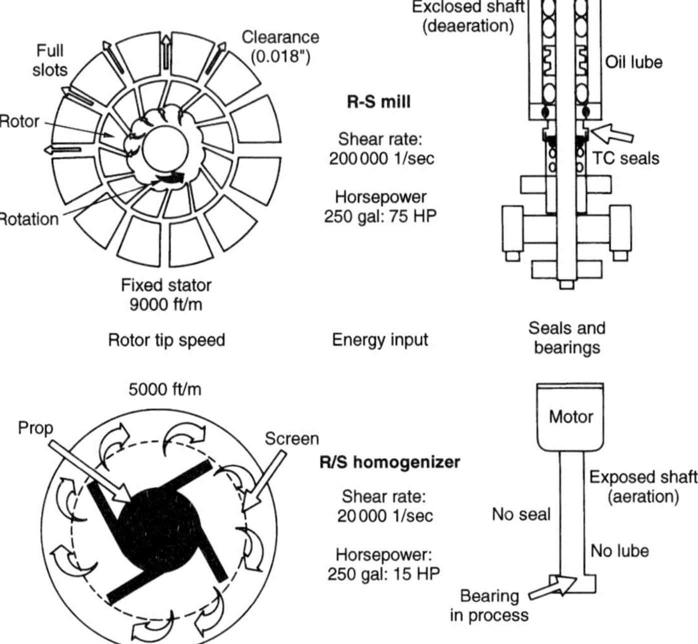

Figure 11.4 R-S mill versus R-S mixer/homogenizer.

3. Viscosity limitations of 10 000–15 000 cps.
4. 30 year old seal technology.

The major difference in the use of the rotor-stator mill is its capability to deaerate. The high speed shaft is sheltered from the process liquid with an outer tube which is sealed by a mechanical seal which separates the process liquid from the lube oil. An exposed high speed shaft will pump air along the shaft and cause excessive turbulence at the surface. Vortexing can be avoided through baffle and rotor design or the use of external sweep or prop to move the batch rather than the rotor-stator head. If no vortexing takes place and no exposed high speed shafts are pumping in air there is a liquid seal that will not allow air in. The only other air with which to be concerned is the air trapped in the pigment particles. The rotor-stator centrifuges the liquid in a very thin film against the stator and separates out the less dense air. One will notice air bubbles that flow to the top and eventually disappear.

Other new developments have allowed these high speed mills to have a wider window of operation to process even organic pigments to a '1–5 NPIRI' in a one-step process:

1. New patent pending rotor/stator design via CFD that actually creates cavitational forces within the stator.

2. Flow assisted designs with sweep arms or paddles to isolate the rotor-stator in performing deagglomeration without moving the total batch. The rotor-stator can now be sized independently to the size of the batch tank (Figure 11.5).
3. The sweep arm allows a larger batch vessel without increasing rotor diameter and horsepower. Less horsepower means less heat per gallon and a larger tank means more cooling jacket area. Heat sensitive materials with a maximum temperature of 110°F can now be processed using the rotor-stator mill design.
4. The sweep arm gives the rotor-stator a wider and higher viscosity capability.
5. Newer and more effective surfactants making it easier to disperse hard to grind pigments.
6. New rotor-stator head designs with flow assist have been able to deaerate these new stronger surfactant (surface active agent) formulas.
7. State-of-the-art positive driven tungsten carbide seal design has allowed the rotor-stator mill to operate at higher rotor tip speeds with minimal maintenance.

A good example of the new flow assisted design is that a major pigment manufacturer needed ten passes in the media mill to make a '3 NPIRI' on an aqueous based dispersion. They had read the 'premilling' article which appeared in the July 1994 issue of PCI magazine and were hoping to cut down a few passes using a rotor-stator mill. A test was run and a final grind of 3 was

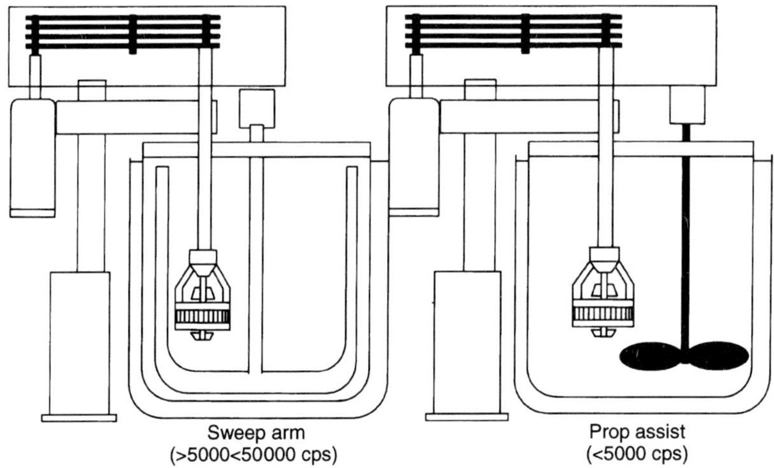

Figure 11.5 Rotor-stator mill with flow assist.

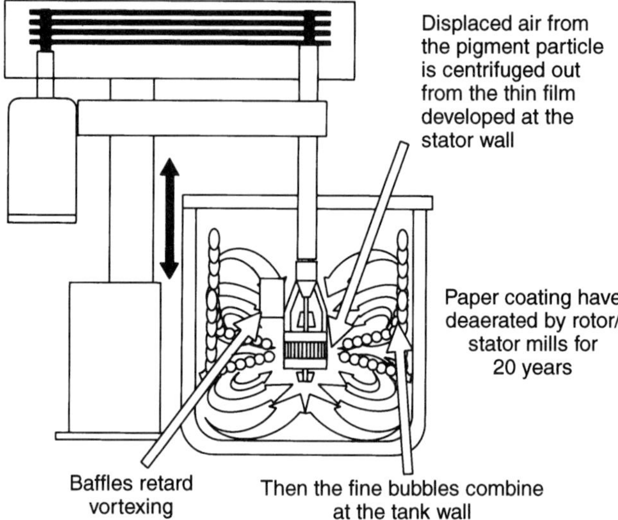

Figure 11.6 Deaeration in a rotor-stator mill.

achieved in 1 h in the one-step operation with a maximum temperature of 140°F In looking at the process, the premixer aerated the batch. This is a major problem for any dispersion device since the air tends to act as a shock absorber and cushions the dispersion. Once the air was removed in the rotor/stator mill the strong effective surfactant made it relatively easy to disperse this organic pigment. The rotor-stator mill at the high tip speeds develops a thin film against the stator and separates out the air bubbles that were generated when air inside the pigment was displaced with vehicle (Figure 11.6). The bubbles then combine and float to the top.

There will be situations where it will not be possible to obtain the final grind in one step processing and these will require secondary processing. A media mill will then be required. Media mills operate at very low capacity and the goal here is to minimize the equipment and maximize throughput.

11.7 Multiple step processes (final media milling)

Media milling is the next step to achieve further size reduction. In order to optimize the performance of the media mill we need to understand its design criteria.

11.7.1 *Small media mill design*

Media mills [1] operate by inducing impact from the collision of media with itself and the vessel walls. Shear is introduced as adjacent media particles pass by one another. They are most effective over a small particle size range reduction. This is due to the important relationship between media and agglomerate size. The rule of thumb is that the most efficient media diameter needs to be about 10 times the size of the initial largest pigment [2].

If the agglomerates are too large, they tend to act as media which effectively eliminate shear as a dispersion tool. The role of impact is also reduced since the media particles have lost some of their size and weight advantage. Attrition now becomes the major contributor to any work being done. Because of its media size and shaft speed the media mill is simply not well suited for attrition. (This is why the basket mill needs a premix.)

If the agglomerates are too small, they tend to get lost in the mill. Impact becomes ineffective as there are not enough media balls to do the job. This is due to an increase in the number of pigment particles during the dispersion process. The number of total particles will be increased by a factor of 8 by grinding its original size in half. In this case attrition is all but eliminated leaving only shear as the means of dispersion.

Advantages of media mill design:

1. Submicronic grind capability.
2. Inline processing.

Limitations of media mill design:

1. Slow production rates.
2. Need a good premix to be efficient.
3. Difficult to clean.
4. Expensive in relation to production rate.
5. Narrow range of particle size reduction.

We have not included the basket mills in the one-step processing because it is a media mill with the same restrictions as we have stated with the standard media mill. The basket mill will not operate efficiently if it is not given a good premix. In most cases it is recommended that a 'sawtooth' mixer be used to make the premix. In the rare cases where a premix is not required the media diameter must be large enough to wet out the pigment which will decrease efficiency and limit the final grind capability. There is also a limitation to the viscosity range it processes, with an additional headache of cleaning. The advantage of this design is to handle low viscosity small fine grind batches of dedicated products that can be processed in the same tank. The new redesigned rotor/stator mill with a sweep arm has the following advantages over a basket mill:

1. Premix and final grind in one-step.
2. Deaeration of the batch.
3. Wider window of operation (higher viscosity).
4. Faster processing.
5. Better suited for premilling.

6. Continuous inline conversions.
7. Easy to clean.

The multiple processing approach should use the following progression to maximize efficiency and minimize testing:

1. 'Sawtooth' blade mixer followed by a small media mill.
2. Rotor-stator mill followed by a micro media mill.
3. Rotor-stator mill followed by small media mill then a micro media mill.
4. Rotor-stator mill with a hi-flow recirculation mill.

11.7.2 *'Sawtooth' blade mixer followed by a small media mill*

This is the most simple two-component system approach and most widely used media mill production setup (Figure 11.7). The 'sawtooth' blade mixer will probably give a premix with a 100 micron particles. The correct media size using the 10:1 ratio should be 1000 microns. If the final particle size is 25 micron (or a 10 NPIRN), you will probably be successful in one pass. (A 4:1 size reduction relationship was stated in the 'premilling' *PCI* article to maintain a single pass.) The problem is if a finer result is desired say 10 microns then it is likely that two media mill passes are required. An interim step (if the viscosity is very low) could be a low speed rotor/stator mixer. Typically the next step would be a rotor-stator mill operating at 9000 ft/m as a premixer and premilling device prior to the media mill.

11.7.3 *Rotor-stator mill followed by a media mill*

The high energy rotor-stator mill offers more flexibility in optimizing media mill design (Figure 11.8). The 'premilling' article described how to

> maintain milling efficiency one must choose the media by the final grind requirements, not the initial particle size. The product then must be 'premilled' to deliver particles within the 10/1 range. A rule of thumb for choosing the media diameter is not well defined but a four fold reduction usually gives an efficient design without sacrificing production rate or requiring multiple passes.

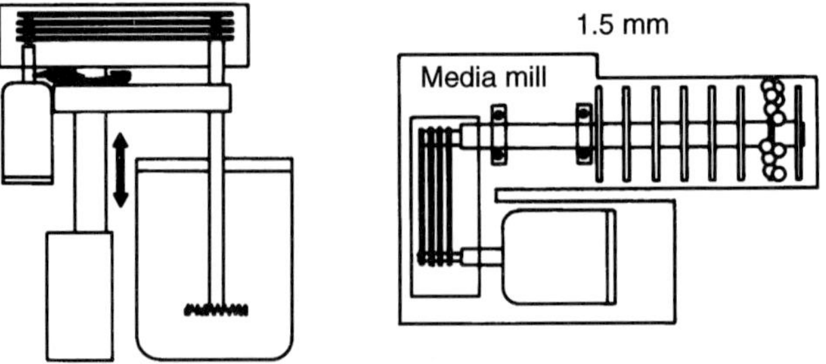

Figure 11.7 High-speed disperser followed by a horizontal media mill (2 + passes).

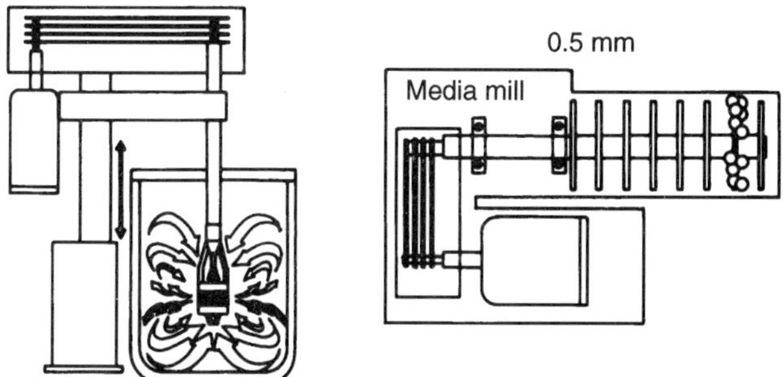

Figure 11.8 Rotor-stator mill (9000 ft/min rotor tip speed) followed by a small media mill.

If we look at an example for obtaining a 10 micron particle we should have a premix of 40 microns to maintain the 4 : 1 particle size reduction ratio. Then using the 10 : 1 ratio of media diameter to the largest agglomerate in the premix one needs to use a 0.5 mm diameter bead.

This micro (0.5 mm and 0.25 mm) media diameter can cause the following problems:

1. The screen plugs easier especially if the 'grit content' (percentage left on a 325 mesh screen) is too high in the premix.
2. The flow rate has to be reduced if plugging occurs.
3. The screen will plug if too much of the pigment settles out in the premix tank.

The rotor/stator mill addresses these problems and has additional advantages over a 'sawtooth' premixer:

1. No settling in the premix tank.
2. Deaeration of foamy formulations.
3. Reduced 'grit content'.
4. The better the premix, the faster the rate through the media mill.

11.7.4 *Rotor-stator mill followed by a small media mill then a micro media mill*

Many of the ink dispersions require a submicron particle size (Figure 11.9). In this case use the smallest media possible which is currently 0.25 mm. If we go back and use our 10 : 1 rule of thumb for premixing and a 4 : 1 size reduction, we will need a premix of 4 microns to get below 1 micron. This may be too much to ask for a premixing device so what we would probably need to use is an intermediate media mill to get to a 4 micron 'premix'. This 'piggyback' or 'cascading' approach would require a premix of 16 microns to get to a 4 micron feed to the final mill. A 0.5 mm bead may be used in the intermediate mill in situations with a slightly larger premix prior to final milling is required.

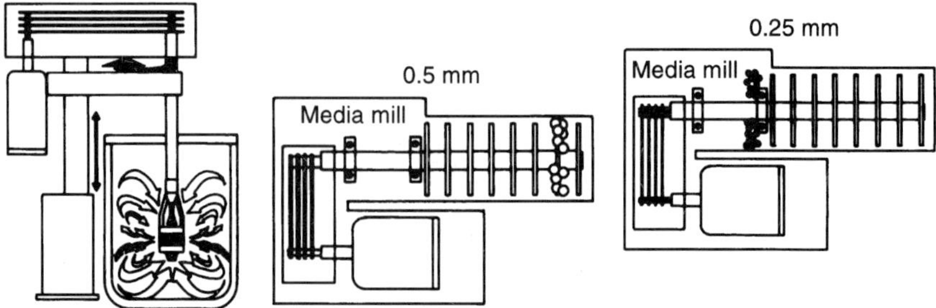

Figure 11.9 Rotor-stator mill followed by a small media mill, followed by a micro media mill.

11.7.5 *Rotor-stator mill with a hi-flow recirculation mill*

A relatively new approach for making submicronic dispersion is to recirculate with many passes but at high flow rates. The problem with using a 'sawtooth' premixer is that the initial passes must be kept at very low flow rates to get the particle size down to an optimum 10 : 1 ratio which defeats the purpose of the hi-flow recirculation. The rotor-stator mill will quickly give a much finer 'premill' faster and let the media mill do what it was designed for. The advantage of this design is that more energy can be put in faster with a more uniform particle size. The disadvantage is that it generates more heat. This is one of the reasons one sees a sweep arm (Figure 11.10) in the rotor-stator mill premix tank. The rotor-stator mill can be turned off after the desired 'premilled' particle size is reached. The sweep arm (which also can be cored for cooling) is kept rotating and the product can be cooled with the jacketed walls of the tank.

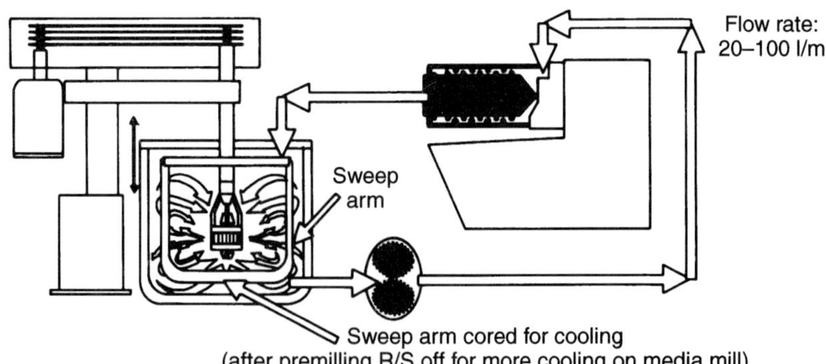

Figure 11.10 Rotor-stator premiller on hi-flow recirculation mill.

11.7.6 *Example of rotor-stator mill with a hi-flow recirculation mill*

Crude blue is one of the most difficult pigments to grind due to the following:

1. Large amounts of surfactant are required for the 0.2 micron final particle size. The formulation generates foam and large amounts of defoamer are needed to wetout in a conventional sawtooth blade mixer.
2. A huge amount of energy must be imparted into the crude blue crystals in order to convert it to the desired phthalo blue.
3. The viscosity is very low initially then it increases in the final stages of processing.
4. The surfactant used can be heat sensitive (130–160°F).

One ink company decided to try a rotor-stator mill to reduce media mill passes in making its crude blue formulation. In a conventional process of using a sawtooth for the premix and using a horizontal small media mill with 0.5 mm, it required at least five passes (50 h for a 1000 gal batch). It also required about 1 gal of defoamer in a 1000 gal batch. The initial loading of the premix (without defoamer) in the rotor-stator mill created some concern since some foam appeared at the top of the batch. In 15 min the foam was completely eliminated due to the proper wetout and the removal of air instead of pigment. In 2 h the grind was off scale on scratches and fine enough so the 0.25 mm high flow recirculation media mill would not clog and could operate at full speed. The rotor-stator motor was turned off while the sweep arm in the same jacketed tank was kept on to continue to cool the product as it was recirculated through the media mill. The processing time was reduced to 12 h instead of 50 h.

11.7.7 *Debottlenecking existing dispersion lines*

There are literally thousands of existing dispersion lines that require inefficient multiple media mill passes. Many manufacturers are saddled with existing equipment and will find it difficult to replace these systems due to operating budgets or even convincing management to scrap 'perfectly good' equipment. There is hope. A new approach which was hatched in response to the huge interest in the 'premilling concept' was an inline continuous rotor-stator mill unit on an existing sawtooth mixing tank (Figure 11.11).

11.7.8 *Continuous inline rotor-stator mill*

The use of inline rotor/stator mixer (5000 ft/m rotor tip speeds) have been used for making emulsions for years but they lack the power to have a great effect on fine dispersions. Inline continuous rotor-stator mills have been sold for 15 years in niche applications and were not generally thought to be as effective as batch units and were not marketed to the ink industry. Recent developments in the rotor-stator design via CFD such as the first patent pending rotor-stator that creates cavitation within the stator has given this approach new life as part of the

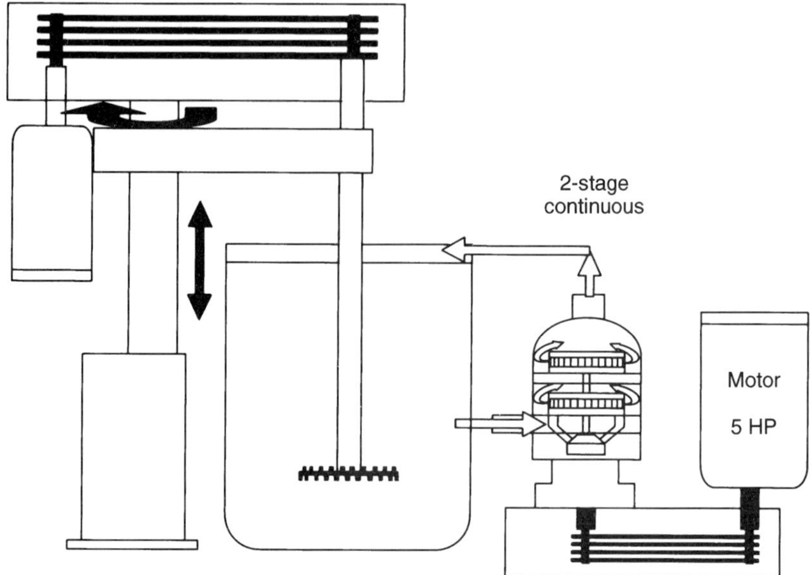

Figure 11.11 Continuous rotor-stator mill on a 'sawtooth'.

'premilling' revolution. The idea is to convert an existing mixer whether it be a 'sawtooth' or a paddle mixer to a rotor-stator mill.

The flow arrangement is depicted in (Figure 11.12). The batch is pumped through the continuous inline unit at a high rate (40–100 gpm on a 1000 gal tank). In some arrangements a pump is not required to push the material through the rotor stator but it can give more flexibility to control flow at an optimum rate

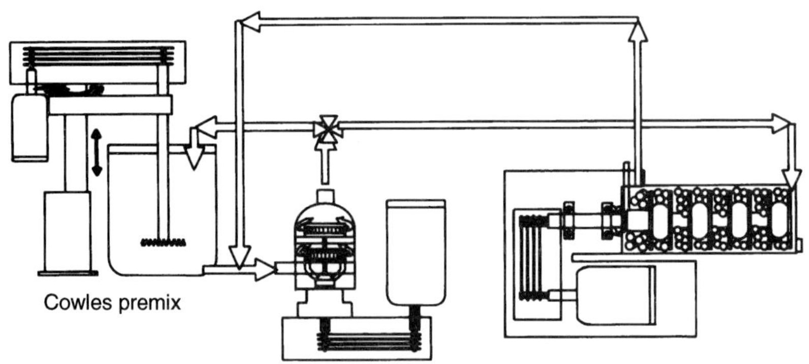

- Deaeration
- No settling
- Convert existing sawtooth to a mill
- Less heat/time

Figure 11.12 Inline premiller for retrofit and higher viscosity.

through the continuous unit. Multiple passes are required to achieve a 3 or 5 NPIRI on organic pigments. This will take 1–3 h of recirculation dependent on the size of inline unit and the flow rate. Flow rates of 900 gpm are being used in the paper coatings industry to deaerate and deagglomerate.

The advantages of the continuous inline unit are:

1. Utilize and optimize existing equipment.
2. Possible final grind in the same premix tank.
3. Portability. The unit can be put on wheels for multiple tank processing.
4. Recovering of bad batches.
5. Wide viscosity range. If you can pump it you can get it through the continuous unit.
6. Unit can be sized independent of tank size. There are some 30 HP units operating on 2500 gal tanks and have doubled the rate through the sand mills.
7. Less heat is generated with smaller units (will take longer).
8. Deaeration can be achieved if the tank is vented and no vortexing is taking place in the premix tank.

The disadvantages are:

1. The standard batch rotor-stator mill is still preferred since it is one-step processing and can handle multiple tanks and there are less components to clean.
2. Most 'sawtooth' mixers do not have jacketed tanks and the high energy mill will raise the temperature.

11.8 Summary

The trend to aqueous based formulations has burdened existing dispersion lines due to formulation, aeration and difficulty to deagglomerate. The use of vacuum to deaerate is cumbersome, time consuming, and expensive, the conversion of existing equipment is difficult. The proper use of a rotor/stator mill without the use of vacuum will deaerate almost all aqueous based formulations. This can make an enormous impact on production rates, downtime, cleanup and environmental impact. Testing is recommended to determine the impact with various formulations.

References

1. Excerpt taken from 'Premilling can optimize your dispersion process.' Published by *PCI* magazine, July, 1994.
2. Dombrowski, L. (1982) Optimizing small media mills. *American Paint & Coatings Journal* 29 November and 6 December issues.

Appendix A: Trouble shooting guide

Part 1: Manufacturing

Defects	How recognized	Probable cause	Suggested remedy
Adhesion, poor ink	Ink fails to adhere to material.	1. Improper ink formula.	1. Make certain you have the correct ink for the kind and grade of stock being run.
	Fails tape test.	2. Ink thinned too much, binder destroyed.	2. Restore and hold ink viscosity at optimum point. Add fresh ink or resin binder.
	Fails crinkle test.	3. Insufficient heat; too low web temperature.	3. Increase heat and/or air volume.
	Fails scuff test.	4. Lack of treatment of some materials.	4. Check surface of the material for adequate treatment.
		5. Surface of stock may be contaminated.	5. Check with supplier of stock; check advisability and effectiveness of applying wash-coat before printing.
Bleed	An under color wetting into an over color. Diffused or migrating colors.	1. Under color drying too slowly or over color drying too fast.	1. Use faster or slower solvent as required. (Preceding colors must be dry enough to receive subsequent colors laid down.)
		2. Effect of some plasticizers of some stocks or materials on dye stuffs.	2. Avoid use of dye colorants when unknown plasticizers are likely to be involved.
Blocking	Undesired adhesion between two web surfaces.	1. Improper ink drying.	1. Proper solvent balance.
		2. Trapped solvents.	2. Effective drying system or solvent balance.
		3. Excessive pressure in rewind.	3. Reduce rewind tension.
		4. Softening of preapplied coatings.	4. Use solvents that do not attack prior coatings.
		5. Web rewound too warm.	5. Reduce web temperature by chilling within 10°F of room temperature or reducing dryer temperature.
		6. Web rewound with excess surface moisture.	6. Avoid rewinding excess surface moisture into finished roll. Avoid over-chilling that condenses moisture on web prior to rewind

Defects	How recognized	Probable cause	Suggested remedy
Brittleness	Stock or material that breaks when flexed.	Excess heat in drying system causing release of moisture and plasticizer from paper or film.	1. Control web or film temperature. 2. Introduce moisture into 'after' drying chamber or tunnel. 3. Direct web through remoisturing chamber. 4. Reduce heat and increase volume of air through drying chamber.
Chatter	Parallel lines of misprint.	Gear bottoming off pitch line.	Adjust plate thickness to pitch line.
Drying too fast	Ink drying on plates and/or roller and failing to transfer to stock or materials.	1. Improper use of solvents.	1. Proper solvent selection.
		2. Uncontrolled or unrestricted air movement in the vicinity of plates and rollers.	2. Properly balanced between color dryers. No excessive air movement in vicinity of plates or rolls by fans or open windows, doors, etc.
		3. Failure to use fountain covers.	3. Use fountain covers.
		4. Improper ink for material.	4. Proper selections of ink for material being printed.
		5. Dried ink on plates from start up.	5. Wash plates after color OK and before full speed run.
Drying too slow	One color bleeding into another (improper trapping).	1. Use of improper solvents.	1. Use of proper solvents.
	Ink pick off or transfer to press rollers and/or plates.	2. Ink viscosity too heavy.	2. Controlled ink viscosities. Make viscosity cup checks each 20 and 30 min.
	Ink offset or blocking.	3. Inadequate or unbalanced drying system.	3. Adequate and balanced drying system to accommodate press speeds.
	Ink penetration of paper.	4. Improper ink selection.	4. Proper ink selection. One having more hold-out capability.
	Tacky surface of overprint.	5. Lack of slip compound in ink.	5. Add wax to ink.
Feathering	Irregular string-like edges around print, often on trailing edges.	1. Improper pressure between ink transfer roll and plate	1. Set proper pressure between ink transfer roll and plate.
		2. Ink drying on the plate or anilox.	2. Use proper solvents and fountain covers.
		3. Improper setting of ink roller nip.	3. Control pressure setting of ink roller nip.
		4. Improper ink rollers.	4. Select proper ink rollers for design and material being printed.
		5. Uncontrolled ink viscosity.	5. Maintain proper ink viscosity. Check it more often.
		6. Lint being picked up from material on plate.	6. Clean or dust material when necessary and avoid having ink too tacky.
		7. Static electricity.	7. Use static eliminator bars.

Defects	How recognized	Probable cause	Suggested remedy
Fill-in	Piling or specking of excess ink on and around printing surface of plates, especially with relation to small type and screen dots.	1. Specks of ink pigment. 2. Foreign matter in ink. 3. Excessively fine type, design, or anilox screen selection. 4. Souring or precipitation of ink vehicle solids caused by moisture absorption by the solvent in wet ink film. 5. Too much pressure. 6. Rubber roller too soft. 7. Excessive plate inking. 8. Ink drying too fast.	1. Use well dispersed inks. 2. Strain ink. 3. Check design for suitability for flexo work, i.e. sufficient open areas, proper typeface, etc. 4. See section under 'Souring'. 5. Check pressure settings. 6. Use 10 points higher durometer. 7. Select finer metering roll or improve wipe. 8. Use slower solvent. Speed up press.
Film curl	Curl over on edges of coated cellophane	1. Curl toward coated side. 2. Curl away from coated side.	1. Excessive moisture. Introduce heat to edges. Use heat guns. 2. Lack of moisture. Introduce moisture to edges. Use moist sponge.
Flat process printing	Lack of contrast. Hazy, milky	1. Poor molds and plates. 2. Excessive plate pressure. 3. Viscosity too low. 4. Contamination of primary colors by any other color.	1. Prepare molds to proper floor. 2. Print with kiss impression. Use compressible sticky back. 3. Raise viscosity. 4. Replace with pure transparent process colors.
Ghosting	A faint image of a plate or plate portion within a printed solid	1. Two in-line plates being inked from same area of the metering roll before re-inking or the metering roll can take place. 2. Improperly cleaned metering roll.	1. Change diameter of plate cylinder or change speed differential of fountain roll to metering roll. 2. Thorough cleaning with proper brush.
	A hazy, colorless image of the print on inline unprinted areas of film.	Moisture or solvent trapped in rewound roll and migrated onto the back of the wrap of film.	Adequate solvent removal, drying and reduced tension.
Halo	An unwanted line surrounding a printed image.	1. Excess pressure between plate and material. 2. Poor plates or makeready.	1. Reset plate pressure and ink viscosity. 2. Perform necessary makeready.
	Double edge printing.	3. Cupped edges of plate. 4. Plate cylinder running out of round. 5. Plate durometer too hard. 6. Stickyback too thin and too firm.	3. Make new plates and new plate mold, avoid cupping. 4. Check plate cylinder, shafts and journals and gears for run-out. 5. Use recommended rubber plates for printing given materials. 6. Use compressible stickyback.

Defects	How recognized	Probable cause	Suggested remedy
Ink heavy/ strong	Excess ink on anilox roll or plates.	1. Insufficient solvent and/or high viscosity.	1. Reduce ink to proper viscosity.
		2. Improper setting of ink roller nip.	2. Properly adjust roller nip. Increase or decrease nip pressure.
		3. Improper ink rollers for ink transfer.	3. Consult roller manufacturer for proper specification for printing on given materials and type of printing to be processed.
		4. Solvent imbalance.	4. Consult your ink formulator for optimum 'make-up' solvent mixture.
		5. Anilox doctor blade improperly set.	5. Adjust doctor blade to shave metering roll more cleanly.
		6. Improper anilox.	6. Select anilox roll with lower cell volume or ink carrying capacity.
Ink weak/ thin	Lacking color strength.	1. Too much solvent or viscosity too low.	1. Add fresh ink and bring to proper viscosity.
		2. Improper setting of ink roller nip.	2. Adjust roller nip to allow more ink on metering roll.
		3. Improper ink rollers for ink transfer.	3. Consult roller manufacturer for proper specification for printing on given materials.
		4. Worn anilox roll.	4. Replace with new or re-engraved roll.
		5. Ink settled.	5. Thoroughly mix ink from container before adding to fountain.
		6. Excess extender varnish.	6. Add pigment concentrate (toner)
Ink mottling	Spotted or speckled appearance of solid print. Dark or colored intermittent streaks, usually in a light color.	1. Ink too weak or thin, and loss of tack.	1. Add fresh ink and bring to proper viscosity. Add varnish to increase tack.
		2. Surface of plates uneven.	2. Remake plates if mottled appearance is evident on surface. Also examine plate mold for same mottle pattern.
		3. Anilox roll cell walls too wide.	3. Check wear condition of anilox lands. Replace anilox roll.
		4. Foreign matter on surface of plates.	4. Wash plates thoroughly.
		5. Dirty or pitted impression cylinders.	5. Clean impression cylinders thoroughly of all foreign matter, inks, waxes, etc.
		6. Unevenly absorptive materials.	6. Softer plates may sometimes help; more opaque ink may help.
		7. Uneven surface materials.	7. Softer plates should be tried.
		8. Ink starvation from anilox.	8. Adjust with richer or slower drying solvent.
		9. Contaminated ink.	9. Clean and strain ink or replace with fresh ink.
		10. Ink lacks proper flow.	10. Consult ink supplier for reformulation.

Defects	How recognized	Probable cause	Suggested remedy
Misregister	One part of design not correctly positioned with another	1. Incorrect plate-up; plates not mounted in register.	1. Study plate-up, mounting and makeready procedures.
		2. Incorrect web tension.	2. Study procedures involving auxiliary equipment, P.I.V. speed variators, unwind and rewind tension controls.
		3. Incorrect draw roll adjustment.	3. Check draw roll parallel constant side to side pressure, center wear condition.
		4. Excessive web temperatures.	4. Reduce dryer temperatures.
		5. Failure to center press register compensators before putting job in press.	5. Center forward, backward and side register compensators before manually keying in job register.
		6. Idle rolls dragging or running intermittently.	6. Free all idle rolls so that they turn freely mechanically. Drive at web speed if necessary.
		7. Press out of line.	7. Check and record press alignment.
		8. Gauge variations in stock.	
Moire	Undesirable dot pattern in process printing.	1. Anilox screen count too similar to plate screen count. For instance, a 165 line anilox used with an 85 line screen plate can cause moire, because of harmonic $85 \times 2 = 170$, too close to 165 line.	1. Proper solvent selection.
		2. Improper screen angles.	2. Refer to information on process printing for proper screen angles.
Odor	Undesirable odor in printed substrate	Retained solvents.	1. Balance solvents and check for proper solvent.
			2. Check dryer efficiency and temperatures.
			3. Increase between-color drying.
			4. Reduce line speed.
Offset or set-off	Transfer of ink to opposite side of material than that on which it is printed.	1. Ink not dried at rewind.	1. Use faster solvents. Check between color and final dryers for obstructions and adequate air and temperature. Reduce ink viscosity to minimum acceptable; increase wipe between fountain and anilox rolls.
		2. Trapped solvents.	2. Same as above. Reduce ink film to minimum acceptable.
		3. Excess pressure in roll.	3. Reduce rewind tension.
		4. Film treated on both sides for printing so that ink contacts opposite treated surface, such as some polyolefin films.	4. Avoid excess pressure in rewound rolls. Apply offset powder to web before rewinding; overprint with nonblocking varnish if necessary.
		5. Film plasticizers subject to migration, such as some vinyl film.	5. Same as remedy number 4.

Defects	How recognized	Probable cause	Suggested remedy
Pick-off/ ink	First down ink transferring to subsequent plate	1. First color too slow.	1. Add fast solvent to first color; increase between-color dryer temperature.
		2. Second color too fast.	2. Add slow solvent to overprinting color; check dryer.
		3. Second-down color too high in viscosity.	3. Increase press operating speed and/or reduce viscosity of second-down color.
Pinholing	Small holes in solid print.	1. Physical surface of some materials.	1. Consult your material supplier.
		2. Failure of ink to form continuous ink-film laydown	2. Increase ink film thickness. Check impression, anilox roll to plate and plate to stock; add impression if setting is too light. Add pinhole compound to ink. If 2% or 3% fails to correct condition, call your inkmaker.
		3. Dirt on impression cylinder.	3. Clean impression cylinders.
		4. Ink drying too fast.	4. Use slower solvent or speed up press.
		5. Worn anilox.	5. Examine wear condition of anilox lands. Replace with new or re-engraved roll.
Plate swelling	Dimensionally larger and softer than when mounted on cylinder.	Ink or solvents not compatible with printing plates.	Check ink and solvents being used with your inkmaker and plate supplier.
Poor ink transfer	Insufficient ink being applied to stock	1. Ink reduced too much.	1. Rebuild tack and color strength addition of fresh ink or extender or both.
		2. Ink drying on plates.	2. See defect 'Drying too fast'. Retard drying with slower drying solvent or thinner; consult inkmaker.
		3. Surface of stock not receptive to ink.	3. Verify ink formulation with ink supplier for given material; check for 'souring'; add 'anti-pinhole compound'; check stock for treatment.
		4. Improper pressure among ink roller, transfer roller and plate.	4. Adjust pressure.
Screening	A definite pattern of small holes appearing in surface of solid print.	1. Ink drying too fast on anilox roll and failing to transfer properly to printing plate.	1. Use fountain covers. Retard drying speed of ink. Adjust exhaust on between-color dryers. Add flow compound to ink.
		2. Worn anilox roll.	2. Examine wear condition of anilox lands. Replace with new or re-engraved anilox.

Defects	How recognized	Probable cause	Suggested remedy
Skip print	Areas of plate failing to print.	1. Poor plates or makeready.	1. Check impressions to determine accuracy of plates. Secure new plates or do necessary makeready.
		2. Impression set too light.	2. Readjust impression settings.
		3. Failure to lock down deck.	3. Be sure printing sections are locked firmly in place when impression is properly set.
		4. Plate cylinder 'bounce'.	4. Check concentricity of plate cylinder and gears; plate cylinder shaft for flexing; journals and bearings for excessive wear. Consider nature of printing plate design. Check for dried ink in gear teeth.
Souring ink	Thixotropic. Loss of flow. Curdling.	Excessive moisture in ink due to humidity.	Keep fountains and reservoirs covered. Add *n*-propyl acetate.
Separation or 'kick-out'	Curdling and thixotropic (much like souring).	Presence of wrong solvent or excess of diluent.	Add rich true solvent to return ink to proper balance.
Streaks or smears	Blobs of lighter or darker color in unwanted areas on web.	1. Gel particles are undissolved.	1. Filler particles out in circulation system and clean plates and anilox.
		2. Uneven plates.	2. Check plates.
		3. Foam bubbles being printed.	3. Add antifoam to ink.
		4. Ink splash dripping into web.	4. Check fountain guards, ink sling rings and pans for overflow, leaks.
Streaks in web direction	Continuous dark lines through print.	1. Damaged fountain or anilox roll.	1. Sharp shot on anilox can cut fountain roll. Polish anilox and grind streak out of fountain roll.
		2. Nick in doctor blade.	2. Hone or replace doctor blade.
Web weave	Web failing to track or follow a true course through the press	1. Press out of alignment.	1. Check press alignment.
		2. A roller out of alignment.	2. Check individual rolls for alignment.
		3. Build up of ink, tape or foreign matter on press rollers.	3. Clean all rolls of foreign matter.
		4. Too much heat on web, such as for polyethylene.	4. Reduce web temperature.
		5. Web guides not operating or set properly.	5. Check and clean web guides regularly per manufacturer's instructions. Set web guides and position cores so that web will unwind and rewind at an equal measurement from the end of the respective shafts.
		6. 'Gauge' problems with stock.	6. Replace roll of stock with new roll.
Wrinkling	Wrinkles in substrate.	1. Baggy substrate.	1. Tape rollers at web edge to draw out wrinkles.
		2. Equipment misalignment.	2. Adjust all roller alignment.

Part 2: Flexographic packaging printing

Defects	How recognized	Probable cause	Suggested remedy
Ink mottling	Spotted or speckled appearance of solid print. Dark or colored intermittent streaks, usually in a light color.	1. Ink too weak, or too thin, and loss of tack.	1. Add fresh ink and bring to proper viscosity. Add varnish to increase tack.
		2. Surface of plates uneven.	2. Remake plates if mottled appearance is evident on surface. Also examine plate mold for like pattern of mottle.
		3. Anilox roll cell walls too wide.	3. Check wear condition of anilox lands. Replace anilox roll.
		4. Foreign matter on surface of plates.	4. Wash plates thoroughly.
		5. Dirty or pitted impression cylinders.	5. Clean impression cylinders thoroughly of all foreign matter, inks, waxes, etc.
		6. Unevenly absorptive materials.	6. Softer plates may sometimes help; more opaque ink may help.
		7. Uneven surface materials.	7. Softer plates should be tried.
		8. Ink starvation from anilox.	8. Adjust with richer or slower drying solvent.
		9. Contaminated ink.	9. Clean and strain ink or replace with fresh ink.
		10. Ink lacks proper flow.	10. Reformulate ink.
Drying too fast	Ink drying on plates and/or roller and failing to transfer to stock or materials.	1. Improper use of solvents.	1. Proper solvent selection.
		2. Uncontrolled or unrestricted air movement in the vicinity of plates and rollers.	2. Properly balanced between color dryers. No excessive air movement in vicinity of plates or rolls by fans or open windows, doors, etc.
		3. Failure to use fountain covers.	3. Use fountain covers.
		4. Improper ink for material.	4. Proper selections of ink for material being run.
		5. Dried ink on plates from start up.	5. Wash plates after color OK and before full speed run.
Blocking	Undesired adhesion between two web surfaces.	1. Improper ink drying.	1. Proper solvent balance.
		2. Trapped solvents.	2. Effective drying system or solvent balance.
		3. Excessive pressure in rewind.	3. Reduce rewind tension.
		4. Softening of preapplied coatings.	4. Use solvents that do not attack prior coatings.
		5. Web rewound too warm.	5. Reduce web temperature by chilling within 10°F of room temperature or reducing dryer temperature.
		6. Web rewound with excess surface moisture.	6. Avoid rewinding excess surface moisture into finished roll. Avoid over-chilling that condenses moisture on web prior to rewind.

Defects	How recognized	Probable cause	Suggested remedy
Skip print	Areas of plate failing to print.	1. Poor plates or makeready.	1. Check impressions to determine accuracy of plates. Secure new plates or do necessary makeready.
		2. Impression set too light.	2. Readjust impression settings.
		3. Failure to lock down deck.	3. Be sure printing sections are locked firmly in place when impression is properly set.
		4. Plate cylinder 'bounce'.	4. Check concentricity of plate cylinder and gears; plate cylinder shaft for flexing; journals and bearings for excessive wear. Consider nature of printing plate design. Check for dried ink in gears.
Web weave	Web failing to track or follow a true course through the press.	1. Press out of alignment.	1. Check press alignment.
		2. A roller out of alignment.	2. Check individual rolls for alignment.
		3. Build up of ink, tape or foreign matter on press rollers.	3. Clean all rolls of foreign matter.
		4. Too much heat on web, such as for polyethylene.	4. Reduce web temperature.
		5. Web guides not operating or set properly.	5. Check and clean web guides regularly per manufacturer's instructions. Set web guides and position cores so that web will unwind and rewind at an equal measurement from the end of the respective shafts.
		6. 'Gauge' problems with stock.	6. Replace roll of stock with new roll.
Brittleness	Stock or material that breaks when flexed.	Excess heat in drying system causing release of moisture and plasticizer from paper or film.	1. Control web or film temperature. 2. Introduce moisture into 'after' drying chamber or tunnel. 3. Direct web through remoisturing chamber. 4. Reduce heat and increase volume of air through drying chamber.
Ink weak/ thin	Lacking color strength.	1. Too much solvent or viscosity too low.	1. Add fresh ink and bring to proper viscosity.
		2. Improper setting of ink roller nip.	2. Adjust nip to allow more ink on metering roll.
		3. Improper ink rollers for ink transfer.	3. Consult roller manufacturer for proper specification for printing on given materials.
		4. Worn anilox rolls.	4. Replace with new or re-engraved roll.
		5. Ink settled.	5. Thoroughly mix ink from container before adding to fountain.
		6. Excess extender varnish.	6. Add pigment concentrate (toner)

Defects	How recognized	Probable cause	Suggested remedy
Streaks or smears	Blobs of lighter or darker color in unwanted areas on web.	1. Gel particles are undissolved. 2. Uneven plates. 3. Foam bubbles being printed. 4. Ink splash, dripping onto web.	1. Filler particles out in circulation system and clean up plates and anilox. 2. Check plates. 3. Add antifoam to ink. 4. Check fountain guards, ink sling rings and pans for overflow, leaks.
Film curl	Curl over on edges of coated cello.	1. Curl toward coated side. 2. Curl away from coated side.	1. Excessive moisture. Introduce heat to edges. Use heat guns. 2. Lack of moisture. Introduce moisture to edges. Use moist sponge.
Screening	A definite pattern of small holes appearing in surface of solid print.	1. Ink drying too fast on anilox roll and failing to transfer properly to printing plate. 2. Worn anilox roll.	1. Use fountain covers. Retard drying speed of ink. Adjust exhaust on between-color dryers. Add flow compound to ink. 2. Examine wear condition of anilox lands. Replace with new or re-engraved anilox.
Plate swelling	Dimensionally larger and softer than when mounted on cylinder.	Ink or solvents not compatible with printing plates.	Check ink and solvents being used.
Offset or set-off	Transfer of ink to opposite side of material than that on which it was printed.	1. Ink not dried at rewind. 2. Trapped solvents. 3. Excess pressure in roll. 4. Film treated on both sides for printing so that ink contacts opposite treated surface, such as, some polyolefin. 5. Film with plasticizers subject to migration, such as some vinyl film.	1. Use faster solvents. Check between color and final dryers for obstructions and adequate air and temperature. Reduce ink viscosity to minimum acceptable; increase wipe between fountain and anilox rolls. 2. Same as above. Reduce ink film to minimum acceptable. 3. Reduce rewind tension. 4. Avoid excess pressure in rewound rolls. Apply offset powder to web before rewinding or overprint with nonblocking varnish. 5. Same as remedy number 4.

Defects	How recognized	Probable cause	Suggested remedy
Drying too slow	One color bleeding into another (improper trapping).	1. Use of improper solvents.	1. Use of proper solvents.
	Ink pick off or transfer to press rollers and/or plates.	2. Ink viscosity too heavy.	2. Controlled ink viscosities. Make viscosity cup checks each 20 to 30 min.
	Ink offset or blocking.	3. Inadequate or unbalanced drying system.	3. Adequate and balanced drying system to accommodate press speeds.
	Ink penetration of paper.	4. Improper ink selection.	4. Choose an ink with more hold-out capability.
	Tacky surface of overprint.	5. Lack of slip compound in ink.	5. Add wax to ink.
Fill-in	Piling or specking of excess ink on and around printing surface of plates, especially with relation to small type and screen dots.	1. Specks of ink pigment.	1. Use well dispersed inks.
		2. Foreign matter in ink.	2. Strain ink.
		3. Excessively fine type, design, or anilox screen selection.	3. Check design for suitability for flexo work, i.e. sufficient open areas, proper typeface, etc.
		4. Souring or precipitation of ink vehicle solids caused by moisture absorption by the solvent in wet ink film.	4. See section under Souring.
		5. Too much pressure.	5. Check pressure settings.
		6. Too soft rubber roller.	6. Use 10 points higher durometer.
		7. Excessive plate inking.	7. Select finer metering roll or improve wipe.
		8. Ink drying too fast.	8. Use slower solvent. Speed up press.
Ghosting	A faint image of a plate or plate portion within a printed solid.	1. Two in-line plates being inked from same area of the metering roll before re-inking of the metering roll can take place.	1. Change diameter of plate cylinder or change speed differential of fountain roll to metering roll.
		2. Improperly cleaned metering roll.	2. Thorough cleaning with proper brush.
	A hazy, colorless image of the print on inline unprinted areas of film.	Moisture or solvent trapped in rewound roll and migrated onto the back of the wrap of film.	Adequate solvent removal, drying and reduced tension.
Pinholing	Small holes in solid print.	1. Physical surface of some materials.	1. Consult your material supplier.
		2. Failure of ink to form continuous film.	2. Increase ink film thickness. Check impression, anilox roll to plate and plate to stock; add impression if setting is too light. Add pinhole compound to ink. If 2% or 3% fails to correct condition, call your inkmaker.

Defects	How recognized	Probable cause	Suggested remedy
Pinholing *continued*		3. Dirt on impression cylinder.	3. Clean impression cylinders.
		4. Ink drying too fast.	4. Use slower solvent or speed up press.
		5. Worn anilox.	5. Examine wear condition of anilox lands. Replace with new or re-engraved roll.
Pick-off/ ink	First down ink transferring to subsequent plates.	1. First color too slow.	1. Add fast solvent to first color; increase between-color dryer temperature.
		2. Second color too fast.	2. Add slow solvent to overprinting color; check dryer.
		3. Second down color too high in viscosity.	3. Increase press operating speed and/or reduce viscosity of second down color.
Poor ink transfer	Insufficient ink being applied to stock	1. Ink reduced too much.	1. Rebuild tack and color strength by addition of fresh ink or extender or both.
		2. Ink drying on plates.	2. See defect 'Drying too fast'. Retard drying with slower drying solvent or thinner; consult your ink manufacturer.
		3. Surface of stock not receptive to ink.	3. Verify ink formulation for given material; check for 'souring'; add 'anti-pinhole compound'; check stock for treatment.
		4. Improper pressure among ink roller, transfer roller and plate.	4. Adjust pressure.
Bleed	An under color wetting into an over color.	1. Under color drying too slowly or over color drying too fast.	1. Use faster or slower solvent as required. (Preceding colors must be dry enough to receive subsequent colors laid down.)
	Diffused or migrating colors.	2. Effect of some plasticizers of some stocks or materials on dye stuffs.	2. Avoid use of dye colorants when unknown plasticizers are likely to be involved.
Souring ink	Thixotropic. Loss of flow. Curdling.	Excessive moisture in ink due to humidity.	Keep fountains and reservoirs covered. Add *n*-propyl acetate.
Streaks in web direction	Continuous dark lines through print.	1. Damaged fountain or anilox roll.	1. Sharp spot on anilox can cut fountain roll. Polish anilox and grind streak out of fountain roll.
		2. Nick in doctor blade.	2. Hone or replace doctor blade.
Wrinkling	Wrinkles in substrate.	1. Baggy substrate.	1. Tape rollers at web edge to draw out wrinkles.
		2. Equipment misalignment.	2. Adjust all roller alignment.

Defects	How recognized	Probable cause	Suggested remedy
Misregister	One part of design not correctly positioned with another.	1. Incorrect plate-up; plates not mounted in register.	1. Study plate-up, mounting and makeready procedures.
		2. Incorrect web tensions.	2. Study procedures involving auxiliary equipment, P.I.V. speed variators, unwind and rewind tension controls.
		3. Incorrect draw-rolls adjustment.	3. Check draw-roll parallel constant side-to-side pressure, center wear condition.
		4. Excessive web temperatures.	4. Reduce dryer temperatures.
		5. Failure to center press register compensators before putting job in press.	5. Center forward, backward and side register compensators before manually keying in job register.
		6. Idle rolls dragging or running intermittently.	6. Free all idle rolls so that they turn freely, mechanically. Drive at web speed if necessary.
		7. Press out of line.	7. Check and record press alignment.
		8. Gauge variations in stock.	8. Change rolls of stock.
Moire	Undesirable dot pattern in process printing.	1. Anilox screen count too similar to plate screen count. For instance, a 165 line anilox used with an 85 line screen plate can cause moire, because of harmonic $85 \times 2 = 170$, too close to 165 line.	1. Change anilox. Select a screen count appropriate for tone work.
		2. Improper screen angles.	2. Check screen angles.
Ink heavy/ strong	Excess ink on anilox roll or plates.	1. Insufficient solvent and/or high viscosity.	1. Reduce ink to proper viscosity.
		2. Improper setting of ink roller nip.	2. Properly adjust roller nip. Increase or decrease nip pressure.
		3. Improper ink rollers for ink transfer.	3. Consult roller manufacturer for proper specification for printing on given materials and type of printing to be processed.
		4. Solvent 'imbalance'.	4. Consult your ink manufacturer for optimum 'make-up' solvent mixture.
		5. Anilox doctor blade improperly set.	5. Adjust doctor blade, to shave metering roll more cleanly.
		6. Improper anilox.	6. Select anilox roll with lower cell volume or ink carrying capacity.

Defects	How recognized	Probable cause	Suggested remedy
Feathering	Irregular string-like edges around print, often on trailing edges.	1. Improper pressure between ink transfer roll and plate	1. Set proper pressure between ink transfer roll and plate.
		2. Ink drying on the plate or anilox.	2. Use proper solvents and fountain covers.
		3. Improper setting of ink roller nip.	3. Control pressure setting of ink roller nip.
		4. Improper ink rollers.	4. Select proper ink rollers for design and material being printed.
		5. Uncontrolled ink viscosity.	5. Maintain proper ink viscosity.
		6. Lint being picked up from material on plate.	6. Clean or dust material when necessary and avoid having ink too tacky.
		7. Static electricity.	7. Use static eliminator bars.
Halo	An unwanted line surrounding a printed image.	1. Excess pressure between plate and material.	1. Reset plate pressure and ink viscosity.
		2. Poor plates or makeready.	2. Perform necessary makeready.
	Double edge printing.	3. Cupped edges of plate.	3. Make new plates and new plate mold or matrix, avoid cupping.
		4. Plate cylinder running out of round.	4. Check plate cylinder, shafts or journals and gears for run-out.
		5. Plate durometer too hard.	5. Use recommended rubber plates for printing given materials.
		6. Stickyback too thin and too firm.	6. Use compressible stickyback.
Flat process printing	Lack of contrast. Hazy, milky.	1. Poor molds and plates.	1. Prepare molds to proper floor.
		2. Excessive plate pressure.	2. Print with kiss impression. Use compressible stickyback.
		3. Viscosity too low.	3. Raise viscosity.
		4. Contamination of primary colors by any other colors.	4. Replace with unadulterated transparent process colors.
Adhesion, poor ink	Ink fails to adhere to material.	1. Improper ink formula.	1. Make certain you have the correct ink for the kind and grade of stock being run.
	Fails tape test.	2. Ink thinned too much, binder destroyed.	2. Restore and hold ink viscosity at optimum point. Add fresh ink or resin binder.
	Fails crinkle test.	3. Insufficient heat; too low web temperature.	3. Increase heat and/or air volume.
	Fails scuff test.	4. Lack of treatment of some materials.	4. Check surface of the material for adequate treatment.
		5. Surface of stock may be contaminated.	5. Check with supplier of stock; check advisability and effectiveness of applying wash-coat before printing.
Odor	Undesirable odor in printed substrate.	Retained solvents.	1. Balance solvents and check for proper solvent.
			2. Check dryer efficiency and temperatures.
			3. Increase between-color drying.
			4. Reduce line speed.

Defects	How recognized	Probable cause	Suggested remedy
Separation or kick-out	Curdling and thixotropic (much like souring).	Presence of wrong solvent or excess of diluent.	Add rich true solvent to return ink to proper balance.
Chatter	Parallel lines of misprint.	Gear bottoming off pitch line.	Adjust plate thickness to pitch line.

Part 3: Packaging, gravure printing

Defects	How recognized	Probable cause	Suggested remedy
Abrasion	Cylinder wear.	1. Unground pigment. 2. Inherent character of pigment. 3. Solvent too fast (dry wipe). 4. Poor chrome. 5. Cylinder design. 6. No blade oscillation. 7. Excessive doctor blade pressure. 8. Porcelain in ink.	1. Better grind. 2. Pigment selectivity. 3. Balanced solvents. 4. Rechrome. 5. Remake. 6. Adjust. 7. Adjust. 8. Destroy ink and remake.
Picking	Imperfection in print with noticeable areas with no ink.	1. Slow drying ink. 2. Too little heat. 3. Too heavy an ink.	1. Add faster solvent. 2. Increase heat and air velocity. 3. Reduce viscosity with extender varnish.
Cobwebs	Filmy, web-like build up on doctor blade, impression roll, engraving, or press frame.	1. Applicator. 2. Air drafts at nip. 3. Dried ink. 4. Drying rate. 5. High viscosity.	1. Repair defect. 2. Excessive air movement must be stopped. 3. Check ovens for down draft on cylinder. 4. Check solvent mix. 5. Reduce viscosity to normal level.
Bleed	One color into another or a tint or stain of the color on the printed substrate.	1. Ink too slow. 2. Not enough heat. 3. Insufficient air. 4. Wrong ink.	1. Better solvent choice. 2. Check oven temperature. 3. Clock air velocity. 4. Reformulate.
Haze	1. Appearance of slight opacity in clear film. 2. Foggy appearance in ink film.	1. Slight roughness in cylinder. 2. Poor wipe. 3. Over-pigmented ink. 4. Poor chrome job. 5. Humidity too high. 6. Poor solvent comb. 7. Improper tension.	1. Polish cylinder. 2. Check doctor blade set up. 3. Add clear extender. 4. Remake cylinder. 5. Slow down ink. 6. Change solvents. 7. Check tension on stock.

Defects	How recognized	Probable cause	Suggested remedy
Snowflakes	Random, generally minute, spots of unprinted areas showing through a deposited ink.	1. Rough stock. 2. Too high viscosity. 3. Ink too fast drying.	1. Electrostatic assist (use polar solvents). 2. Lower viscosity. 3. Slow down ink.
Crawling (mottle)	Poor lay of ink.	1. Inks' inability to set substrate evenly. 2. Ink too thin. 3. Poor pigment selection. 4. Press speed too slow. 5. Cylinder too deep. 6. Too much impression.	1. Change blade angle, change solvents. 2. Increase viscosity. 3. Reformulate. 4. Increase speed. 5. Re-etch. 6. Adjust pressure.
Pin holes	Appearance of small holes in the printed area.	1. Ink fails to form a complete film. 2. Imperfections in stock. 3. Wrong ink for stock. 4. Stock too rough.	1. Adjustment of vehicle to reduce viscosity, use active solvent. 2. Adjust blade angle. 3. Reformulate. 4. Electrostatic assist may help.
Blocking	Undesired adhesion between surfaces.	1. Improper drying. 2. Trapped solvent. 3. Web rewound too warm. 4. Excessive pressure in rewind. 5. Over-plasticized ink or varnish. 6. Low melting point binders. 7. Too little heat.	1. Proper solvent balance. 2. Proper solvent balance. 3. Use of chill rollers. 4. Reduce tension. 5. Check on ink and varnish formulas. 6. Same as 5. 7. Check oven temperature.
Foaming	Small bubbles in ink. (Surface tension problem.)	1. Internal friction. 2. Ink pump too efficient. 3. Poor formulation.	1. Use anti-foam. 2. Reduce ink flow. 3. Use anti-foam.
Static	Fuzz hairs.	1. In web. 2. Low moisture content. 3. Low viscosity ink.	1. Tinsel. 2. Increase humidity. 3. Increase viscosity use pigmented extender. 4. Use polar solvents, if possible. 5. Add steam vapor to web.
Adhesion	Ink and\or varnish do not bind to each other or to substrate.	1. Surface will not accept ink. 2. Not enough plasticizer. 3. Too much 'non-stick'.	1. Reformulate ink. 2. Add judiciously. 3. Same as 2.
Drying-in	Weak print.	1. High viscosity ink. 2. Short body ink. 3. Air drafts at nip. 4. Too fast drying. 5. Wiping. 6. Too much impression. 7. Press speed too fast.	1. Reduce viscosity. 2. Add clear extender. 3. Check oven and fountain. 4. Slow down ink. 5. Adjust doctor blade. 6. Check impression dial. 7. Slow down press.

Defects	How recognized	Probable cause	Suggested remedy
Screening	Improper print showing a screen pattern.	1. High viscosity ink. 2. fast drying ink. 3. Too sharp a blade angle.	1. Lower viscosity. 2. Slow down ink (avoid picking). 3. Flatten wipe of blade.
Skips	Engraving dots that have not printed.	1. Rough printing surface. 2. Lack of impression. 3. Drying in. 4. Ink not on cylinder. 5. Ink too heavy. 6. Poor ink flow.	1. Use primer coats. 2. Check impression. 3. Slower solvent. 4. Check applicator. 5. Check viscosity. 6. Check circulation pump.
Grainy print	Print not smooth.	1. High viscosity ink. 2. Press too slow. 3. Stock too dry.	1. Reduce with slowest active solvent. 2. Increase speed. 3. Introduce moisture or humid air. 4. Vary temperatures of stock and fountain ink.
Drag-out (slur)	A bead of excessive ink that appears at trailing edge of print.	1. Ink too thin. 2. Wavy doctor blade. 3. Blade angle. 4. Blade make-up. 5. Poor tension control. 6. Ink drying too slow.	1. Increase viscosity (add clay extender). 2. Align blade properly. 3. Adjust and clean blade. 4. Check back-up blade, move closer to doctor blade edge. 5. Check tension. 6. Speed up drying.
Doughnut	Prints only circumference of screen dot.	1 Too fast drying. 2. Improper solvent balance. 3. Board receptivity.	1. Slower solvents. 2. Balanced solvents. 3. Change specific gravity of ink.
Comets and darts	Ink deposited in shape of comets and darts.	1. Cylinder 2. Foreign substance under doctor blade. 3. Poor chrome job. 4. Ink too dry.	1. Polish cylinder. 2. Strain ink. 3. Check for loose or flaky chrome. 4. Lubrication to add to ink. 5. Check engraver's proofs.
Offset	Transfer of the printed matter to the reverse side of sheet or web.	1. Press problem. 2. Wet or tacky ink film. 3. Improper rewind tension. 4. Ink too slow drying.	1. Check oven and air exhaust. 2. Solvent comb. 3. Check rewind. 4. Speed up drying rate.
Railroads	Continuous line showing in the unprinted areas.	1. Scratches on cylinder. 2. Scratches on blade. 3. Particles lodged under doctor blade (shows in oscillation of blade). 4. High spot on cylinder or chrome deposits.	1. Polish or remake cylinder. 2. Replace doctor blade. 3. Filter ink. 4. Remove burr and polish cylinder.
Brittleness	Substrate breaks when flexed.	1. Excessive heat in drying system. 2. Loss of moisture in substrate. 2. Plasticizer migration. 4. Adding 'non-stick'.	1. Control web temperature. 2. Introduce moisture. 3. May need reformulating. 4. Add plasticizer reasonably.

Defects	How recognized	Probable cause	Suggested remedy
Picking in multicolor work	Previous ink picks off sheet or on roller.	1. First down ink too slow. 2. Increase viscosity of second down ink. 3. Check for best wipe on offending cylinder.	1. Faster solvents. 2. Use extender. 3. Doctor blade adjustment. 4. Use air blast on offending cylinder.

Part 4: Viscosity conversion guide

Centipose	Zahn #1	Zahn #2	Zahn #3	Zahn #4	Zahn #5	Ford #3	Ford #4	Gardner Holt	Shell #1	Shell #2	Shell #3	Shell #4
1.00									17			
2.00									21.5			
3.00									26			
5.00									35			
7.50	30.5								46	18.3		
10.00	32							A-4	57	22		
15.00	35							A-3		30.4		
20.00	38	18								39		
25.00	42	19						A-2		47	18.6	
30.00	45	20						A-1		56	22	
40.00	52	22.5						A			28.6	
50.00	60	25									35	
60.00	68	28						B			42	18
70.00		30				32	20				48	21
80.00		34				37	22	C			55	24
90.00		38				40	25					27
100		43	17			43	28	D				30
125		53	19.5			53	35	E				37
150		63	22			63	40	G				45
175		72	24.5	17		73	48					52
200			27	18		83	55	H				60
225			30	19.5		93	62	I				
250			32	21		103	70	J				
275			35.5	22.5		113	76					
300			38	24	17	120	82	L				
325			41	26	18.5	130	90	M				
350			43	28	20		98					
375			45	30	21		103					
400			48.5	32	22.5		110	P				
500			58	38	27		138	S				
600			70	45	31			U				
700				51	35							
800				59	40							
900				65	45			V				
1000				70	49			W				
1250					60			X				
1500					73							
1750								Y				

Part 5: Ink mileage estimation

If there is no previous history, Table A.1 is a guide that can be used as a reasonable starting point for mileage estimation. To estimate the number of pounds of ink for each color needed for the job:

1. Determine the amount of square inch coverage for each color in a single repeat of the design.
2. Multiply the number of square inches of each color in a design repeat length by the number of impressions to be printed.
3. Divide by the appropriate square inches yield per gallon indicated in Table A.1.
4. Add the amount of ink reduced to press viscosity needed to supply the fountain in ink circulating system.

To determine the actual ink mileage on a job the exact amount of ink and solvent actually used must be known. To calculate this, a record must be kept of the amount of ink brought to the press, the amount of solvent used in reduction of the ink and from this must be subtracted the amount of ink left, including the proportioned ink and solvent in the circulating system when the job is completed. The actual square inches or coverage must be known – (square inches per impression × the number of impressions). With the foregoing information along with the cost of ink and solvent it is relatively simple to calculate mileage (square inches per gallon or per pound) and ink cost of the job, value analysis of the ink (square inches per dollar) or any computation that may be relevant to the operation in regard to ink mileage.

The simple criteria of percent solvent reduction has been used to make a value analysis of an ink. This simple approach can be misleading. The inherent tack or transfer of inks can vary. An ink with more transfer can print a greater volume even though it is reduced more and is at lower viscosity than an ink with less tack or transfer. A highly pigmented ink may require less solvent to reduce its viscosity than a low strength ink containing a lot of tacky resin. It is the volume applied that is the important factor not just the amount of reducing solvent that can be used. An ink mileage test is a more accurate approach to making a value analysis.

Table A.1 A guide for estimating ink mileage

Type of ink wt/gal lbs/gal range	Substrate	Square inches, yield per gallon
Pigmented solvent 7.9–12.0	Non-porous film foil	1 000 000
	Porous paper board	750 000
Pigmented water 8.3–13.5	Paper and board	700 000
	Very absorbent paper and board	600 000
100% dye solvent 7.0–8.0	Paper and board	1 300 000

Part 6: Viscosity (poises)

Poises	0.10	0.20	0.50	1	2	5	10	20	50	100	200	500	1000	2000	5000
American Can—sec. XD					57	143	286	572	1428	2860	5720	14 280			
ASTM.07—sec. XD				72	143	357	715	1430	3570	7150	14 300	35 700			
ASTM.10—sec. XD					42	104	208	417	1041	2080	4170	10 410			
ASTM.15—sec. XD						24	48	95	238	476	953	2380	4760	9530	23 800
ASTM.20—sec. XD						8	16	33	82	164	328	820	1640	3280	8200
ASTM.25—sec. XD							7	14	36	72	143	357	715	1430	3570
Ault & Wiborg B—sec. XD					11	27	54	108	270	540	1080	2700	5400	18 800	27 000
Ault & Wiborg crucible—sec. XD					17	43	86	171	427	855	1710	4270	8550	17 100	42 700
Barbey—units XD	740	320	130	65	32	13	6	3							
Braebender—units						120	260	480	1020						
Brookfield—cps	10	20	50	100	200	500	1000	2000	5000	10 000	20 000	50 000	100 000	200 000	500 000
Casper's tin plate—sec. XD					55	139	278	555	1390	2780	5550	13 900	27 800	55 500	
Continental can—sec. XD					60	151	303	606	1515	3030	6060	15 150	30 300	60 600	
Crown cork & seal—sec. XD					60	151	303	606	1515	3030	6060	15 150	30 300	60 600	
Demmler #1—sec. X factor, D				32	63	156	312								
Demmler #10—sec. X factor, D				3	6	15	31								
Engler—degrees XD				14	27	68	137	274	685	1370	2740	6850	13 700	27 400	
Engler—sec. XD				690	1300	3460	7000	14 000	34 800	70 500					
Fenske—sec. X factor, D	10	20	50	100	200	500	1000	2000	5000	10 000	20 000	50 000	100 000	200 000	
Ford #3—sec. XD				42	84	208	416	834	2081	4160	8340	20 810	41 600	83 400	
Ford #4—sec. XD				30	55	135	270	540	1350	2700	5400	13 500	27 000	54 000	
Fisher #1—sec. XD	20	30													
Fisher #2—sec. XD		15	24	50											
Fisher electro—cps.	10	20	50	100	200	500	1000	2000	5000	10 000	20 000	50 000			
Gardner Holdt—units XD	A-3	A-2	A	D	H	S	W	Y-Z	Z3	Z5	Z6-Z7	Z7-Z8	Z10		
Gardner Holdt—sec. XD						5	10	20	50	100	200	500	1000		
Gardner vertical—sec. XD						5	10	20	50	100	200	500	1000	2000	5000
Hercules—sec. X factor, D							3	5	13	27	53	133	265	530	1330
Hoeppler—sec. X factor, D	10	20	50	100	200	500	1000	2000	5000	10 000	20 000	50 000	100 000		
Interchemical 8°—sec. XD				10	20	50	100	200	500	1000					
Krebs-Stormer—units (KU)						67	85	105	140						
Litho var.—units XD					000	00	0	2-	3-4	5-	6		8-		
Mac Michael (ASTM)—degree #30							18	36	89						

Mobilometer 100 g, 20 cm—units				12	20	45	90	190	500						
Oko Oil—sec. XD								S37	S70	M2½	M7½	M17-M25	M37		
Ostwald—sec. X factor, D	10	20	50	100	200	500	1000	2000	5000	10 000	20 000	50 000	100 000	200 000	
Parlin 7—sec. XD				77	154	385	770	1540	3850	7700	15 400	38 500	77 000		
Parlin 10—sec. XD				21	42	104	208	416	1040	2080	4160	10 400	20 800	41 600	
Parlin 15—sec. XD					10	25	47	93	232	465	930	2320	4650	9300	
Parlin 20—sec. XD						8	17	33	83	167	333	833	1667	3330	8330
Parlin 25—sec. XD								15	36	72	143	357	715	1430	3570
Parlin 30—sec. XD									19	38	77	192	384	770	1920
Pratt & Lambert A—sec. XD			82	164	328	820	1640	3280	8200	16 400	32 800	82 000			
Pratt & Lambert B—sec. XD				82	164	410	820	1640	4100	8200	16 400	41 000			
Pratt & Lambert C—sec. XD				41	82	206	411	823	2060	4110	8230	20 600	41 100	82 300	
Pratt & Lambert D—sec. XD					41	103	205	410	1027	2050	4100	10 270	20 500	41 000	
Pratt & Lambert E—sec. XD					20	51	103	205	513	1027	2050	5130	10 270	20 500	51 300
Pratt & Lambert F—sec. XD					10	26	51	102	256	513	1025	2560	5130	10 250	25 600
Pratt & Lambert G—sec. XD						13	26	52	131	263	526	1315	2630	5260	13 150
Pratt & Lambert H—sec. XD							13	26	66	131	264	658	1315	2640	6580
Pratt & Lambert I—sec. XD							7	13	33	66	131	329	660	1315	3290
Saybolt Furol—sec. XD			24	48	96	238	476	954	2380	4760	9540	23 800	47 600	95 400	
Saybolt Universal (SUS)—sec. XD		96	238	476	954	2380	4760	9540	23 800	47 600	95 400				
Sears—sec. XD			20	27	44										
Shell #1—sec. XD	57														
Shell #2—sec. XD		39	93												
Shell #3—sec. XD			15	35	68										
Shell #4—sec. XD			7	15	30	60									
Shell #5—sec. XD			4	8	16	32	79								
Shell #6—sec. XD					7	13	32	62							
Stormer, cyl (150 gr)—sec.		10	16	27	49	114	223	440	1090	2180					
Ubbelohde—sec. X factor, D	10	20	50	100	200	500	1000	2000	5000	10 000	20 000	50 000	100 000	200 000	
Westinghouse—sec. XD				30	59	147	294	588	1470	2940	5880	14 700	29 400	58 800	
Zahn G 1—sec. XD		38	60	100	267	667	1332	2670	6670	13 320	26 700	66 700			
Zahn G 2—sec. XD		16	24	42	82	161	323	645	1610	3230	6450	16 100	32 300	64 500	
Zahn G 3—sec. XD					27	58	113	204	510	1020	2040	5100	10 200	20 400	51 000
Zahn G 4—sec. XD					19	38	71	160	400	800	1600	4000	8000	16 000	40 000
Zahn G 5—sec. XD					13	27	50	97	212	424	848	2120	4240	8480	21 200

D—Specific gravity. Factor—Instrument factor. Sec.—Reading in seconds. Cps.—Centipoises. Temperature 77°F. Accuracy in most cases: 10%

Appendix B: Typical starting formulations for water based inks

E.J. FLICK

Part 1: General inks

Ink (acrylic)

Raw materials	% by weight
1. Elftex 8 carbon black	13.00
2. Huber 80 kaolin pigment	6.00
3. MP-22 wax	1.00
4. Colloid 675 defoamer	1.00
5. *Iso*-propyl alcohol	3.00
6. Gro-Rez 2050 acrylic resin solution	35.00
7. Ammonia (28%)	0.50
8. Water	39.50
9. Transaid 1280 polymeric material	1.00
	100.00

Key properties
Low cost GCMI #90 black ink

Source: Grow polymer: Technical data sheet: Starting formulation

Ink (acrylic)

Raw materials	% by weight
1. Gro-Rez 6064 acrylic resin solution	29.00
2. Water	35.00
3. Colloid 675 defoamer	0.80
4. SWS-213 silicone compound	0.20
5. *Iso*-propanol	2.00
6. Ammonia (28%)	0.42
7. Elftex 8 carbon black	15.00
8. MP-22 wax	1.70
9. Colloid 675 defoamer	0.20
10. Water	14.68
11. Transaid 1280 polymeric material	1.00
	100.00

Physical constants
Viscosity-Zahn #2: 29 seconds
pH: 8.30

Key properties
GCMI 90 black ink

Source: Grow polymer: Technical bulletin: Starting formulation

Ink (acrylic)

Raw materials	% by weight
1. Nacrylic 78-6175 acrylic copolymer emulsion	74.00
2. Titanium dioxide	21.00
3. Water	5.00
	100.00

Physical constants
  Viscosity-Zahn #2: 23–28 seconds
  Nonvolatile content: 61.2%
  Pigment/binder ratio: 68/32

Key properties
  White ink with improved gloss and block resistance

Source: National Starch and Chemical Corp.: Technical bulletin: Nacrylic polymers for aqueous inks: Formula 4327-56A

Ink (acrylic)

Raw materials	% by weight
1. Nacrylic 78-6175 acrylic copolymer resin	60.00
2. Phthalo blue	30.00
3. Water	10.00
	100.00

Physical constants
  Viscosity-Zahn #2: 25–30 seconds
  Nonvolatile content: 44.7%
  Pigment/binder ratio: 50/50
Key properties
  Blue ink with improved gloss and block resistance
Source: National Starch and Chemical Corp.: Technical bulletin: Nacrylic polymers for aqueous inks: Formula 4327-56B

Ink (acrylic)

Raw materials	% by weight
1. Joncryl 99 acrylic solution polymer	50.00
2. Water	34.80
3. Defoamer	0.20
4. Barium lithol red	15.00
	100.00

Physical constants
  Total solids content: 33.5%
  Viscosity-Zahn #2: 28.0 seconds

Key properties
  Red printing ink, using a sole vehicle
  Very flat viscosity profile, with good color strength development and good aging characteristics
Source: S.C. Johnson & Son, Inc.: Technical service information: Joncryl 99 universal corrugated solution: Suggested formulation

Ink (acrylic)

Raw materials	% by weight
1. Joncryl 99 acrylic solution polymer	50.00
2. Water	39.80
3. Defoamer	0.20
4. Barium lithol red	15.00
5. ASP 352 aluminum silicate	20.00
	100.00

Physical constants
Viscosity-Zahn #2: 30.0 seconds

Key properties
Red printing ink, using a sole vehicle
Very flat viscosity profile, with good color strength development and good aging characteristics
Source: S.C. Johnson & Son, Inc.: Technical service information: Joncryl 99 universal corrugated solution: Suggested formulation

Ink (acrylic)

Raw materials	% by weight
1. Joncryl 99 acrylic solution polymer	50.00
2. Water	34.80
3. Defoamer	0.20
4. Raven 890 carbon black	15.00
	100.00

Physical constants
Total solids content: 33.50%
Viscosity-Zahn #2: 38.0 seconds

Key properties
Black printing ink, using a sole vehicle
Very flat viscosity profile, with good color strength development and good aging characteristics
Source: S.C. Johnson & Son, Inc.: Technical service information: Joncryl 99 universal corrugated solution: Suggested formulation

Ink (acrylic)

Raw materials	% by weight
1. Joncryl 99 acrylic solution polymer	45.00
2. Water	34.80
3. Defoamer	0.20
4. Raven 890 carbon black	15.00
5. Aluminum silicate clay	5.00
	100.00

Physical constants
Viscosity-Zahn #2: 40.0 seconds

Key properties
Black printing ink, using a sole vehicle
Very flat viscosity profile, with good color strength development and good aging characteristics

Source: S.C. Johnson & Son, Inc.: Technical service information: Joncryl 99 universal corrugated solution: Suggested formulation

Ink (acrylic)

Raw materials	% by weight
1. Joncryl 99 acrylic solution polymer	53.00
2. Water	31.80
3. Defoamer	0.20
4. Phthalo blue	15.00
	100.00

Physical constants
Total solids content: 33.5%
Viscosity-Zahn #2: 22.0 seconds

Key properties
Blue printing ink, using a sole vehicle
Very flat viscosity profile, with good color strength development and good aging characteristics

Source: S.C. Johnson & Son, Inc.: Technical service information: Joncryl 99 universal corrugated solution: Suggested formulation

Ink (acrylic)

Raw materials	% by weight
1. Joncryl 99 acrylic solution polymer	30.00
2. Water	34.80
3. Defoamer	0.20
4. Phthalo blue	15.00
5. Aluminum silicate clay	20.00
	100.00

Physical constants
Viscosity-Zahn #2: 28.0 seconds

Key properties
Blue printing ink, using a sole vehicle
Very flat viscosity profile, with good color strength development and good aging characteristics

Source: S.C. Johnson & Son, Inc.: Technical service information: Joncryl 99 universal corrugated solution: Suggested formulation

Ink (acrylic)

Raw materials	% by weight
1. Joncryl 99 acrylic solution polymer	52.00
2. Water	32.80
3. Defoamer	0.20
4. Diarylide yellow 1270	15.00
	100.00

Physical constants
Total solids content: 33.5%
Viscosity-Zahn #2: 21.5 seconds

Key properties
Yellow printing ink, using a sole vehicle
Very flat viscosity profile, with good color strength development and good aging characteristics

Source: S.C. Johnson & Son, Inc.: Technical service information: Joncryl 99 universal corrugated solution: Suggested formulation

Ink (acrylic)

Raw materials	% by weight
1. Joncryl 99 acrylic solution polymer	30.00
2. Water	34.80
3. Defoamer	0.20
4. Diarylide yellow 1270	15.00
5. ASP 352 aluminum silicate	20.00
	100.00

Physical constants
Viscosity-Zahn #2: 29.0 seconds

Key properties
Yellow printing ink, using a sole vehicle
Very flat viscosity profile, with good color strength development and good aging characteristics

Source: S.C. Johnson & Son, Inc.: Technical service information: Joncryl 99 universal corrugated solution: Suggested formulation

Ink (acrylic)

Raw materials	% by weight
1. Pigment dispersion	26.69
2. Carboset *XL*-37 resin	54.96
3. Silane A-1120 adhesion promoter	1.28
4. Ammonium stearate (33% solids)	6.40
5. Water	10.67
	100.00

Key properties
Water based ink, with good adhesion to treated and untreated polyethylene
Source: B.F. Goodrich Co.: Technical data sheet CR-79-7: Carboset resins: Suggested formulation

Ink (acrylic)

Raw materials	% by weight
1. Blue pigment	20.00
2. MPP-123 polyethylene wax	0.50
3. *Iso*-propyl alcohol	6.00
4. Grocryl 6057 modified acrylic copolymer	40.00
5. Water	32.20
6. Ammonia (28%)	1.30
	100.00

Key properties
Blue film/foil ink for high speed disperser

Source: Grow polymer: Technical data sheet: Starting formulation

Ink (acrylic/polyethylene)

Raw materials	% by weight
1. Flexiverse dispersion	40.00
2. Gro-Rez 2020 acrylic resin solution	49.80
3. Growax 35 polyethylene emulsion	5.00
4. Defoamer	0.20
5. Transaid 1280 polymeric material	1.00
6. Water	4.00
	100.00

Source: Grow polymer: Technical data sheet: Starting formulation

Ink (acrylic/polyethylene)

Raw materials	% by weight
1. Flexiverse dispersion	40.00
2. Gro-Rez 2050 acrylic resin solution	15.50
3. Groplex 6066 vehicle	30.80
4. Water	7.00
5. Defoamer	0.20
6. Transaid 1280 polymeric material	1.00
7. Growax 35 polyethylene emulsion	5.00
8. Ammonia (28%)	0.50
	100.00

Source: Grow polymer: Technical data sheet: Starting formulation

Ink (acrylic/polyethylene)

Raw materials	% by weight
1. Flexiverse dispersion	40.00
2. Gro-Rez 2020 acrylic resin solution	20.00
3. Groplex 6066 vehicle	33.80
4. Defoamer	0.20
5. Growax 35 polyethylene emulsion	5.00
6. Transaid 1280 polymeric material	1.00
	100.00

Source: Grow polymer: Technical data sheet: Starting formulation

Ink (acrylic/polyethylene)

Raw materials	% by weight
1. Flexiverse dispersion	40.00
2. Gro-Rez 2050 acrylic resin solution	37.30
3. Growax 35 polyethylene emulsion	5.00
4. Water	16.00
5. Defoamer	0.20
6. Transaid 1280 polymeric material	1.00
7. Ammonia (28%)	0.50
	100.00

Source: Grow polymer: Technical data sheet: Starting formulation

Ink (acrylic/wax)

Raw materials	% by weight
1. Red lake C acroverse chip	16.45
2. Water	15.05
3. *Iso*-propyl alcohol	2.10
4. Ammonia (28%)	0.70
5. Morpholine	0.70
6. Grocryl 6057 modified acrylic copolymer	57.00
7. Growax 35 polyethylene emulsion	4.00
8. *Iso*-propyl alcohol	4.00
	100.00

Key properties
Red flexo/foil ink

Source: Grow polymer: Technical data sheet: Starting formulation

Ink (resin)

Raw materials	% by weight
1. Filtrez 5014 resin	25.00
2. Water	7.00
3. Titanium dioxide	50.00
4. Wax (finely powdered)	1.00
5. Styrenated shellac (40% solids)	17.00
	100.00

Formulation notes
1. Primrose, lemon, chrome yellow and molybdate orange may also be used as pigments
2. Solvents may be used to speed drying
3. Ethylene or propylene glycol may be used to slow the drying for sharper print detail

Key properties
Standard guide formula for white ink

Source: FRP Co.: Technical bulletin: Suggested formulation

Ink (resin)

Raw materials	% by weight
1. Filtrez 5012 resin	50.50
2. Water	24.00
3. Antifoam emulsion	0.50
4. Red pigment	25.00
	100.00

Key properties
Flexo water type cal. litho red base ink guide formula

Source: FRP Co.: Technical bulletin: Suggested formulation

Ink (varnish)

Raw materials	% by weight
1. Filtrez 5001 varnish	56.50
2. Water	18.00
3. Antifoam	0.50
4. Barium lithol red pigment	25.00
	100.00

Key properties
Water-type barium lithol red base guide formula

Source: FRP Co.: Technical bulletin: Suggested formulation

Ink (varnish)

Raw materials	% by weight
1. Water	51.50
2. Ammonia (28%)	8.00
3. Antifoam emulsion	0.50
4. Filtrez 5001 varnish	40.00

Key properties
Varnish guide formula

Source: FRP Co.: Technical bulletin: Suggested formulation

Ink (varnish)

Raw materials	% by weight
1. *Iso*-propanol	3.00
2. Ammonia (28%)	4.50
3. Water	11.00
4. Filtrez 5001 varnish	21.00
5. GS presscake @ 52% solids	60.50
	100.00

Key properties
Blue guide formula

Source: FRP Co.: Technical bulletin: Suggested formulation

Ink (varnish)

Raw materials	% by weight
1. Orange pigment @ 39.5% solids	66.00
2. *Iso*-propanol	3.40
3. Ammonia (28%)	5.10
4. Water	1.70
5. Filtrez 5001 varnish	23.80
	100.00

Key properties
Orange guide formula

Source: FRP Co.: Technical bulletin: Suggested formulation

Part 2: Flexo/gravure inks

Flexo/gravure ink (acrylic)

Raw materials	% by weight
1. Joncryl 142 acrylic polymer emulsion	75.00
2. Balab 748 defoamer	0.50
3. *Iso*-propanol	5.00
4. Water	18.20
5. Ammonia (28%)	1.30
	100.00

Physical constants
Brookfield viscosity: 30 000 cp
Nonvolatile content: 30%
pH: 7.2

Key properties
Flexo/gravure low solids water based system
Vehicle which can be used to formulate low solids, pigmented water ink systems without sacrificing printability or performance
Can lower ink costs when economics in formulating are essential

Source: S.C. Johnson & Son, Inc.: Technical service information: Joncryl 142: Formula I

Flexo/gravure ink (acrylic)

Raw materials	% by weight
1. Joncryl 142 acrylic polymer emulsion	30.00
2. Balab 748 defoamer	0.20
3. *Iso*-propanol	12.00
4. Water	42.28
5. Ammonia (28%)	0.52
6. Balithiol 20-4509 pigment	15.00
	100.00

Physical constants
Viscosity-Zahn #2 (initial): 20 seconds
Viscosity–Zahn #2 (overnight): 22 seconds

Key properties
Balithiol flexo/gravure low solids water-based system
Low solids, pigmented water ink system with good printability and performance
Can lower ink costs when economics in formulating are essential

Source: S.C. Johnson & Son, Inc.: Technical service information: Joncryl 142: Suggested formulation

Flexo/gravure ink (acrylic)

Raw materials	% by weight
1. Joncryl 142 acrylic polymer emulsion	30.00
2. Balab 748 defoamer	0.20
3. *Iso*-propanol	12.00
4. Water	42.28
5. Ammonia (28%)	0.52
6. Phthalocyanine blue	15.00
	100.00

Physical constants
Viscosity-Zahn #2 (initial): 19 seconds
Viscosity-Zahn #2 (overnight): 21 seconds

Key properties
Blue flexo/gravure low solids water based system
Low solids, pigmented water ink system with good printability and performance
Can lower ink costs when economics in formulating are essential

Source: S.C. Johnson & Son, Inc.: Technical service information: Joncryl 142: Suggested formulation

Flexo/gravure ink (acrylic)

Raw materials	% by weight
1. Joncryl 142 acrylic polymer emulsion	30.00
2. Balab 748 defoamer	0.20
3. *Iso*-propanol	12.00
4. Water	42.28
5. Ammonia (28%)	0.52
6. Raven 890 carbon black	15.00
	100.00

Physical constants
Viscosity-Zahn #2 (initial): 26 seconds
Viscosity-Zahn #2 (overnight): 37 seconds

Key properties
Black flexo/gravure low solids water based system
Low solids, pigmented water ink system with good printability and performance
Can lower ink costs when economics in formulating are essential

Source: S.C. Johnson & Son, Inc.: Technical service information: Joncryl 142: Suggested formulation

Flexo/gravure ink (acrylic)

Raw materials	% by weight
1. Joncryl 142 acrylic polymer emulsion	30.00
2. Balab 748 defoamer	0.20
3. *Iso*-propanol	2.00

4. Water	27.28
5. Ammonia (28%)	0.52
6. Harshaw 2737 chrome yellow	40.00
	100.00

Physical constants
 Viscosity-Zahn #2: 25 seconds
 Viscosity-Zahn #2: 30 seconds

Key properties
 Yellow flexo/gravure low solids water based system
 Low solids, pigmented water ink system with good printability and performance
 Can lower ink costs when economics in formulating are essential

Source: S.C. Johnson & Son, Inc.: Technical service information: Joncryl 142: Suggested formulation

Flexo/gravure ink (acrylic)

Raw materials	% by weight
1. Joncryl 142 acrylic polymer emulsion	50.00
2. Balab 748 defoamer	0.50
3. *Iso*-propanol	2.00
4. Water	30.00
5. Ammonia (28%)	17.50
	100.00

Physical constants
 Brookfield viscosity: 850 cp
 Nonvolatile content: 26%
 pH: 7.2

Key properties
 Flexo/gravure low solids water based system
 Vehicle which can be used to formulate low solids, pigmented water ink systems without sacrificing printability or performance
 Can lower ink costs when economics in formulating are essential

Source: S.C. Johnson & Son, Inc.: Technical service information: Joncryl 142: Formula II

Flexo/gravure ink (acrylics/polyethylene)

Raw materials	% by weight
1. Joncryl 87 styrenated acrylic dispersion	20.00
2. Jonwax 22 microcrystalline wax emulsion	5.00
3. Joncryl 67 acrylic resin	12.00
4. Ammonia (28%)	1.60
5. Morpholine	1.00
6. *Iso*-propanol	4.00
7. Dibutyl phthalate	1.20
8. Ethylene glycol monoethyl ether	1.20
9. Water	39.80
10. Sag 471 antifoam	0.20
11. Organic pigment	14.00
	100.00

Key properties
Flexographic or gravure ink, with fast drying, good finish, water resistance and printability

Source: S.C. Johnson & Son, Inc.: Technical service information: Joncryl 67: Suggested formulation

Flexo/gravure ink (acrylics/polyethylene)

Raw materials	% by weight
1. Joncryl 87 styrenated acrylic dispersion	20.00
2. Jonwax 22 microcrystalline wax emulsion	5.00
3. Joncryl 67 acrylic resin	6.00
4. Ammonia (28%)	0.80
5. Morpholine	0.50
6. *Iso*-propanol	2.00
7. Dibutyl phthalate	0.60
8. Ethylene glycol monoethyl ether	0.60
9. Water	24.30
10. Sag 471 antifoam	0.20
11. Inorganic pigment	40.00
	100.00

Key properties
Flexographic or gravure ink, with fast drying, good finish, water resistance and printability

Source: S.C. Johnson & Son, Inc.: Technical service information: Joncryl 67: Suggested formulation

Flexo/gravure ink (acrylics/polyethylene)

Raw materials	% by weight
1. Joncryl 67 acrylic resin	6.25
2. Ammonia (28%)	0.33
3. Monoethanolamine	0.50
4. Diethylaminoethanol	0.75
5. *Iso*-propanol	2.50
6. Water	24.67
7. Organic pigment (high gloss)	15.00
8. Joncryl 87 styrenated acrylic dispersion	45.00
9. Jonwax 22 microcrystalline wax emulsion	5.00
	100.00

Key properties
Flexo/gravure ink which provides optimum balance of low cost, high plasticizing effect and good press resolubility

Source: S.C. Johnson & Son, Inc.: Technical service information: Joncryl 87: Suggested formulation

Flexo/gravure ink (acrylics/polyethylene)

Raw materials	% by weight
1. Joncryl 67 acrylic resin	12.00
2. Ammonia (28%)	1.60
3. Morpholine	1.00
4. *Iso*-propanol	4.00
5. Dibutyl phthalate	1.20
6. Cellosolve solvent	1.20
7. Water	40.00
8. Organic pigment (high rub)	14.00
9. Joncryl 87 styrenated acrylic dispersion	20.00
10. Jonwax 22 microcrystalline wax emulsion	5.00
	100.00

Key properties
Flexo/gravure ink with optimum balance of water resistance and wash-up

Source: S.C. Johnson & Son, Inc.: Technical service information: Joncryl 87: Suggested formulation

Flexo/gravure ink (acrylics/polyethylene)

Raw materials	% by weight
1. Joncryl 67 acrylic resin	6.00
2. Ammonia (28%)	0.31
3. Monoethanolamine	0.48
4. Diethanolaminoethanol	0.72
5. *Iso*-propanol	2.40
6. Water	14.09
7. Inorganic pigment (high gloss)	36.00
8. Joncryl 87 styrenated acrylic dispersion	35.00
9. Jonwax 22 microcrystalline wax emulsion	5.00
	100.00

Key properties
Flexo/gravure ink which provides optimum balance of low cost, high plasticizing effect and good press resolubility

Source: S.C. Johnson & Son, Inc.: Technical service information: Joncryl 87: Suggested formulation

Flexo/gravure ink (acrylics/polyethylene)

Raw materials	% by weight
1. Joncryl 67 acrylic resin	6.00
2. Ammonia (28%)	0.80
3. Morpholine	0.50
4. *Iso*-propanol	2.00
5. Dibutyl phthalate	0.60
6. Cellulose solvent	0.60
7. Water	24.50
8. Inorganic pigment (high rub)	40.00
9. Joncryl 87 styrenated acrylic dispersion	20.00
10. Jonwax 22 microcrystalline wax emulsion	5.00
	100.00

Key properties
Flexo/gravure ink with optimum balance of water resistance and wash-up

Source: S.C. Johnson & Son, Inc.: Technical service information: Joncryl 87: Suggested formulation

Flexo/gravure ink (acrylics/polyethylene)

Raw materials	% by weight
1. Joncryl 67 acrylic resin	9.00
2. Ammonia (28%)	1.20
3. Morpholine	1.00
4. Cellosolve solvent	0.90
5. *Iso*-propanol	3.00
6. Water	33.50
7. Organic pigment	15.00
8. Joncryl 87 styrenated acrylic dispersion	22.00
9. Jonwax 22 microcrystalline wax emulsion or Jonwax 26 polyethylene wax emulsion	5.00
10. Joncryl 74F acrylic polymer solution or Joncryl 77 acrylic polymer solution	8.50
11. Dibutyl phthalate	0.90
	100.00

Key properties
Flexo/gravure ink with high wet-rub and gloss, excellent film formation, printability, strength and viscosity stability

Source: S.C. Johnson & Son, Inc.: Technical service information: Joncryl 87: Suggested formulation

Flexo/gravure ink (acrylics/polyethylene)

Raw materials	% by weight
1. Joncryl 67 acrylic resin	6.00
2. Ammonia (28%)	0.80
3. Morpholine	0.75
4. Cellosolve solvent	0.60
5. *Iso*-propanol	2.00
6. Water	19.50
7. Inorganic pigment	40.00
8. Joncryl 87 styrenated acrylic dispersion	18.00
9. Jonwax 22 microcrystalline wax emulsion or Jonwax 26 polyethylene wax emulsion	4.75
10. Joncryl 74F acrylic polymer solution or Joncryl 77 acrylic polymer solution	7.00
11. Dibutyl phthalate	0.60
	100.00

Key properties
Flexo/gravure ink with high wet-rub and gloss, excellent film formation, printability, strength and viscosity stability

Source: S.C. Johnson & Son, Inc.: Technical service information: Joncryl 87: Suggested formulation

Flexo/gravure ink (acrylics/polyethylene)

Raw materials	% by weight
1. Barium litho pigment	15.00
2. Joncryl 61LV acrylic resin solution or Joncryl 678 acrylic resin	20.00
3. Water	15.00
4. Joncryl 87 styrenated acrylic dispersion	25.00
5. Joncryl 77 acrylic polymer solution	15.00
6. Jonwax 26 microcrystalline wax emulsion	5.00
7. Water	5.00
	100.00

Key properties
Flexo/gravure ink with high wet-rub and gloss, excellent film formation, printability, strength and viscosity stability

Source: S.C. Johnson & Son, Inc.: Technical service information: Joncryl 87: Suggested formulation

Flexo/gravure ink (acrylics/polyethylene)

Raw materials	% by weight
1. Cyan blue pigment	15.00
2. Joncryl 61LV acrylic resin solution or Joncryl 678 acrylic resin	25.00
3. Joncryl 87 styrenated acrylic dispersion	40.00
4. Jonwax 26 polyethylene wax emulsion	5.00
5. Water	15.00
	100.00

Key properties
Flexo/gravure ink with high wet-rub and gloss, excellent film formation, printability, strength and viscosity stability

Source: S.C. Johnson & Son, Inc.: Technical service information: Joncryl 87: Suggested formulation

Flexo/gravure ink (acrylics/polyethylene)

Raw materials	% by weight
1. Raven 500 furnace black	15.00
2. Joncryl 61LV acrylic resin solution or Joncryl 678 acrylic resin	35.00
3. Water	20.00
4. Joncryl 87 styrenated acrylic dispersion	25.00
5. Jonwax 26 microcrystalline wax emulsion	5.00
	100.00

Key properties
Flexo/gravure ink with high wet-rub and gloss, excellent film formation, printability, strength and viscosity stability

Source: S.C. Johnson & Son, Inc.: Technical service information: Joncryl 87: Suggested formulation

Flexo/gravure ink (acrylics/polyethylene)

Raw materials	% by weight
1. Raven 890 furnace black	15.00
2. Joncryl 61LV acrylic resin solution	25.00
3. Water	19.00
4. Joncryl 138 acrylic polymer dispersion	31.00
5. Jonwax 26 polyethylene wax emulsion	10.00
	100.00

Physical constants
Viscosity (Zahn #2), initial: 25.2 seconds
Viscosity (Zahn #2), 24 hours: 29.0 seconds

Key properties
Flexo/gravure ink, especially useful in high speed operations
Keeps current EPA solvent restrictions in mind

Source: S.C. Johnson & Son, Inc.: Technical service information: Joncryl 138: Formula #1-F

Flexo/gravure ink (acrylics/polyethylene)

Raw materials	% by weight
1. Cyan blue	15.00
2. Joncryl 61LV acrylic resin solution	25.00
3. Water	20.00
4. Joncryl 138 acrylic polymer dispersion	30.00
5. Jonwax 26 polyethylene wax emulsion	10.00
	100.00

Physical constants
Viscosity (Zahn #2), initial: 27.0 seconds
Viscosity (Zahn #2), 24 hours: 33.4 seconds

Key properties
Flexo/gravure ink, especially useful in high speed operations
Keeps current EPA solvent restrictions in mind

Source: S.C. Johnson & Son, Inc.: Technical service information: Joncryl 138: Formula #2-F

Flexo/gravure ink (acrylics/polyethylene)

Raw materials	% by weight
1. Joncryl 138 acrylic polymer dispersion	59.00
2. Joncryl 61LV acrylic resin solution	25.00
3. Zinc oxide solution #1	6.00
4. Jonwax 26 polyethylene wax emulsion	10.00
	100.00

Physical constants
Viscosity (Brookfield), initial: 680 cp
Viscosity (Brookfield), one week: 720 cp

Key properties
Flexo/gravure ink, a typical overprint formula, especially useful in high speed operations
Keeps current EPA solvent restrictions in mind

Source: S.C. Johnson & Son, Inc.: Technical service information: Joncryl 138: Formula 1

Flexo/gravure ink (acrylics/polyethylene)

Raw materials	% by weight
1. Joncryl 138 acrylic polymer dispersion	65.00
2. Joncryl 61LV acrylic resin solution	17.00
3. Jonwax 26 polyethylene wax emulsion	5.00
4. Water	13.00
	100.00

Physical constants
Viscosity (Brookfield), initial: 700 cp
Viscosity (Brookfield), one week: 750 cp

Key properties
Flexo/gravure ink, a typical overprint formula, especially useful in high speed operations
Keeps current EPA solvent restrictions in mind

Source: S.C. Johnson & Son, Inc.: Technical service information: Joncryl 138: Formula 2

Flexo/gravure ink (acrylics/polyethylene)

Raw materials	% by weight
1. Joncryl 77 acrylic polymer solution	65.00
2. Joncryl 61 acrylic polymer solution	20.00
3. Poly-Em 40 emulsion	10.00
4. Water	5.00
	100.00

Physical constants
pH: 8.3
Brookfield viscosity (initial): 150 cp
Brookfield viscosity (30 days): 160 cp

Key properties
Represents a combination of ingredients providing high gloss in addition to excellent water, oil and rub resistance

Source: S.C. Johnson & Son, Inc.: Technical service information: Joncryl 77 acrylic polymer solution: Formula #1

Flexo/gravure ink (acrylics/wax)

Raw materials	% by weight
1. Joncryl 74F acrylic polymer solution	65.00
2. Joncryl 61 acrylic polymer solution	20.00
3. Jonwax 26 polyethylene wax emulsion	10.00
4. Water	5.00
	100.00

Physical constants
pH: 8.3
Brookfield viscosity (initial): 150 cp
Brookfield viscosity (30 days): 160 cp

Key properties
Combination of ingredients providing high gloss, in addition to excellent water, oil and rub resistance

Source: S.C. Johnson & Son, Inc.: Technical service information: Joncryl 74F acrylic polymer solution: Formula #1

Flexo/gravure ink (acrylics/wax)

Raw materials	% by weight
1. Joncryl 74F acrylic polymer solution	65.00
2. Joncryl 61 acrylic polymer solution	20.00
3. Jonwax 26 polyethylene wax emulsion	5.00
4. Zinc oxide solution #1	5.00
5. Water	5.00
	100.00

Physical constants
pH: 9.5
Brookfield viscosity (initial): 500 cp
Brookfield viscosity (30 days): 550 cp

Key properties
Exhibits lower gloss with a corresponding lower oil and water resistance, but with maximum heat resistance

Source: S.C. Johnson & Son, Inc.: Technical service information: Joncryl 74F acrylic polymer solution: Formula #2

Flexo/gravure ink (acrylics/wax)

Raw materials	% by weight
1. Joncryl 77 acrylic polymer solution	65.00
2. Joncryl 61 acrylic polymer solution	20.00
3. Jonwax 26 polyethylene wax emulsion	10.00
4. Zinc oxide solution #1	5.00
	100.00

Physical constants
pH: 9.5
Brookfield viscosity (initial): 500 cp
Brookfield viscosity (30 days): 550 cp

Key properties
Exhibits slightly less gloss with a corresponding lower oil and heat resistance, but with maximum heat resistance

Source: S.C. Johnson & Son, Inc.: Technical service information: Joncryl 77 acrylic polymer solution: Formula #2

Part 3: Flexo/roto inks

Flexo/roto ink (acrylic)

Raw materials	% by weight
1. Joncryl 67 acrylic resin	20.00
2. Ammonia (28%)	4.70
3. Water	75.30
	100.00

Physical constants
Viscosity (Gardner–Holdt): A-4
pH: 8.5
Color: straw

Key property
Popular water ink varnish

Source: S.C. Johnson & Son, Inc.: Technical service information: Joncryl 67 acrylic resin: Resin cut A

Flexo/roto ink (acrylic)

Raw materials	% by weight
1. Joncryl 67 acrylic resin	20.00
2. Morpholine	7.00
3. Water	73.00
	100.00

Physical constants
Viscosity (Gardner–Holdt): A-1
pH: 8.5
Color: straw

Key property
Popular water ink varnish modification

Source: S.C. Johnson & Son, Inc.: Technical service information: Joncryl 67 acrylic resin: Resin cut B

Flexo/roto ink (acrylic)

Raw materials	% by weight
1. Joncryl 67 acrylic resin	30.00
2. Ammonia (28%)	7.00
3. Tall oil fatty acid	5.00
4. Ethylene glycol monoethyl ether	3.00
5. Water	55.00
	100.00

Physical constants
Viscosity (Gardner–Holdt): Z-5
pH: 8.4
Color: brown

Key property
'Let-down' varnish and low-slip flexo/roto overprint coating

Source: S.C. Johnson & Son, Inc.: Technical service information: Joncryl 67 acrylic resin: Resin cut C

Flexo/roto ink (acrylic)

Raw materials	% by weight
1. Joncryl 67 acrylic resin	30.00
2. Ammonia (28%)	4.00
3. Morpholine	2.50
4. *Iso*-propyl alcohol	10.00
5. Dibutyl phthalate	3.00
6. Ethylene glycol monoethyl ether	3.00
7. Water	47.50
	100.00

Physical constants
Viscosity (Gardner–Holdt): T-U
pH: 8.2
Color: straw

Key property
High solids varnish for wet rub resistant flexo and gravure inks

Source: S.C. Johnson & Son, Inc.: Technical service information: Joncryl 67 acrylic resin: Resin cut D

Flexo/roto ink (acrylic)

Raw materials	% by weight
1. Joncryl 67 acrylic resin	25.00
2. Dibutyl phthalate	2.00
3. Acid or basic dye (alcohol soluble)	15.00
4. Ethanol (denatured)	58.00
	100.00

Formulation note
  Adding fugitive amines promotes water dilutability and water wash-up characteristics

Key properties
  Excellent replacement for solvent based type inks
  Good adhesion
  Very strong dye intensity with good dilutability and water resistance

Source: S.C. Johnson & Son, Inc.: Technical service information: Joncryl 67 acrylic resin: Suggested formulation

Flexo/roto ink (acrylic)

Raw materials	% by weight
1. Joncryl 678 acrylic resin	25.00
2. Ammonia (28%)	5.00
3. Water	70.00
	100.00

Physical constants
  Brookfield viscosity: 30 cps
  pH: 8.4

Key property
  Modifications can be added to satisfy specific requirements

Source: S.C. Johnson & Son, Inc.: Technical service information: Joncryl 678 acrylic resin: Formula 4076M95

Flexo/roto ink (acrylic)

Raw materials	% by weight
1. Joncryl 678 acrylic resin	35.00
2. Ammonia (28%)	7.50
3. Ethylene glycol	1.50
4. *Iso*-propyl alcohol	3.00
5. Water	53.00
	100.00

Physical constants
  Brookfield viscosity: 5000 cps
  pH: 8.5

Key property
  Modifications can be added to satisfy specific requirements

Source: S.C. Johnson & Son, Inc.: Technical service information: Joncryl 678 acrylic resin: Formula 3504M148

Flexo/roto ink (acrylic)

Raw materials	% by weight
1. Joncryl 678 acrylic resin	27.50
2. Ammonia (28%)	4.53
3. Ethylene glycol	0.38
4. *Iso*-propyl alcohol	0.75
5. Water	51.84
6. Organic pigment	15.00
	100.00

Key properties
Pigment is efficiently dispersed and suspended
Many pigments may be effectively used
Modifications can be added to satisfy specific requirements

Source: S.C. Johnson & Son, Inc.: Technical service information: Joncryl 678 acrylic resin: Suggested formulation

Flexo/roto ink (acrylic)

Raw materials	% by weight
1. Joncryl 678 acrylic resin	25.00
2. Morpholine	9.00
3. Water	66.00
	100.00

Physical constants
Brookfield viscosity: 100 cp
pH: 8.5

Key property
Modifications can be added to satisfy specific requirements

Source: S.C. Johnson & Son, Inc.: Technical service information: Joncryl 678 acrylic resin: Formula 4325M2

Flexo/roto ink (acrylic)

Raw materials	% by weight
1. Joncryl 678 acrylic resin	16.12
2. Ammonia (28%)	3.35
3. Ethylene glycol	0.38
4. *Iso*-propyl alcohol	0.75
5. Water	39.20
6. Inorganic pigments	40.20
	100.00

Key properties
Pigment is efficiently dispersed and suspended
Many pigments may be effectively used
Modifications can be added to satisfy specific requirements

Source: S.C. Johnson & Son, Inc.: Technical service information: Joncryl 678 acrylic resin: Suggested formulation

Appendix C: Trademarked raw materials for typical water based inks

E.J. FLICK

Raw material	Chemical description	Source
A49-1551 flushed color	Phthalo blue (40% pigment)	Sun Chemical Corp.
Abitol hydrobietyl alcohol	Technical grade of hydrobietyl alcohol, derived from rosin	Hercules, Inc.
A-C 6 polyethylene resin	Polyethylene homopolymer resin. Softening point of 222°F	Allied Chemical Corp.
Acryloid B-72 polymer	Acrylic ester resin	Rohm & Haas Co.
Acryloid NAD-10 polymer	Acrylic ester resin	Rohm & Haas Co.
Aerosil R-972 hydrophobic silica	Hydrophobic silica	Degussa Corp.
Amberol M-82 polymer	Phenolic resin	Rohm & Haas Co.
Arochem 404 resin	Maleic resin	Spencer-Kellogg
Aromatic solvent SC-100	Petroleum solvent with 311°F. IBP	Exxon Chemical
Aromatic solvent SC-150	Petroleum solvent with 362°F. IBP	Exxon Chemical
ASM-5029 Alglos setmaster varnish	Ultra-fast quick set let-down varnish. Modified phenolic/T.S.O.R. in Magie 470	Algan, Inc.
ASP-352 aluminum silicate	Hydrous aluminum silicate, used as pigment and extender	Engelhard Minerals & Chemicals
B19-1750 flushed color	Lithol rubine, B.S. 33% pigment	Sun Chemical
B49-1202 flushed color	Phthalo blue, G.S. 37% pigment	Sun Chemical
B49-1752 flushed color	Phthalo blue, G.S. Pigment color 49	Sun Chemical
B49-2194 flushed color	Carbon black. Pigment class, Black 7	Sun Chemical
B49-2210 flushed color	Phthalo blue, G.S. (36% pigment)	Sun Chemical
B49-2262 flushed color	Phthalo blue, G.S. (40% pigment)	Sun Chemical
B49-2316 flushed color	Phthalo blue, G.S. (34% pigment)	Sun Chemical
Balab 748 defoamer	Organic, nonsilicone proprietary defoamer (100% active)	Witco Chemical
Bartyl F antiskinning agent	Proprietary composition antiskinning agent	Sindar Corp.
Beckamine 21-511 resin	Urea–formaldehyde resin (60% solids in alcohol)	Reichhold
Bentone 38 gelling agent	Organo-clay thixotropic additive	NL Chemicals
Bentone 500 rheological additive	Organo-clay rheological additive	NL Chemicals
Bronze powder XM18G	Gold pigment. 5.5 microns average particle size	Obron Corp.
Butyl cellosolve solvent	Ethylene glycol monobutyl ether acetate solvent	Union Carbide
BYK-301 resin (50%)	Ink resin	Mallinckrodt
CAB-381-0.5 cellulose acetate butyrate	Cellulose acetate buryrate	Eastman Chemical
CAP-482-0.5 cellulose acetate propionate	Cellulose acetate propionate ester	Eastman Chemical
CAP-504-0.2 cellulose acetate propionate	Cellulose acetate propionate ester, alcohol soluble	Eastman Chemical

Raw material	Chemical description	Source
Carbitol solvent	Diethylene glycol monoethyl ether solvent	Union Carbide
Carboset XL-37 resin	Acrylic polymer (35% solids)	B.F. Goodrich
Cellolyn 21 synthetic resin	Dibasic-acid-modified rosin ester	Hercules, Inc.
Cellosolve solvent	Ethylene glycol monoethyl ether solvent	Union Carbide
Chlorafin 40 chlorinated paraffin	Chlorinated paraffin. 40% chlorine content	Hercules, Inc.
Cobalt/manganese drier 2.4	Cobalt/manganese tallate mixture. 2/4% ratio	Shepherd Chemical
Colloid 675 defoamer	Proprietary composition defoamer. 100% active	Colloids, Inc.
Colloid 680 defoamer	Proprietary composition defoamer	Colloids, Inc.
D49-2035 flushed color	Phthalo blue, G.S. (37% pigment)	Sun Chemical
D49-2286 flushed color	Phthalo blue, G.S. (85% pigment)	Sun Chemical
D49-2397 flushed color	Phthalo blue, G.S. Color 49	Sun Chemical
Day-Glo A pigment series	Fluorescent pigment series	Day-Glo Color
Day-Glo AX pigment series	Fluorescent pigment series. Stronger than A line	Day-Glo Color
Day-Glo IRB base color	Fluorescent pigment series	Day-Glo Color
Day-Glo special heatset base	Special heatset pigment base series	Day-Glo Color
Decotherm varnish	Printing ink vehicle, for high gloss (78% solids)	Lawter
Diarylide yellow 1270	Dichlorobenzidine coupled pigment. Pigment Yellow 14. Color Index No. 21095	Harshaw Chemical
Drier #1269 paste	Metal salt of neodecanoate acid (6% cobalt paste)	Shepherd Chemical
Dyall C-124 polyethylene dispersion	Polyethylene dispersion	Lawter
Dyall C-306 wax compound	Wax compound	Lawter
Eastman resin H-130	Ink resin	Eastman Chemical
Ektasolve EB solvent	Ethylene glycol monobutyl ether solvent	Eastman Chemical
Ektasolve EE solvent	Ethylene glycol monoethyl ether solvent	Eastman Chemical
Elftex pellets 115 carbon black	Furnace process carbon black. 27 millimicrons particle size	Cabot Corp.
Elftex 8 carbon black	Furnace process carbon black. 27 millimicrons particle size	Cabot Corp.
Elvacite 2013 resin	Acrylic resin	duPont
Epolene C-10 wax	Polyolefin wax. Softening point of 104°C	Eastman Chemical
Epolene C-13 wax	Polyolefin wax. Softening point of 110°C	Eastman Chemical
Ester gum 8D	Glycerol ester of rosin	Hercules, Inc.
Ethylcellulose	Organosoluble ethyl ether of cellulose	Hercules, Inc.
Ethylhydroxyethylcellulose	Organosoluble ethyl ether of cellulose	Hercules, Inc.
Exkin #2 anti-skinning agent	Anti-skinning agent of the volatile oxime type	Tenneco Chemicals
Filtrez 525 resin	Fumaric resin. Melt point 148°C	FRP Co.
Filtrez 526 resin	Fumaric resin. Melt point 130°C	FRP Co.
Filtrez 530 resin	Fumaric resin. Melt point 150°C	FRP Co.
Filtrez 593A resin	Fumaric resin. Melt point 130°C	FRP Co.
Filtrez 5001 varnish	Varnish for inks	FRP Co.
Filtrez 5008 resin	Fumaric resin	FRP Co.

Raw material	Chemical description	Source
Filtrez 5012 resin	Fumaric resin. Melt point 135°C	FRP Co.
Filtrez 5014 resin	Fumaric resin. Melt point 140°C	FRP Co.
Filtrez 5400 resin	Fumaric resin. Melt point 130°C	FRP Co.
Flexiverse dispersion	Pigment dispersion line	Sun Chemical
Fluo HT dry teflon compound	Micronized PTFE. Melt point 620°F	Micro Powders
Grocryl P-260 polymer emulsion	High solids (48%), low viscosity polymer emulsion	Grow Polymer
Grocryl 6057 modified acrylic	Modified acrylic copolymer (40% solids)	Grow Polymer
Groplex 6066 vehicle	Polymer vehicle for inks	Grow Polymer
Gro-Rez 2020 acrylic resin solution	Modified acrylic resin solution (30% solids)	Grow Polymer
Gro-Rez 2050 acrylic resin solution	Acrylic resin dispersing vehicle solution	Grow Polymer
Gro-Rez 6064 acrylic resin solution	Acrylic resin solution (24.5% solids)	Grow Polymer
Growax 35 polyethylene emulsion	Nonionic emulsion of 275°F melt point polyethylene	Grow Polymer
Gulf 581 naphthenic mineral oil	Naphthenic mineral oil	Gulf Oil
Halex repellant varnish	Water/alcohol repellant varnish	Lawter
Harshaw 2737 chrome yellow	Chrome yellow pigment	Harshaw Chemical
Hercolyn D resin	Hydrogenated methyl ester of rosin	Hercules, Inc.
Huber 80 kaolin pigment	Aluminum silicate extender pigment	J.M. Huber Corp.
Ionol CP phenol compound	2,6-di-tert-butyl-4-methyl-phenol compound	Shell Chemical
IRB base color	Fluorescent pigment base color	Day-Glo Color
Joncryl 61 acrylic polymer solution	Acid functional styrene/acrylic resin (34% solids)	S.C. Johnson
Joncryl 61LV acrylic resin solution	Improved acrylic resin varnish solution (34% solids)	S.C. Johnson
Joncryl 67 acrylic resin	Acrylic resin, versatile and hard, in flake form	S.C. Johnson
Joncryl 74F acrylic polymer solution	Acrylic polymer solution (49% solids)	S.C. Johnson
Joncryl 77 acrylic polymer solution	Acrylic polymer solution gloss vehicle (45% solids)	S.C. Johnson
Joncryl 80 acrylic polymer	Acrylic polymer (49% solids)	S.C. Johnson
Joncryl 87 styrenated acrylic dispersion	Styrenated acrylic dispersion gloss vehicle (49% solids)	S.C. Johnson
Joncryl 89 styrenated acrylic dispersion	Styrenated acrylic dispersion economical gloss (48% solids)	S.C. Johnson
Joncryl 99 acrylic solution polymer	Acrylic solution polymer vehicle (37% solids)	S.C. Johnson
Joncryl 134 acrylic polymer emulsion	Acrylic polymer emulsion for gravure (45% solids)	S.C. Johnson
Joncryl 138 acrylic polymer dispersion	Acrylic water-borne polymer for high gloss systems	S.C. Johnson
Joncryl 142 acrylic polymer emulsion	Acrylic polymer emulsion for flexo/gravure (39% solids)	S.C. Johnson
Joncryl 537 acrylic emulsion polymer	Acrylic detergent-resistant emulsion polymer (46% solids)	S.C. Johnson
Joncryl 678 acrylic resin	Acrylic resin in flake form	S.C. Johnson
Joncryl 682 acrylic oligomer	Solid grade acrylic oligomer for high solids inks	S.C. Johnson

Raw material	Chemical description	Source
Joncryl 1535 acrylic mixing vehicle	Acrylic mixing vehicle for metallic pigments (37% solids)	S.C. Johnson
Jonwax 22 microcrystalline wax emulsion	Microcrystalline wax emulsion (35% solids)	S.C. Johnson
Jonwax 26 polyethylene wax emulsion	Polyethylene wax emulsion (25% solids)	S.C. Johnson
Kodaflex DOP plasticizer	Dioctyl phthalate plasticizer	Eastman Chemical
Kodaflex DBP plasticizer	Dibutyl phthalate plasticizer	Eastman Chemical
Lactol spirits solvent	Aliphatic naphtha in the toluene evaporation range	Union Chemicals
Lewisol 28 synthetic resin	Maleic-modified glycerol ester of rosin	Hercules, Inc.
Lin-All P.I. drier	Printing ink drier. 4.3% manganese metal	Mooney Chemicals
LoCal A-7-T dispersion vehicle	Dispersion vehicle system	Lawter
LoCal FST dispersion vehicle	Dispersion vehicle system, with thixotropy	Lawter
LoCal G-33 dispersion vehicle	Medium heatset gel system	Lawter
Magie #2 oil	Ink oil	Magie
Magie #3 oil	Ink oil	Magie
Magie 415 oil	Ink oil	Magie
Magie 470 oil	Ink oil	Magie
Magie 500 oil	Ink oil	Magie
Magie 535 oil	Ink oil	Magie
Magie 590 oil	Ink oil	Magie
Magiesol 47 oil	Ink oil	Magie
Magruder I.R. color flush	I.R. color flush series	Magruder
Mineral spirits 360	Mineral spirits solvent	Union Chemicals
Mogul L carbon black	Furnace process carbon black	Cabot
MP-22 wax	Micronized synthetic wax. Melt point 219°F	Micro Powders
MPP-123 polyethylene wax	Micronized polyethylene wax. Melt point 233°F	Micro powders
MPP-620 VF polyethylene	Finely micronized polyethylene wax. Melt point 241°F	Micro Powders
Multimix color flush	Color flush series	BASF Wyandotte
Nacrylic 78-6175 acrylic copolymer	Solid alkali soluble acrylic copolymer resin	National Starch
NiPar S-20 solvent	2-Nitropropane solvent	Angus
NiPar S-30 solvent	Mixed nitropropane isomers	Angus
Obron bronze pigment	Bronze pigment series	Obron
Obron XM-18 pigment	Highest quality bronze pigment	Obron
Obron XM-18G pigment	Superfine ink lining	Obron
Parlon S10 chlorinated rubber	Chlorinated rubber viscosity grade	Hercules, Inc.
Parlon S20 chlorinated rubber	Chlorinated rubber viscosity grade	Hercules, Inc.
Pentalyn G synthetic resin	Pentaerythritol ester of rosin	Hercules, Inc.
Pentalyn K synthetic resin	Pentaerythritol ester of resin	Hercules, Inc.
Picco 6140 resin	Proprietary aromatic resin	Hercules, Inc.
Piccotex 120 resin	Thermoplastic copolymer resin. Softening point 120°C	Hercules, Inc.
Pliolite 50 resin	High-styrene/butadiene resin	Goodyear
Poly-Em 40 emulsion	Polyethylene emulsion	Gulf Oil
Pope BW-813 black flush	33% Carbon black in mineral oil	Pope Chemical

Raw material	Chemical description	Source
Pope VWO hydrocarbon/ mineral oil vehicle	Hydrocarbon/mineral oil vehicle	Pope Chemical
Raven 500 furnace black	Industrial furnace black. Mean particle diameter 56 nm	Columbian Chemicals
Raven 890 carbon black	Industrial furnace black. Mean particle diameter 30 nm	Columbian Chemicals
Regal 330R carbon black	Furnace process carbon black. 25 millimicrons particle size	Cabot Corp.
Regal 400R carbon black	Furnace process carbon black. Medium flow	Cabot Corp.
Regal 500 carbon black	Furnace process carbon black. Regular color	Cabot Corp.
Resimene V-980 resin	Ink resin	Monsanto
Rex orange X-1939 pigment	Coprecipitated lead pigment	Hercules, Inc.
RS nitrocellulose, $\frac{1}{2}$ second	RS nitrocellulose, $\frac{1}{2}$ second	Hercules, Inc.
RS nitrocellulose, 5–6 second	RS nitrocellulose, 5–6 second	Hercules, Inc.
S-394 polyethylene wax	Dry polyethylene wax	Shamrock Chemicals
Sag 471 antifoam	Proprietary silicone antifoam	Union Carbide
Silane A-1120 adhesion promoter	Amino organofunctional silane	Union Carbide
Sucrose acetate isobutyrate (SAIB)	Sucrose acetate isobutyrate solvent	Eastman Chemical
Sunprint 996 naphthenic mineral oil	Naphthenic mineral oil	Sun Petroleum
Surfynol 104-H surfactant	Organic surfactant	Air Products
SWS-213 silicone compound	Silicone compound	SWS Silicones
Syloid 308 silica	Micron-sized silica	Davison
Tecsol C solvent	Special industrial solvent	Eastman
Tecsol 3 solvent	Special industrial solvent	Eastman
Telura 797 process oil	Process oil	Exxon Chemical
Tetron 60 heat–set compound	Fluorinated wax blend heat–set compound (60% solids)	Lawter
Thixcin R thixotrope	Powder form thixotrope	NL Chemicals
Ti-Pure R-902 titanium dioxide	Rutile titanium dioxide (99%+ assay)	duPont
Titanox 2090 titanium dioxide	Rutile titanium dioxide	NL Chemicals
Transaid 1280 polymeric material	Proprietary composition polymeric material	Grow Polymer
Trionol No. 7 varnish	Quickset vehicle. #7 litho viscosity	Lawter
TXIB solvent	Proprietary solvent	Eastman
U49-2356 flushed color	Phthalo blue, G.S. 50% pigment	Sun Chemical
Ultrex quickset varnish	Gloss quickset varnish	Lawter
Uni-Rez 304 resin	Maleic resin	Union Camp
Uni-Rez 710 maleic resin	Maleic resin. Softening point 143°C	Union Camp
Uni-Rez 7020 resin	Maleic resin	Union Camp
Uni-Rez 7024 maleic resin	Modified maleic resin. Softening point 118°C	Union Camp
Unitane OR-580 titanium dioxide	Rutile titanium dioxide	American Cyanamid
Varnish 936	Ink varnish	Degen Oil
V-2630 urethane Q.S. varnish	Urethane Q.S. varnish	Superior varnish
Versamid 930 thermoplastic polyamide resin	Thermoplastic polyamide resin. Softening point 110°C	Henkel

Raw material	Chemical description	Source
WD-2507 raw umber	Raw umber pigment dispersion (60% pigment)	Daniel Products
WD-2509 burnt umber	Burnt umber pigment dispersion (40% pigment)	Daniel Products
XJ-12 compound	Anti-offset compound	Lawter
Zinc oxide solution #1	Zinc ammonium crosslinking agent (15% solids)	S.C. Johnson

Appendix D: Addresses of US suppliers of materials used in water based inks

E.J. FLICK

Air Products & Chemicals, Inc.
P.O. Box 538
Allentown, PA 18105

Alcan Ingot and Powders
Alcan Aluminum Corp.
P.O. Box 290
Elizabeth, NJ 07207

Algan, Inc.
10741 West Park Circle
Chagrin Falls, OH 44022

Allied Chemical Corp.
Industrial Chemicals
P.O. Box 1139R
Morristown, NJ 07960

American Cyanamid Co.
Chemical Products Division
One Cyanamid Plaza
Wayne, NJ 07470

Angus Chemical Co.
2211 Sanders Road
P.O. Box 3037
Northbrook, IL 60062

Atlantic Powdered Metals, Inc.
225 Broadway
New York, NY 10007

BASF Wyandotte Corp.
491 Columbia Ave.
Holland, MI 49423

Braznell Co.
5215 Manchester Ave.
St. Louis, MO 63110

Cabot Corp.
Concord Road
Billerica, MA 01821

CasChem, Inc.
40 Avenue A
Bayonne, NJ 07002

Colloids, Inc.
394 Frelinghuysen Ave.
Newark, NJ 07114

Columbian Chemicals Co.
2431 East 61 St.
P.O. Box 37
Tulsa, OK 74102

Daniel Products Co.
400 Claremont Ave.
Jersey City, NJ 07304

Davison Chemical Division
W.R. Grace & Co.
P.O. Box 2117
Baltimore, MD 21203

Day-Glo Color Corp.
4515 St. Clair Ave.
Cleveland, OH 44103

Degen Oil & Chemical Co.
200 Kellogg St.
Jersey City, NJ 07305

Degussa Corp.
Route 46 at Hollister Road
P.O. Box 2004
Teterboro, NJ 07608

E.I. duPont de Nemours & Co.
Wilmington, DE 19898

Eastman Chemical Products
Kingsport, TN 37662

Engelhard Minerals and Chemicals Division
Menlo Park
Edison, NJ 08817

Exxon Chemical U.S.A.
P.O. Box 3272
Houston, TX 77001

FRP Co.
P.O. Box 349
Baxley, GA 31513

B.F. Goodrich Chemical Division
6100 Oak Tree Blvd.
Cleveland, OH 44131

Goodyear Tire & Rubber Co.
1144 East Market St.
Akron, OH 44316

Grow Polymer
1354 Old Post Road
P.O. Box 366
Havre de Grace, MD 21078

Gulf Oil Chemicals Co.
P.O. Box 1563
Houston, TX 77001

Harshaw Chemical Co.
1945 East 97th St.
Cleveland, OH 44106

Henkel Corp.
7900 West 78th St.
Minneapolis, MN 55435

Hercules, Inc.
910 Market St.
Wilmington, DE 19899

Hilton-Davis Chemical Co.
P.O. Box 37869
Cincinnati, OH 45222

J.M. Huber Corp.
P.O. Box 310
Havre de Grace, MD 21078

S.C. Johnson & Son, Inc.
Racine, WI 53403

Lawter International, Inc.
990 Skokie Blvd.
Northbrook, IL 60062

Magie Brothers Oil Co.
9101 Fullerton Ave.
Franklin Park, IL 60131

Magruder Color Co., Inc.
1029 Newark Ave.
Elizabeth, NJ 07201

Mallinckrodt, Inc.
P.O. Box 5439
St. Louis, MO 63147

Micro Powders, Inc.
1730 Central Park Ave.
Yonkers, NY 10710

Monsanto Co.
800 North Lindbergh Blvd.
St. Louis, MO 63166

Mooney Chemicals, Inc.
2301 Scranton Road
Cleveland, OH 44113

NL Chemicals
Wyckoff Mills Road
P.O. Box 700
Hightstown, NJ 08520

National Starch and Chemical
Finderne Ave.
Bridgewater, NJ 08807

Obron Corp.
8 North State St.
Painesville, OH 44077

Pope Chemical Corp.
Patterson, NJ 07500

Rohm & Haas Co.
Independence Mall West
Philadelphia, PA 19105

Shamrock Chemicals Corp.
Foot of Pacific St.
Newark, NJ 07114

Shell Chemical Co.
100 Executive Drive
P.O. Box 600
West Orange, NJ 07052

Shepherd Chemical Co.
4900 Beech St.
Cincinnati, OH 45212

Spencer Kellogg Division
Textron, Inc.
120 Delaware Ave.
P.O. Box 807
Buffalo, NY 14240

Sun Chemical Corp.
Pigments Division
411 Sun Ave.
Cincinnati, OH 45232

Sun Petroleum Products Co.
1608 Walnut St.
Philadelphia, PA 19103

SWS Silicones
Division Stauffer Chemical
Sutton Road
Adrian, MI 49221

Tenneco Chemicals, Inc.
Organics & Polymers Division
P.O. Box 365
Piscataway, NJ 08854

U.S. Bronze Powders, Inc.
Route 202
P.O. Box 31
Flemington, NJ 08822

Union Camp Corp.
Chemical Products Division
P.O. Box 60369
Jacksonville, FL 32236

Union Carbide Corp.
Chemicals and Plastics
Danbury, CT 06817

Union Chemicals Division
Union Oil Co. of California
1345 Meacham Road
Schaumburg, IL 60196

Witco Chemical Corp.
Organics Division
277 Park Ave.
New York, NY 10017

Index

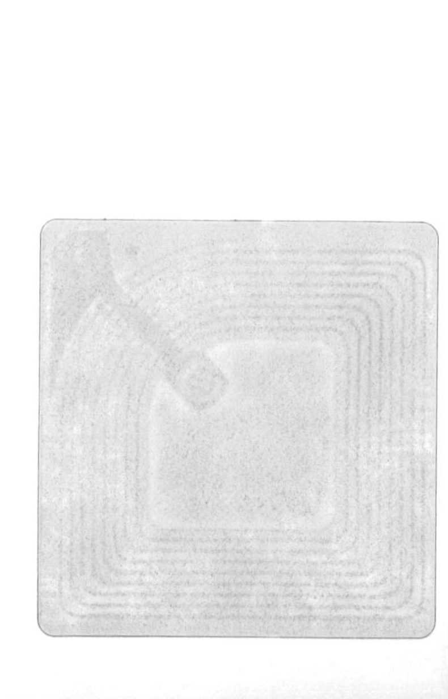

T